Birkhäuser

Trends in Mathematics

Trends in Mathematics is a series devoted to the publication of volumes arising from conferences and lecture series focusing on a particular topic from any area of mathematics. Its aim is to make current developments available to the community as rapidly as possible without compromise to quality and to archive these for reference.

Proposals for volumes can be submitted using the Online Book Project Submission Form at our website www.birkhauser-science.com.

Material submitted for publication must be screened and prepared as follows:

All contributions should undergo a reviewing process similar to that carried out by journals and be checked for correct use of language which, as a rule, is English. Articles without proofs, or which do not contain any significantly new results, should be rejected. High quality survey papers, however, are welcome.

We expect the organizers to deliver manuscripts in a form that is essentially ready for direct reproduction. Any version of TeX is acceptable, but the entire collection of files must be in one particular dialect of TeX and unified according to simple instructions available from Birkhäuser.

Furthermore, in order to guarantee the timely appearance of the proceedings it is essential that the final version of the entire material be submitted no later than one year after the conference.

For further volumes:
http://www.springer.com/series/4961

Harmonic and Complex Analysis and its Applications

Alexander Vasil'ev
Editor

Birkhäuser

Editor
Alexander Vasil'ev
Department of Mathematics
University of Bergen
Bergen
Norway

ISBN 978-3-319-01805-8 ISBN 978-3-319-01806-5 (eBook)
DOI 10.1007/978-3-319-01806-5
Springer Cham Heidelberg New York Dordrecht London

Mathematics Subject Classification (2010): 13P15, 17B68, 17B80, 30C35, 30E05, 31A05, 31B05, 42C40, 46E15, 70H06, 76D27, 81R10

Printed on acid-free paper

Springer is part of Springer Science+Business Media (www.birkhauser-science.com)

Contents

Introduction

"Harmonic and Complex Analysis and its Applications" (HCAA) was a 5-year (2007–2012) European Science Foundation Programme whose aim was to explore and to strengthen the bridge between two scientific communities: analysts with broad backgrounds in complex and harmonic analysis and mathematical physics and specialists in physics and applied sciences. The programme was a multidisciplinary European activity uniting leading European scientists from these communities in 11 countries (Austria, Finland, Germany, Israel, Ireland, Luxembourg, Norway, Spain, Sweden, Switzerland, and UK) in a project of developing a coherent viewpoint on harmonic and complex analysis in the general context of mathematical physics. It coordinated actions for advancing harmonic and complex analysis and for increasing its application to challenging scientific problems. Particular topics considered by this programme included conformal and quasiconformal mappings, potential theory, Banach spaces of analytic functions and their applications to the problems of fluid mechanics, conformal field theory, Hamiltonian and Lagrangian mechanics, and signal processing. The programme had partnerships with other European and non-European networks. More on HCAA one can read at http://org.uib.no/hcaa/. The programme was steered by a committee

- *Alexander Vasil'ev* (Chairman) (University of Bergen, Norway);
- *Zoltan Balogh* (University of Bern, Switzerland);
- *Hans G. Feichtinger* (University of Vienna, Austria);
- *Stephen Gardiner* (University College Dublin, Ireland);
- *Björn Gustafsson* (Kungliga Tekniska Högskolan, Stockholm, Sweden);
- *Ilkka Holopainen and Olli Martio* (University of Helsinki, Finland);
- *John King and Linda Cummings* (University of Nottingham, UK);
- *Fernando Pérez-González* (Universidad de la Laguna, Tenerife, Spain);
- *Dierk Schleicher* (Jacobs University Bremen, Germany);
- *Martin Schlichenmaier* (Université du Luxembourg);
- *Lawrence Zalcman*, (Bar-Ilan University, Ramat-Gan, Israel).

Technical support was provided by an ESF (Physical and Engineering Sciences Unit) officer Mrs. *Catherine Werner* to whom we are all thankful for her kind timely help and assistance. This programme was very successful which is documented by several joint papers in highly ranked journals in line with many other ESF programmes in fundamental research and applications. We accept that it was a very useful and necessary instrument of funding in the European research area especially in the basic research, in particular in mathematics, in which there are not many other sources of external financing. At the beginning of the programme a kick-off conference was organized in Norway in 2007 and a volume of proceedings appeared in [1]. During the programme we started a new journal [2] under the same name that publishes current research results as well as selected high-quality survey articles in real, complex, harmonic, and geometric analysis originating and/or having applications in mathematical physics. The journal promotes dialog among specialists in these areas. After the time period of HCAA had finished, the Steering Committee decided to edit a special volume of surveys reflecting important research lines represented by the participants of the programme. The committee served as the Advisory Board of this volume.

We would like to acknowledge the efforts of all participants of the programme HCAA, in particular, the authors of this volume. We hope that it will be interesting and useful for professionals and novices in analysis and mathematical physics. We strongly believe that graduate students will be one of the target audiences of this book. Browsing the volume, the reader undoubtedly notices that the scope of the programme being rather broad exhibits many interrelations between the various contributions, which can be regarded as different facets of a common theme. We hope that this volume will enrich the further development in analysis and its prosperous interaction with mathematical physics and applied sciences.

Bergen, Norway Alexander Vasil'ev

References

1. B. Gustafsson, A. Vasil'ev (eds.), *Analysis and Mathematical Physics*, Trends in Mathematics (Birkhäuser Verlag, Boston, 2009)
2. B. Gustafsson, A. Vasil'ev (eds.), *Analysis and Mathematical Physics* (Birkhäuser Verlag, Boston). ISSN:1664–2368

Function Spaces of Polyanalytic Functions

Luis Daniel Abreu and Hans G. Feichtinger

Abstract This article is meant as both an introduction and a review of some of the recent developments on Fock and Bergman spaces of polyanalytic functions. The study of polyanalytic functions is a classic topic in complex analysis. However, thanks to the interdisciplinary transference of knowledge promoted within the activities of HCAA network it has benefited from a cross-fertilization with ideas from signal analysis, quantum physics, and random matrices. We provide a brief introduction to those ideas and describe some of the results of the mentioned cross-fertilization. The departure point of our investigations is a thought experiment related to a classical problem of multiplexing of signals, in order words, how to send several signals simultaneously using a single channel.

Keywords Gabor frames • Landau levels • Polyanalytic spaces

L.D. Abreu (✉)
Acoustic Research Institute, Austrian Academy of Sciences, Wohllebengasse 12–14, Wien A-1040, Austria

CMUC, Department of Mathematics, University of Coimbra, Portugal
e-mail: daniel@mat.uc.pt

H.G. Feichtinger
NuHAG, Faculty of Mathematics, University of Vienna, Nordbergstraße 15, A-1090 Wien, Austria
e-mail: hans.feichtinger@univie.ac.at

A. Vasil'ev (ed.), *Harmonic and Complex Analysis and its Applications*, Trends in Mathematics, DOI 10.1007/978-3-319-01806-5_1,

1 Introduction

1.1 Definition of a Polyanalytic Function

Among the most widely studied mathematical objects are the solutions of the Cauchy–Riemann equation

$$\partial_{\overline{z}}F(z) = \frac{1}{2}\left(\frac{\partial}{\partial x} + i\frac{\partial}{\partial \xi}\right) F(x+i\xi) = 0,$$

known as *analytic functions*. The properties of analytic functions are so remarkable that, at a first encounter, they are often perceived as "magic." However, the analyticity restriction is so strong that it created a prejudice against non-analytic functions, which are often perceived as unstructured and bad behaved objects and therefore not worthy of further study. Nevertheless, there are non-analytic functions with significant structure and with properties reminiscent of those satisfied by analytic functions.

Such nice non-analytic functions are called *polyanalytic functions.*

A function $F(z,\overline{z})$, defined on a subset of $\mathbb{C}$, and satisfying the generalized Cauchy–Riemann equations

$$(\partial_{\overline{z}})^n \, F(z,\overline{z}) = \frac{1}{2^n}\left(\frac{\partial}{\partial x} + i\frac{\partial}{\partial \xi}\right)^n F(x+i\xi, x-i\xi) = 0, \tag{1}$$

is said to be *polyanalytic of order* $n-1$. It is clear from (1) that the following polynomial of order $n-1$ in $\overline{z}$

$$F(z,\overline{z}) = \sum_{k=0}^{n-1} \overline{z}^k \varphi_k(z), \tag{2}$$

where the coefficients $\{\varphi_k(z)\}_{k=0}^{n-1}$ are analytic functions is a polyanalytic function of order $n-1$. By solving $\partial_{\overline{z}}F(z,\overline{z}) = 0$, an iteration argument shows that every $F(z,\overline{z})$ satisfying (1) is indeed of the form (2). Some fundamental properties of analytic functions cease to be true for polyanalytic functions. For instance, a simple polyanalytic function of order 1 is

$$F(z,\overline{z}) = 1 - |z|^2 = 1 - z\overline{z}.$$

Since

$$\partial_{\overline{z}}F(z,\overline{z}) = -z \text{ and } (\partial_{\overline{z}})^2 \, F(z,\overline{z}) = 0,$$

the function $F(z, \overline{z})$ is not analytic in z, but is *polyanalytic of order one*. This simple example already highlights one of the reasons why the properties of polyanalytic functions can be different of those enjoyed by analytic functions: they can vanish on closed curves without vanishing identically, while analytic functions cannot even vanish on an accumulation set of the complex plane! Still, many properties of analytic functions have found an extension to polyanalytic functions, often in a nontrivial form, as we shall see later on in a few examples.

1.2 What Are Polyanalytic Functions Good for?

Imagine some application of analytic functions. By definition, they allow to represent the objects of our application as a function of z (because the function is analytic). We may want to represent the object to obtain a nice theory, we may want to store the information contained in the object and send it to someone. Whatever we want to do, we will always end up with a representation involving powers of z (because the functions are analytic). Not that bad, since we have an infinite number of them. However, several applications of mathematics, like quantum mechanics and signal analysis, require infinite dimensions for their theoretical formulation. And when we build a model in the complex plane using analytic functions, all the powers of z are taken.

What if we want to build several models simultaneously for the same plane? Introducing an extra complex variable will bring us the complications related to the study of analytic functions in several complex variables. If $\mathbb{C}$ is not enough for some models, $\mathbb{C}^2$ may seem too much to handle if we want to keep the mathematical problems within a tangible range. One is tempted to ask if there is something in between, but it may seem hard to believe that it is possible to "store" more information in a complex plane without introducing an extra independent variable.

Enter the world of polyanalytic functions!

We are now allowed to use powers of z and $\overline{z}$. This introduces an enormous flexibility. Consider the Hilbert space $\mathcal{L}_2(\mathbb{C})$ of all measurable functions equipped with the norm

$$\|F\|^2_{\mathcal{L}_2(\mathbb{C})} = \int_{\mathbb{C}} |F(z)|^2 \, e^{-\pi|z|^2} \, d\mu(z). \tag{3}$$

It is relatively easy to observe, using integration by parts (see formula (6) below) that, given an analytic function $F(z) \in \mathcal{L}_2(\mathbb{C})$, the function

$$F'(z) - \pi \overline{z} F(z)$$

is orthogonal to $F(z)$. We can create several subspaces of $\mathcal{L}_2(\mathbb{C})$ by multiplying elements of the Fock space of analytic functions by a power of $\overline{z}$. If we consider the sum of all such spaces, we obtain the whole $\mathcal{L}_2(\mathbb{C})$. We can do even better: by proper combination of the powers of z and $\overline{z}$ we can obtain an orthogonal decomposition of $\mathcal{L}_2(\mathbb{C})$! This fact, first observed by Vasilevski [76], is due to the following: the polynomials

$$e_{k,j}(z,\overline{z}) = e^{\pi|z|^2}(\partial_z)^k\left[e^{-\pi|z|^2}z^j\right]$$

are orthogonal in both the index j and k and they span the whole space $\mathcal{L}_2(\mathbb{C})$ of square integrable functions in the plane weighted by a gaussian $e^{-\pi|z|^2}$. For every k, we have thus a "copy" of the space of analytic functions which is orthogonal to any of the other copies. It is a remarkable fact that every polyanalytic function of order n can be expressed as a combination of the polynomials $\{e_{k,j}(z,\overline{z})\}_{k<n,j\in\mathbb{N}}$. Thus, we can work simultaneously in n planes keeping the number of degrees of freedom of each one intact. We will see in this paper how this fact can be put in good use, notably in the analysis of the higher Landau levels and in the multiplexing of signals (analysis of several signals simultaneously).

1.3 Some Historical Remarks

Polyanalytic functions were for the first time considered in [55] by the Russian mathematician G. V. Kolossov (1867–1935) in connection with his research on elasticity. This line of research has been developed by his student Muskhelishvili and the applications of polyanalytic functions to problems in elasticity are well documented in his book [63].

Polyanalytic function theory has been investigated intensively, notably by the Russian school led by Balk [15]. More recently the subject gained a renewed interest within operator theory and some interesting properties of the function spaces whose elements are polyanalytic functions have been derived [16, 17]. A new characterization of polyanalytic functions has been obtained by Agranovsky [9]. The two visionary papers of Vasilevski [76] and [75] had a profound influence in what we will describe. In the beginning, our investigations in the topic were motivated by applications in signal analysis, in particular by the results of Gröchenig and Lyubarskii on Gabor frames with Hermite functions [44, 45], but soon it was clear that Hilbert spaces of polyanalytic functions lie at the heart of several interesting mathematical topics. Remarkably, they provide explicit representation formulas for the functions in the eigenspaces of the Euclidean Laplacian with a magnetic field, the so-called *Landau levels*. It is also worth of historical remark that, in 1951, Richard Feynman has obtained formulas very similar to those involving

Gabor transforms with Hermite functions in his work on quantum electrodynamics [37]. Precisely the same functions have been used by Daubechies and Klauder in their formulation of Feynman integrals [25]. The historically conscious reader may recognize in polyanalytic function theory some of the eclectic flavor emblematic of the mathematics oriented to signal analysis and quantum mechanics, something particularly notorious in the body of mathematics which became known as the Bell papers of the 1960s (see the review [73]) and in the advent of wavelets and coherent states [10, 24].

The following are the main books that we used as sources in important topics which are briefly outlined in this survey. Balk's book [15] is still an authoritative reference for the function theoretical aspects of polyanalytic functions. Standard references for Fock and Bergman spaces are [29, 51, 79] and for the applications of such spaces in sampling we have followed [72]. Our notations and essential facts about Gabor analysis are extracted from Gröchenig's book [42]. We used [10] as a reference for coherent states and Daubechies classical monograph [24] for wavelets and general frame theory. An introduction to Determinantal Point Processes can be found in [18].

1.4 Outline

Our main goal is to highlight the connections between different topics. We try to present the material in a such a way that the reader can gain from the time-frequency point of view, even if it is the case of a reader who is not familiar with the basic concepts of time-frequency analysis. Thus, the basic concepts are presented and a special attention is given to those particular regions of knowledge where two different mathematical topics intersect.

We have organized the paper as follows. We start with a section on the Hilbert space theory of polyanalytic Fock spaces. This includes the description of the theoretical multiplexing device which is the basic signal analytic model for our viewpoint. The third section explains how the topic connects to time-frequency analysis, more precisely, to the theory of Gabor frames with Hermite functions. Then we make a review of the basic facts of the L^p theory of polyanalytic Fock spaces for $p \neq 2$ and of the polyanalytic Bargmann transform in modulation spaces. We review some physical applications in Sect. 4, namely the interpretation of the so-called true polyanalytic Fock spaces as the eigenspaces of the Euclidean Landau Hamiltonian with a constant magnetic field. In Sect. 6 we provide a brief introduction to the polyanalytic Ginibre ensemble. The last section is devoted to hyperbolic analogues of the theory, where wavelets and the Maass Laplacian play a fundamental role.

2 Fock Spaces of Polyanalytic Functions

2.1 The Orthogonal Decomposition and the Polyanalytic Hierarchy

Recall that $\mathcal{L}_2(\mathbb{C})$ denotes the Hilbert space of all measurable functions equipped with the norm

$$\|F\|_{\mathcal{L}_2(\mathbb{C})}^2 = \int_{\mathbb{C}} |F(z)|^2 e^{-\pi|z|^2} d\mu(z),$$

where $d\mu(z)$ stands for area measure on $\mathbb{C}$. If we require the elements of the space to be analytic, we are lead to the Fock space $\mathcal{F}_2(\mathbb{C})$. Polyanalytic Fock spaces $\mathbf{F}_2^n(\mathbb{C})$ arise in an analogous manner, by requiring its elements to be polyanalytic of order $n-1$. They seem to have been first considered by Balk [15, p. 170] and, more recently, by Vasilevski [76], who obtained the following decompositions in terms of spaces $\mathcal{F}_2^n(\mathbb{C})$ which he called *true poly-Fock spaces*:

$$\mathbf{F}_2^n(\mathbb{C}) = \mathcal{F}_2^1(\mathbb{C}) \oplus \ldots \oplus \mathcal{F}_2^n(\mathbb{C}). \tag{4}$$

$$\mathcal{L}_2(\mathbb{C}) = \bigoplus_{n=1}^{\infty} \mathcal{F}_2^n(\mathbb{C}).$$

We will use the following definition of $\mathcal{F}_2^n(\mathbb{C})$ which is equivalent to the one given by Vasilevski: a function F belongs to the *true polyanalytic Fock space* $\mathcal{F}_2^{n+1}(\mathbb{C})$ if $\|F\|_{\mathcal{L}_2(\mathbb{C})} < \infty$ and there exists an entire function H such that

$$F(z) = \left(\frac{\pi^n}{n!}\right)^{\frac{1}{2}} e^{\pi|z|^2} (\partial_z)^n \left[e^{-\pi|z|^2} H(z)\right]. \tag{5}$$

With this definition it is easy to verify that the spaces $\mathcal{F}_2^n(\mathbb{C})$ are orthogonal using Green´s formula:

$$\int_{\mathbf{D}_r} f(z)\partial_{\bar{z}}\overline{g(z)}dz = -\int_{\mathbf{D}_r} \partial_{\bar{z}} f(z)\overline{g(z)}dz + \frac{1}{i}\int_{\delta\mathbf{D}_r} f(z)\overline{g(z)}dz. \tag{6}$$

and its higher order version obtained by iterating (6):

$$\int_{\mathbf{D}_r} f(z)\left(\frac{d}{d\bar{z}}\right)^n \overline{g(z)}dz = (-1)^n \int_{\mathbf{D}_r} \left(\frac{d}{d\bar{z}}\right)^n f(z)\overline{g(z)}dz$$
$$+\frac{1}{i}\sum_{j=0}^{n-1}(-1)^j \int_{\delta\mathbf{D}_r} \left(\frac{d}{d\bar{z}}\right)^j f(z)\left(\frac{d}{d\bar{z}}\right)^{n-j-1}\overline{g(z)}dz. \tag{7}$$

A visual image of the polyanalytic hierarchy is the following decomposition of $\mathcal{L}_2(\mathbb{C})$ (an orthogonal decomposition in the spaces $\{\mathcal{F}_2^n(\mathbb{C})\}_{k=1}^{\infty}$ and a union of the nested spaces $\{\mathbf{F}_2^n(\mathbb{C})\}_{k=1}^{\infty}$).

$$\begin{array}{lll}
\mathcal{F}_2^1(\mathbb{C}) & = \mathbf{F}_2^1(\mathbb{C}) & = \mathcal{F}_2(\mathbb{C}) \\
\mathcal{F}_2^1(\mathbb{C}) \oplus \mathcal{F}_2^2(\mathbb{C}) & = \mathbf{F}_2^2(\mathbb{C}) & \\
\quad .. & & \\
\quad \dots & & \\
\mathcal{F}_2^1(\mathbb{C}) \oplus \dots\dots \oplus \mathcal{F}_2^n(\mathbb{C}) & = \mathbf{F}_2^n(\mathbb{C}) & \\
\quad \dots. & & \\
\mathcal{F}_2^1(\mathbb{C}) \oplus \dots\dots \mathcal{F}_2^n(\mathbb{C}) \oplus \mathcal{F}_2^{n+1}(\mathbb{C}) \oplus \dots & = \bigoplus_{n=1}^{\infty} \mathcal{F}_2^n(\mathbb{C}) & = \mathcal{L}_2(\mathbb{C})
\end{array}$$

2.2 *Reproducing Kernels of the Polyanalytic Fock Spaces*

The reproducing kernels of the polyanalytic Fock spaces have been computed using several different methods: invariance properties of the Landau Laplacian [13], composition of unitary operators [76], Gabor transforms with Hermite functions [2], and the expansion in the kernel basis functions [47]. Nice formulas are obtained using the Laguerre polynomials

$$L_k^{\alpha}(x) = \sum_{i=0}^{k} (-1)^i \binom{k+\alpha}{k-i} \frac{x^i}{i!}.$$

The reproducing kernel of the space $\mathcal{F}_2^n(\mathbb{C})$, $\mathcal{K}^n(z,w)$, can be written as

$$\mathcal{K}^n(z,w) = \pi L_{n-1}^0(\pi\,|z-w|^2)e^{\pi z\overline{w}}.$$

This gives the explicit formula for the orthogonal projection P^n required at the step (5) of our theoretical multiplexing device in the next section:

$$(P^n F)(w) = \int_{\mathbb{C}} F(z)\pi L_{n-1}^0(\pi\,|z-w|^2)e^{\pi z(\overline{w}-\overline{z})}d\mu(z). \tag{8}$$

The reproducing kernel of the space $\mathbf{F}_2^n(\mathbb{C})$ is denoted by $\mathbf{K}^n(z,w)$. Using the formula $\sum_{k=0}^{n-1} L_k^{\alpha} = L_{n-1}^{\alpha+1}$, (4) gives

$$\mathbf{K}^n(z,w) = \pi L_{n-1}^1(\pi\,|z-w|^2)e^{\pi z\overline{w}}.$$

2.3 A Thought Experiment: Multiplexing of Signals

A classical problem in the Theory of Signals is the one of Multiplexing, that is, transmitting several signals over a single channel in such a way that it is possible to recover the original signal at the receiver [14]. We will use the multiplexing idea as a thought experiment providing intuition about our ideas, later to be developed rigorously. We suggest the reading of [54] for considerations regarding the role of these kinds of experiences, ubiquitous in Theoretical Physics, in the modern mathematical landscape.

The center of our attention is now the orthogonal decomposition (4). Assume that we can somehow (we will do it in the next section) construct a map $\mathcal{B}^n$ sending an arbitrary $f \in L^2(\mathbb{R})$ to the space $\mathcal{F}_2^n(\mathbb{C})$. Then we can, at least theoretically, proceed as follows.

1. Given n signals $f_1, \ldots, f_n$, with finite energy ($f_k \in L^2(\mathbb{R})$ for every k), process each individual signal by evaluating $\mathcal{B}^k f_k$. This encodes each signal into one of the n orthogonal spaces $\mathcal{F}^1(\mathbb{C}), \ldots, \mathcal{F}^n(\mathbb{C})$.
2. Construct a new signal $F = \mathbf{Bf} = \mathcal{B}^1 f_1 + \ldots + \mathcal{B}^n f_n$ as a superposition of the n processed signals.
3. Sample, transmit, or process F.
4. Let P^k denote the orthogonal projection from $\mathbf{F}_2^n(\mathbb{C})$ onto $\mathcal{F}^k(\mathbb{C})$, then $P^k(F) = \mathcal{B}^k f_k$ by virtue of (4).
5. Finally, after inverting each of the transforms $\mathcal{B}^k$, we recover each component f_k in its original form.

The combination of n independent signals into a single signal $\mathbf{B}^n\mathbf{f}$ and the subsequent processing provides our multiplexing device. With two signals this can be outlined in the following scheme.

$$
\begin{array}{ccccc}
f_1 \to \mathcal{B} f_1 & & & & \mathcal{B} f_1 \to f_1 \\
& \searrow & & P^1 \nearrow & \\
& & \mathcal{B} f_1 + \mathcal{B}^2 f_2 = \mathbf{Bf} & & \\
& \nearrow & & P^2 \searrow & \\
f_2 \to \mathcal{B}^2 f_2 & & & & \mathcal{B}^2 f_2 \to f_2
\end{array}
$$

We believe that the above device has practical applications, but we will pursue another direction in our reasoning: we will use the above scheme as a source of mathematical ideas—with some poetic license, we may say that we apply signal analysis to mathematics.

2.4 The Polyanalytic Bargmann Transform

The construction of the map $\mathcal{B}^k$ above can be done as follows. To map the first signal $f_1 \in L^2(\mathbb{R})$ to the space $\mathcal{F}_2^1(\mathbb{C}) = \mathcal{F}_2(\mathbb{C})$ we can of course use the good old

Bargmann transform $\mathcal{B}$, where

$$\mathcal{B}f(z) = 2^{\frac{1}{4}} \int_{\mathbb{R}} f(t) e^{2\pi t z - \pi z^2 - \frac{\pi}{2} t^2} dt.$$

The remaining signals are mapped using

$$\mathcal{B}^{k+1} f(z) = \left(\frac{\pi^k}{k!} \right)^{\frac{1}{2}} e^{\pi |z|^2} (\partial_z)^k \left[e^{-\pi |z|^2} \mathcal{B} f(z) \right]$$

It can be proved that $\mathcal{B}^k : L^2(\mathbb{R}) \to \mathcal{F}_2^k(\mathbb{C})$ is a Hilbert space isomorphism, by observing that the Hermite functions are mapped to the orthogonal basis $\{e_{k,n}(z, \overline{z}) : n \geq 0\}$ of $\mathcal{F}_2^k(\mathbb{C})$, where

$$e_{k,n}(z) = \left(\frac{\pi^k}{k!} \right)^{\frac{1}{2}} e^{\pi |z|^2} (\partial_z)^k \left[e^{-\pi |z|^2} e_n(z) \right] \tag{9}$$

and

$$e_n(z) = \left(\frac{\pi^n}{n!} \right)^{\frac{1}{2}} z^n$$

is the orthogonal monomial basis of the Fock space. We can now define a transform $\mathbf{B}^n : L^2(\mathbb{R}, \mathbb{C}^n) \to \mathbf{F}^n(\mathbb{C})$ by mapping each vector $\mathbf{f} = (f_1, \ldots, f_n) \in L^2(\mathbb{R}, \mathbb{C}^n)$ to the following polyanalytic function of order n:

$$\mathbf{B}^n \mathbf{f} = \mathcal{B}^1 f_1 + \ldots + \mathcal{B}^n f_n . \tag{10}$$

This map is again a Hilbert space isomorphism and is called the *polyanalytic Bargmann transform* [1].

2.5 A Polyanalytic Weierstrass Function

Our construction of the previous section is done at a purely theoretical level, since we have no way of storing and processing a continuous signal. However, we can construct a discrete counterpart of the theory. Following [3], an analogue of the Whittaker–Shannon–Kotel´nikov sampling theorem can be constructed using a polyanalytic version of the Weierstrass sigma function. Let σ be the Weierstrass sigma function corresponding to Λ defined by

$$\sigma_\Lambda(z) = z \prod_{\lambda \in \Lambda \setminus \{0\}} \left(1 - \frac{z}{\lambda} \right) e^{\frac{z}{\lambda} + \frac{z^2}{2\lambda^2}} ,$$

To simplify our notations we will write the results in terms of the square lattice, $\Lambda = \alpha(\mathbb{Z}+i\mathbb{Z})$ consisting of the points $\lambda = \alpha l + i\alpha m, k, m \in \mathbb{Z}$, but most of what we will say is also true for general lattices. To write down our explicit sampling formulas, the following polyanalytic extension of the Weierstrass sigma function is required:

$$S_{\Lambda}^{n+1}(z) = \left(\frac{\pi^n}{n!}\right)^{\frac{1}{2}} e^{\pi|z|^2} (\partial_z)^n \left[e^{-\pi|z|^2} \frac{(\sigma_\Lambda(z))^{n+1}}{n!z} \right].$$

Clearly, $S_{\Lambda}^{1}(z) = \sigma_\Lambda(z)/z$. Let $\sigma_{\Lambda^\circ}(z)$ be the Weierstrass sigma function associated with the adjoint lattice $\Lambda^\circ = \alpha^{-1}(\mathbb{Z} + i\mathbb{Z})$ of Λ and consider the corresponding polyanalytic Weierstrass function $S_{\Lambda^0}^n(z)$. With this terminology we have:

Theorem 1 ([3]). *If* $\alpha^2 < \frac{1}{n+1}$, *every* $F \in \mathcal{F}_2^{n+1}(\mathbb{C})$ *can be written as:*

$$F(z) = \sum_{\lambda \in \alpha(\mathbb{Z}+i\mathbb{Z})} F(\lambda) e^{\pi\overline{\lambda}z - \pi|\lambda|^2} S_{\Lambda^0}^n(z), \tag{11}$$

Remark 1. If $\alpha^2 > n+1$, then Λ is an interpolating sequence for $\mathcal{F}^{n+1}(\mathbb{C})$. Moreover, the interpolation problem is solved by [3]

$$F(z) = \sum_{\lambda \in \Lambda} a_\lambda e^{\pi\overline{\lambda}z - \pi|\lambda|^2/2} S_{\Lambda}^n(z - \lambda). \tag{12}$$

Observe the duality between formulas (11) and (12).

3 Time-Frequency Analysis of Polyanalytic Functions

3.1 The Gabor Transform

The study of polyanalytic Fock spaces can be significantly enriched via a connection to time-frequency (Gabor) analysis. Recall that the Gabor or Short-time Fourier transform (STFT) of a function or distribution f with respect to a window function g is defined to be

$$V_g f(x,\xi) = \int_{\mathbb{R}} f(t)\overline{g(t-x)} e^{-2\pi i \xi t} dt. \tag{13}$$

There is a very important property enjoyed by inner products of this transforms. The following relations are usually called *the orthogonal relations for the short-time Fourier transform.* Let $f_1, f_2, g_1, g_2 \in L^2(\mathbb{R})$. Then $V_{g_1} f_1, V_{g_2} f_2 \in L^2(\mathbb{R}^2)$ and

$$\langle V_{g_1} f_1, V_{g_2} f_2 \rangle_{L^2(\mathbb{R}^2)} = \langle f_1, f_2 \rangle_{L^2(\mathbb{R})} \overline{\langle g_1, g_2 \rangle}_{L^2(\mathbb{R})}. \tag{14}$$

If $\|g\|_{L^2(\mathbb{R})} = 1$, the Gabor transform provides an isometry

$$V_g : L^2(\mathbb{R}) \to L^2(\mathbb{R}^2),$$

since, if $f, g \in L^2(\mathbb{R}^d)$, then

$$\| V_g f \|_{L^2(\mathbb{R}^2)} = \| f \|_{L^2(\mathbb{R})} \|g\|_{L^2(\mathbb{R})} . \tag{15}$$

In many applications it is required that the window g is in the Feichtinger algebra S_0. Equivalently one is requiring that $V_g g \in L^1(\mathbb{R}^2)$, resp., that g is satisfying some mild conditions concerning smoothness and decay (as typically Fourier summability kernels are supposed to have them). In particular such an assumption implies robustness of the transform and also the guarantee that samples of $V_g f$ over a grid are in $l^2(\Lambda)$. Given a point $\lambda = (\lambda_1, \lambda_2)$ in phase-space $\mathbb{R}^2$, the corresponding *time-frequency shift* is

$$\pi_\lambda f(t) = e^{2\pi i \lambda_2 t} f(t - \lambda_1), \quad t \in \mathbb{R}.$$

Using this notation, the Gabor transform of a function f with respect to the window g can be written as

$$V_g f(\lambda) = \langle f, \pi_\lambda g \rangle_{L^2(\mathbb{R})} .$$

In analogy with the time-frequency shifts π_λ, there are *Bargmann–Fock shifts* β_λ defined for functions on $\mathbb{C}$ by

$$\beta_\lambda F(z) = e^{\pi i \lambda_1 \lambda_2} e^{\pi \bar{\lambda} z} F(z - \lambda) \, e^{-\pi |\lambda|^2/2} . \tag{16}$$

We observe that the true polyanalytic Bargmann transform intertwines the time-frequency shifts π_λ and the Fock representation β_λ on $\mathcal{F}_2^n(\mathbb{C})$ by a calculation similar to [42, p. 185]:

$$\mathcal{B}^n (\pi_\lambda \gamma)(z) = \beta_\lambda \mathcal{B}^n \gamma(z) , \tag{17}$$

for $\gamma \in L^2(\mathbb{R})$. If we choose the Gaussian function $h_0(t) = 2^{\frac{1}{4}} e^{-\pi t^2}$ as a window in (13), then a simple calculation shows that the Bargmann transform is related to these special Gabor transforms as follows:

$$\mathcal{B} f(z) = e^{-i\pi x \xi + \pi \frac{|z|^2}{2}} V_{h_0} f(x, -\xi). \tag{18}$$

This is a well-known fact and the details of the calculation can be found in standard textbooks on time-frequency analysis, see, for instance, [42, p. 53]. It has been proved in [12] that the only possible choice of g yielding spaces of analytic functions is indeed this one. One may be led to think that this means the end of the story concerning Gabor and complex analysis. The next paragraphs show that this conclusion is premature.

The key step now is the choice of higher order Hermite functions as windows in (13). In Fig. 1 one can observe the interesting patterns of the time-frequency concentration when higher Hermite functions are used. We begin by choosing the first Hermite function $h_1(t) = 2^{\frac{1}{4}} 2\pi t e^{-\pi t^2}$. First observe that

$$V_{h_1} f(x, -\xi) = 2\pi V_{h_0}[(.)f(.)](x, -\xi) - 2\pi x V_{h_0} f(x, -\xi)$$

and that

$$2\pi \mathcal{B}[(.)f(.)](z) = \partial_z Bf(z) + \pi \overline{z} Bf(z).$$

Thus, using (18)

$$e^{-i\pi x\xi + \pi \frac{|z|^2}{2}} V_{h_1} f(x, -\xi) = \partial_z Bf(z) - \pi \overline{z} Bf(z) = \mathcal{B}^2 f(z).$$

With a bit more effort, we can choose the *nth Hermite function*

$$h_n(t) = c_n e^{\pi t^2} \left(\frac{d}{dt}\right)^n \left(e^{-2\pi t^2}\right),$$

where c_n is chosen so that $\|h_n\|_2 = 1$, as a special window in (13), and find an important and useful relation between Gabor transforms with Hermite functions and true polyanalytic Bargmann transforms of general order n:

$$e^{-i\pi x\xi + \pi \frac{|z|^2}{2}} V_{h_n} f(x, -\xi) = \mathcal{B}^{n+1} f(z). \tag{19}$$

This simple observation made in [1] connects polyanalytic Fock spaces with Gabor analysis. The important fact to retain is that the multiplier $e^{-i\pi x\xi + \pi \frac{|z|^2}{2}}$ is the same for every n. This leads us to the next observation. We can define a vector-valued Gabor transform

$$\mathbf{V}_{\mathbf{h}_{n-1}} \mathbf{f}(\lambda) = V_{(h_0, \ldots, h_{n-1})}(f_0, \ldots f_{n-1})(\lambda)$$

for the purpose of processing simultaneously n signals using a vectorial window constituted by the first n Hermite functions. Since the windows are orthogonal to each other, we can do this by simple superposition—we know that the transformed

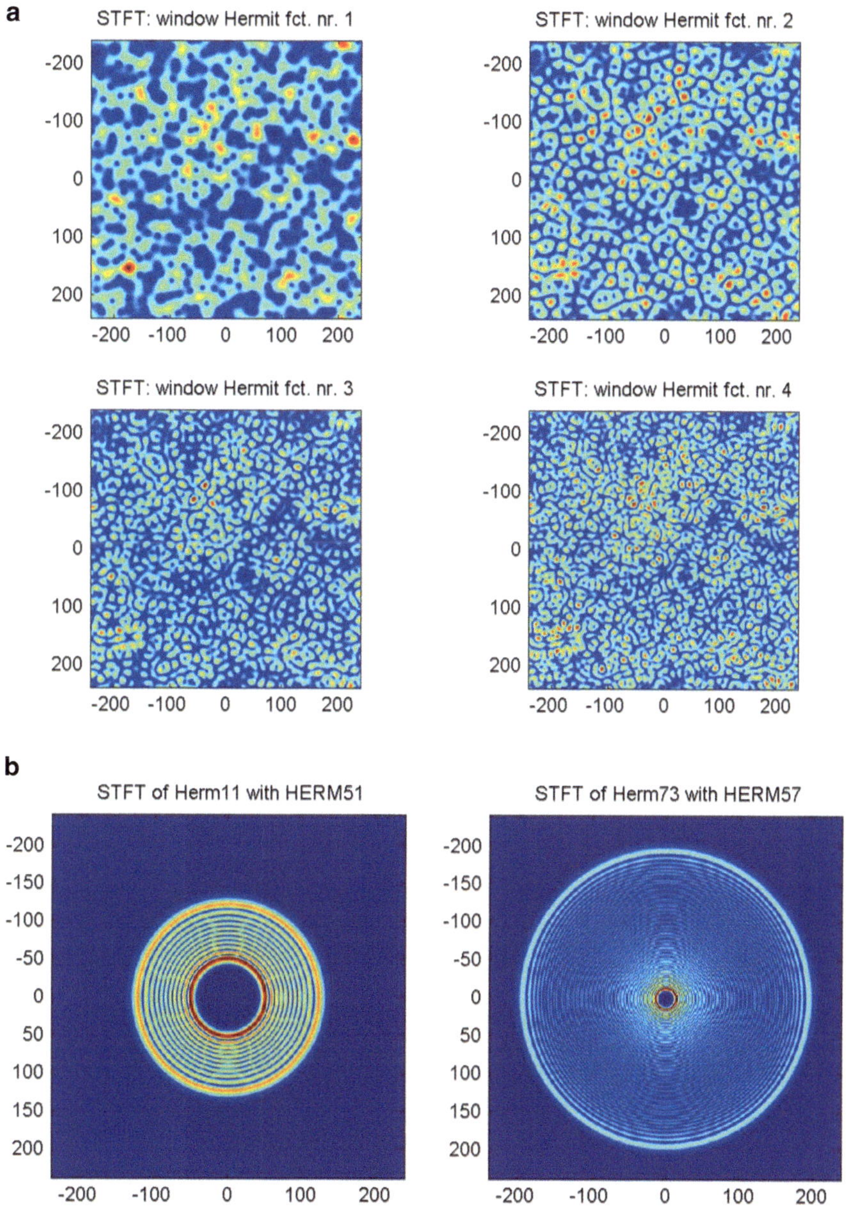

Fig. 1 Short-time Fourier transforms with higher order Hermite windows. (**a**) The image of a function in Gabor spaces generated by different Hermite windows. (**b**) Hermite functions in time-frequency plane using different Hermite windows

signals will live in mutually orthogonal function spaces because of the orthogonality conditions (14). Define

$$\mathbf{V}_{\mathbf{h}_{n-1}}\mathbf{f}(\lambda) = \sum_{k=0}^{n-1} V_{h_k} f_k(\lambda).$$

If follows now from (19) and from (10) that

$$\mathbf{B}^n \mathbf{f} = e^{-i\pi x\xi + \pi \frac{|z|^2}{2}} \mathbf{V}_{\mathbf{h}_{n-1}}\mathbf{f}(\lambda). \tag{20}$$

Formula (20) is the key for a real variable treatment of polyanalytic Fock spaces. This approach already led to the proof of results that seemed hopeless using only complex variables. For instance, it was possible to prove that the sampling and interpolation *lattices* of $\mathbf{F}_2^n(\mathbb{C})$ can be characterized by their density [1]. Previously, this result was known only for $n = 1$ and the proofs [59, 71] strongly depend on tools like Jensen´s formula, which are not available in the polyanalytic case. The above connection to Gabor analysis solved the problem, by means of a remarkable duality result from time-frequency analysis [69], which turned the polyanalytic problem into a Hermite interpolation (multi-sampling) problem in spaces of analytic functions.

On the one side, the connection between Gabor analysis and polyanalytic functions has been used to prove new results about polyanalytic function spaces, as outlined in the above paragraph. On the other side, the same connection offers a new technical ammunition to time-frequency analysis and it has already been used in [28, Lemma 1] in a key step of the proof of the main result, where it is shown that a certain time-frequency localization has infinite rank.

3.2 Gabor Spaces

The *Gabor space* $\mathcal{G}_g$ is the subspace of $L^2(\mathbb{R}^2)$ which is the image of $L^2(\mathbb{R})$ under the Gabor transform with the window g,

$$\mathcal{G}_g = \left\{ V_g f : f \in L^2(\mathbb{R}) \right\}.$$

The spaces $\mathcal{G}_g$ are called *model spaces* in [12]. It is well known (see [24]) that Gabor spaces have a reproducing kernel given by

$$k(z,w) = \langle \pi_z g, \pi_w g \rangle_{L^2(\mathbb{R})} \tag{21}$$

For instance, if we consider the Gaussian window $g(x) = 2^{\frac{1}{4}}e^{-\pi t^2}$, using the notation $z = x + i\xi$ and $w = u + i\eta$, a calculation (see [42, Lemma 1.5.2]) shows that the reproducing kernel of $\mathcal{G}_g$ is

$$k^0(x, \xi, u, \eta) = e^{\pi i(u+x)(\xi-\eta)-\frac{\pi(u-x)^2-\pi(\eta-\xi)^2}{2}}.$$

This reproducing kernel can be related to the reproducing kernel of the Fock space:

$$k^0(z, w) = e^{-i\pi(u\eta-x\xi)-\pi\frac{|z|^2+|w|^2}{2}}e^{\pi\overline{w}z}. \tag{22}$$

Similar calculations can be done for $g = h_{n-1}$ and we obtain

$$k^n(z, w) = e^{-i\pi(u\eta-x\xi)-\pi\frac{|z|^2+|w|^2}{2}}\pi L^0_{n-1}(\pi|z-w|^2)e^{\pi z\overline{w}}$$

for the reproducing kernel of $\mathcal{G}_{h_{n-1}}$. Thus, considering the operator E which maps f to Mf, where

$$M(z) = e^{\pi\frac{|z|^2}{2}-i\pi x\xi},$$

it is clear that E is an isometric isomorphism,

$$E : \mathcal{G}_{h_{n-1}} \to \mathcal{F}^n(\mathbb{C}^d).$$

A similar construction is valid for the vector-valued spaces. See [2] for more details.

3.3 Gabor Expansions with Hermite Functions

To give a more concrete idea of what we are talking about, let us see what Theorem 1 tells about Gabor expansions, more precisely about the required size of the square lattice. From Theorem 1, if $\alpha^2 < \frac{1}{n+1}$, then every $F \in \mathcal{F}_2(\mathbb{C})$ can be written in the form (11). Now, applying the inverse Bargmann transform and doing some calculations involving the intertwining property between the time-frequency shifts and the Fock shifts (see [3] for the details), one can see that this expansion is exactly equivalent to the Gabor expansion of an $L^2(\mathbb{R})$ function. More precisely, if $\alpha^2 < \frac{1}{n+1}$, every $f \in L^2(\mathbb{R})$ admits the following representation as a Gabor series

$$f(t) = \sum_{l,k\in\mathbb{Z}} c_{k,l}e^{2\pi i\alpha lt}h_n(t-\alpha k), \tag{23}$$

with

$$\|c\|_{l^2} \approx \|f\|_{L^2(\mathbb{R})}$$

Stable Gabor expansions of the form (23) can be obtained via frame theory. For a countable subset $\Lambda \in \mathbb{R}^2$, one says that the Gabor system $\mathcal{G}(h_n, \Lambda) = \{\pi_\lambda h_n : \lambda \in \Lambda\}$ is a *Gabor frame* or *Weyl–Heisenberg frame* in $L^2(\mathbb{R})$, whenever there exist constants $A, B > 0$ such that, for all $f \in L^2(\mathbb{R})$,

$$A \|f\|^2_{L^2(\mathbb{R})} \leq \sum_{\lambda \in \Lambda} \left| \langle f, \pi_\lambda h_n \rangle_{L^2(\mathbb{R})} \right|^2 \leq B \|f\|^2_{L^2(\mathbb{R})} . \tag{24}$$

This sort of expansions have been used before for practical purposes, for instance, in image analysis [39]. Their mathematical study [1, 3, 35, 38, 41, 44, 45, 58] used a blend of ideas from signal, harmonic and complex analysis, providing this research field with a nice interdisciplinary flavor.

The problem of finding conditions on the set Λ that yield Gabor frames is known as the density of Gabor frames. See [50] for a survey on the topic and the recent paper [46] for the solution of the problem for a large class of windows.

3.4 Fock Expansions

Observe that, using the Bargmann transform and the intertwining property (17) with $n = 0$, we can write the expansion (23) in the Fock space as follows. Writing $\lambda_{l,k} = \alpha k + i\alpha l$ and using (18) we can expand every $F \in \mathcal{F}_2(\mathbb{C})$ in the form

$$F(z) = \sum_{l,k \in \mathbb{Z}} c_{k,l} \beta_{\lambda_{l,k}} e_n(z),$$

where β_λ is the Bargmann–Fock shift (16) and the system $\{\beta_{\lambda_{l,k}} e_n(z)\}_{l,k \in \mathbb{Z}}$ is *a frame in the Fock space* $\mathcal{F}_2(\mathbb{C})$ of analytic functions. On the other side, if we apply the true polyanalytic Bargmann transform of order $n + 1$ together with (17) with the same $n + 1$, we can use (19) and expand every $F \in \mathcal{F}_2^{n+1}(\mathbb{C})$ in the form

$$F(z) = \sum_{l,k \in \mathbb{Z}} c_{k,l} \beta_{\lambda_{l,k}} e_{n,n}(z) \tag{25}$$

and the system $\{\beta_{\lambda_{l,k}} e_{n,n}(z)\}_{l,k \in \mathbb{Z}}$ is *a frame in the true polyanalytic space* $\mathcal{F}_2^{n+1}(\mathbb{C})$. To see that this is an expansion of a distinguished sort, recall that using (19) and the fact that the reproducing kernel of the Gabor space generated by h_n is given by $\langle \pi_z h_n, \pi_w h_n \rangle_{L^2(\mathbb{R})}$, one can express the reproducing kernel of the true polyanalytic space as

$$e^{-i\pi x\xi+\frac{\pi}{2}|z|^2}\langle \pi_z h_n, \pi_w h_n\rangle_{L^2(\mathbb{R})} = \mathcal{B}^{n+1}(\pi_w h_n)(z) = \beta_w e_{n,n}(z).$$

Thus, (25) is an expansion of reproducing kernels and, applying the reproducing formula, the frame inequality for $\{\beta_{\lambda_{l,k}} e_{n,n}(z)\}_{l,k\in\mathbb{Z}}$,

$$A\,\|F\|^2_{\mathcal{L}_2(\mathbb{C})} \leq \sum_{l,k\in\mathbb{Z}} \left|\left\langle F, \beta_{\lambda_{l,k}} e_{n,n}(z)\right\rangle_{L^2(\mathbb{R})}\right|^2 \leq B\,\|F\|^2_{\mathcal{L}_2(\mathbb{C})}$$

can be written as

$$A\,\|F\|^2_{\mathcal{L}_2(\mathbb{C})} \leq \sum_{l,k\in\mathbb{Z}} |F(\lambda_{l,k})|^2 \leq B\,\|F\|^2_{\mathcal{L}_2(\mathbb{C})}.$$

Thus, the lattice $\{\lambda_{l,k}\}_{l,k\in\mathbb{Z}}$ is a sampling sequence for the true polyanalytic Fock space. Thus, sampling in $\mathcal{F}_2^{n+1}(\mathbb{C})$ is equivalent to (analytic) Fock frames of the form $\{\beta_{\lambda_{l,k}} e_n(z)\}_{l,k\in\mathbb{Z}}$ and both are equivalent to the formulation of Gabor frames with Hermite functions. On the other side, one can construct completely different frames in $\mathcal{F}_2^{n+1}(\mathbb{C})$ by applying $\mathcal{B}^{n+1}$ to the frames $\mathcal{G}(h_m, \Lambda)$ with $m < n$. This means that we have a different representation of the frame $\mathcal{G}(h_m, \Lambda)$ in each space $\mathcal{F}_2^{n+1}(\mathbb{C})$!

An Open Problem

The first proof of the sufficiency of the condition $\alpha^2 < \frac{1}{n+1}$ for the expansion (23) is due to Gröchenig and Lyubarskii [44]. In the same paper, the authors provide some evidence to support the conjecture that the condition may even be sharp (it is known from a general result of Ramanathan and Steger [66] that $\alpha^2 < 1$ is necessary), a statement which would be surprising, since $\alpha^2 < \frac{1}{n+1}$ is exactly the sampling rate necessary and sufficient for the expansion of n functions using the superframe (the superframe [45] is a vectorial version of frame which has be seen to be equivalent to sampling in the polyanalytic space [1]). The following problem seems to be quite hard.

Problem 1 ([44]). Find the exact range of α such that $\mathcal{G}(h_n, \alpha(\mathbb{Z}+i\mathbb{Z}))$ is a frame.

Recently, Lyubarskii and Nes [58] found that $\alpha^2 = \frac{3}{5} > \frac{1}{2}$ is a sufficient condition for the case $n = 1$. They also proved that, if $\alpha^2 = 1 - \frac{1}{j}$, no odd function in the Feichtinger algebra [33] generates a Gabor frame. In [58], supported by their results and by some numerical evidence, the authors formulated a conjecture.

Conjecture 1 ([58]). If $\alpha^2 < 1$ and $\alpha^2 \neq 1 - \frac{1}{j}$, then $\mathcal{G}(h_1, \alpha(\mathbb{Z}+i\mathbb{Z}))$ is a frame.

3.5 Sampling and Interpolation in $\mathbf{F}_2^n(\mathbb{C})$

We say that a set Λ is a set of sampling for $\mathbf{F}_2^n(\mathbb{C})$ if there exist constants $A, B > 0$ such that, for all $F \in \mathbf{F}_2^n(\mathbb{C})$,

$$A \left\| F \right\|^2_{\mathbf{F}_2^n(\mathbb{C})} \leq \sum_{\lambda \in \Lambda} \left| F(\lambda) \right|^2 e^{-\pi|\lambda|^2} \leq B \left\| F \right\|^2_{\mathbf{F}_2^n(\mathbb{C})}.$$

A set Λ is a set of interpolation for $\mathbf{F}_2^n(\mathbb{C})$ if for every sequence $\{a_{i(\lambda)}\} \in l^2$, we can find a function $F \in \mathbf{F}_2^n(\mathbb{C})$ such that

$$e^{i\pi\lambda_1\lambda_2 - \frac{\pi}{2}|\lambda|^2} F(\lambda) = a_{i(\lambda)},$$

for every $\lambda \in \Lambda$. The sampling and interpolation lattices of $\mathbf{F}_2^n(\mathbb{C})$ can be characterized by their density. For the square lattice the results are as follows.

Theorem 2. *The lattice $\alpha(\mathbb{Z} + i\mathbb{Z})$ is a set of sampling for $\mathbf{F}_2^n(\mathbb{C})$ if and only if $\alpha^2 < \frac{1}{n+1}$.*

Theorem 3. *The lattice $\alpha(\mathbb{Z} + i\mathbb{Z})$ is a set of interpolation for $\mathbf{F}_2^n(\mathbb{C})$ if and only if $\alpha^2 > \frac{1}{n+1}$.*

So far, there is no proof of these results using only complex variables. The proof in [1] is based on the observation that the polyanalytic Bargmann transform is an isometric isomorphism

$$\mathbf{B}^n : \mathcal{H} \to \mathbf{F}^n(\mathbb{C}^d).$$

and the sampling problem can be transformed in a problem about Gabor superframes with Hermite functions. Consider the Hilbert space $\mathcal{H} = L^2(\mathbb{R}, \mathbb{C}^n)$ consisting of vector-valued functions $\mathbf{f} = (f_0, \ldots, f_{n-1})$ with the inner product with the inner product

$$\langle \mathbf{f}, \mathbf{g} \rangle_{\mathcal{H}} = \sum_{0 \leq k \leq n-1} \langle f_k, g_k \rangle_{L^2(\mathbb{R}^d)}. \tag{26}$$

The time-frequency shifts π_λ act coordinate-wise in an obvious way. The vector-valued system $\mathcal{G}(\mathbf{g}, \Lambda) = \{\pi_\lambda \mathbf{g}\}_{(x,w)\in\Lambda}$ is a *Gabor superframe* for $\mathcal{H}$ if there exist constants A and B such that, for every $\mathbf{f} \in \mathcal{H}$,

$$A \left\| \mathbf{f} \right\|^2_{\mathcal{H}} \leq \sum_{\lambda \in \Lambda} \left| \langle \mathbf{f}, \pi_\lambda \mathbf{g} \rangle_{\mathcal{H}} \right|^2 \leq B \left\| \mathbf{f} \right\|^2_{\mathcal{H}}. \tag{27}$$

Then the above sampling theorem is equivalent to the following statement about Gabor superframes with Hermite functions:

Theorem 4 ([45]). *Let $h_n = (h_0, \ldots, h_{n-1})$ be the vector of the first n Hermite functions. Then $G(h_n, \alpha(Z+iZ))$ is a frame for $L^2(\mathbb{R}, \mathbb{C}^n)$, if and only if $\alpha^2 < \frac{1}{n+1}$.*

Superframes were introduced in a more abstract form in [49] and in the context of "multiplexing" in [14]. An important result about the structure of Gabor frames, the so-called *Ron–Shen duality* [69] transforms the superframe problem into a problem about multiple Riesz sequences, which can be further transformed in a problem about multiple interpolation in the Fock space (in other words, in a problem of sampling a function of the Fock space using samples of the function and of its derivatives). The solution of the multiple interpolation problem can be obtained as a special case of the results of [20]. The dual of this argument proves the second theorem. The characterization of the lattices yielding Gabor superframes with Hermite functions had been previously obtained by Gröchenig and Lyubarskii in [45], using the Wexler–Rax orthogonality relations and solving the resulting interpolation problem in the Fock space. An approach using group theoretic methods has been considered by Führ [38]. The hard part of the above results is the sufficiency of the condition in Theorem 2. This would follow from the explicit formula in Theorem 1, if a complex variables proof was available (the case $n = 0$ is a simple consequence of the Cauchy formula).

Problem 2. Find a proof of Theorem 1 without using the structure of Gabor frames (in particular, one using only complex variables).

4 The L^p Theory and Modulation Spaces

4.1 Banach Fock Spaces of Polyanalytic Functions

The L^p version of the polyanalytic Bargmann–Fock spaces has been introduced in [3], where the link to Gabor analysis has been particularly useful. For $p \in [1, \infty[$ write $\mathcal{L}_p(\mathbb{C})$ to denote the Banach space of all measurable functions equipped with the norm

$$\|F\|_{\mathcal{L}_p(\mathbb{C})} = \left(\int_{\mathbb{C}} |F(z)|^p \, e^{-\pi p \frac{|z|^2}{2}} \, dz \right)^{1/p}.$$

For $p = \infty$, we have $\|F\|_{\mathcal{L}_\infty(\mathbb{C})} = \sup_{z \in \mathbb{C}} |F(z)| \, e^{-\pi \frac{|z|^2}{2}}$.

Definition 1. We say that a function F belongs to the *polyanalytic Fock space* $\mathbf{F}_p^{n+1}(\mathbb{C})$, if $\|F\|_{\mathcal{L}_p(\mathbb{C})} < \infty$ and F is polyanalytic of order n.

Definition 2. We say that a function F belongs to the *true polyanalytic Fock space* $\mathcal{F}_p^{n+1}(\mathbb{C})$ if $\|F\|_{\mathcal{L}_p(\mathbb{C})} < \infty$ and there exists an entire function H such that

$$F(z) = \left(\frac{\pi^n}{n!} \right)^{\frac{1}{2}} e^{\pi |z|^2} (\partial_z)^n \left[e^{-\pi |z|^2} H(z) \right].$$

Clearly, $\mathcal{F}_p^1(\mathbb{C}) = \mathcal{F}_p(\mathbb{C})$ is the standard Bargmann–Fock space. The space $\mathcal{F}_1^1(\mathbb{C})$ is the Bargmann–Fock image of the Feichtinger algebra.

The orthogonal decomposition (4) extends to the p-norm setting. Similar results appear in [67] for the unit disk case. For $1 < p < \infty$:

$$\mathbf{F}_p^n(\mathbb{C}) = \mathcal{F}_p^1(\mathbb{C}) \oplus \ldots \oplus \mathcal{F}_p^n(\mathbb{C}).$$

$$\mathcal{L}_p(\mathbb{C}) = \bigoplus_{n=1}^{\infty} \mathcal{F}_p^n(\mathbb{C}).$$

This decomposition has been used in [8] as an essential ingredient in the proof of a result relating localization operators to Toeplitz operators.

4.2 Polyanalytic Bargmann Transforms in Modulation Spaces

For the investigation of the mapping properties of the true polyanalytic Bargmann transform $\mathcal{F}_p^n(\mathbb{C})$ we need the concept of *modulation space*. Following [42], the modulation space $M^p(\mathbb{R})$, $1 \leq p \leq \infty$, consists of all tempered distributions f such that $V_{h_0} f \in L^p(\mathbb{R}^2)$ equipped with the norm

$$\|f\|_{M^p(\mathbb{R})} = \|V_{h_0} f\|_{L^p(\mathbb{R}^2)} .$$

Modulation spaces were introduced in [32]. The case $p = 1$ is the Feichtinger Algebra, which has been mentioned previously. It would probably be hard to prove the following statement directly, but the Modulation space theory provides a simple proof.

Theorem 5. *Given $F \in \mathcal{F}_p^n(\mathbb{C})$ there exists $f \in M^p(\mathbb{R})$ such that $F = \mathcal{B}^n f$. Moreover, there exist constants C, D such that:*

$$C \|F\|_{\mathcal{L}_p(\mathbb{C})} \leq \|\mathcal{B} f\|_{\mathcal{L}_p(\mathbb{C})} \leq D \|F\|_{\mathcal{L}_p(\mathbb{C})} . \quad (28)$$

The key observation leading to the proof of this result is the following: since the definition of Modulation space is independent of the particular window chosen [42, Proposition 11.3.1], then the norms

$$\|f\|'_{M^p(\mathbb{R}^2)} = \|V_{h_n} f\|_{L^p(\mathbb{R}^2)}$$

and

$$\|f\|_{M^p(\mathbb{R}^2)} = \|V_{h_0} f\|_{L^p(\mathbb{R}^2)} ,$$

are equivalent. Then, using the relations between the true polyanalytic Bargmann transform and the Gabor transform with Hermite functions (19) provides a norm equivalence which can be transferred to the whole $\mathcal{F}_p^n(\mathbb{C})$ due to the mapping properties of the true polyanalytic Bargmann transform [3].

The properties of the polyanalytic projection are kept intact. Indeed, we have the following result.

Proposition 1. *The operator P^n is bounded from $\mathcal{L}_p(\mathbb{C})$ to $\mathcal{F}_p^{n+1}$ for $1 \leq p \leq \infty$. Moreover, if $F \in \mathcal{F}_p^{n+1}$ then $P^n F = F$.*

Combining the above L^p theory of the polyanalytic Bargmann–Fock spaces, a result from localization [43] and coorbit theory [34], with the estimates from [44], provides a far reaching generalization of Corollary 1.

Theorem 6. *Assume that $\Lambda \subseteq \mathbb{R}^2$ is a lattice and $\alpha^2 < \frac{1}{n+1}$*

(i) Then F belongs to the true poly-Fock space $\mathcal{F}_p^{n+1}$, if and only if the sequence with entries $e^{-\pi|\lambda|^2/2}F(\lambda)$ belongs to $\ell^p(\Lambda)$, with the norm equivalence

$$\|F\|_{\mathcal{F}_p^{n+1}} \asymp \left(\sum_{\lambda\in\Lambda} |F(\lambda)|^p e^{-\pi p|\lambda|^2/2}\right)^{1/p} .$$

(ii) Let $S^n_{\Lambda^0}(z)$ be the polyanalytic Weierstrass function on the adjoint lattice Λ^0. Then every $F \in \mathcal{F}_p^{n+1}(\mathbb{C})$ can be written as

$$F(z) = \sum_{\lambda\in\Lambda} F(\lambda)e^{\pi\overline{\lambda}z-\pi|\lambda|^2} S^n_{\Lambda^0}(z-\lambda) . \tag{29}$$

The sampling expansion converges in the norm of $\mathcal{F}_p^{n+1}$ for $1 \leq p < \infty$ and pointwise for $p = \infty$.

5 The Landau Levels and Displaced Fock States

In addition to time-frequency analysis, polyanalytic Fock spaces also appear in several topics in quantum physics. We will now describe how the polyanalytic structure shows up in the Landau levels associated with a single particle within an Euclidean plane in the presence of an uniform magnetic field perpendicular to the plane and also a connection to displaced Fock states.

5.1 The Euclidean Laplacian with a Magnetic Field

Consider a single charged particle moving on a complex plane with an uniform magnetic field perpendicular to the plane. Its motion is described by the Schrödinger operator

$$H_B = -\frac{1}{4}\left((\partial_x + iBy)^2 + \left(\partial_y - iBx\right)^2\right) - \frac{1}{2}$$

acting on $L^2(\mathbb{C})$. Here $B > 0$ is the strength of the magnetic field. Writing

$$\widetilde{\Delta_z} = e^{\frac{B}{2}|z|^2} H_B e^{-\frac{B}{2}|z|^2}$$

we obtain the following Laplacian on $\mathbb{C}$

$$\widetilde{\Delta_z} = -\partial_z\partial_{\overline{z}} + B\overline{z}\partial_{\overline{z}}. \tag{30}$$

This Laplacian is a positive and self-adjoint operator in the Hilbert space $\mathcal{L}_2(\mathbb{C})$ and the set $\{n, n \in \mathbb{Z}^+\}$ can be shown to be the pure point spectrum of $\widetilde{\Delta_z}$ in $\mathcal{L}_2(\mathbb{C})$. There are other Laplacians in the literature related to this one [26, 27, 74]. The eigenspaces of $\widetilde{\Delta_z}$ are known as the *Landau levels*. In [13] the authors consider

$$A^2_{n,B}(\mathbb{C}) = \{F \in \mathcal{L}_2(\mathbb{C}) : \widetilde{\Delta_{z,B}} F = nf\},$$

and obtain an orthogonal basis for the spaces $A^2_{n,B}$. When $B = \pi$ we can use the results in [13] (comparing either the orthogonal basis or the reproducing kernels of both spaces) to see that

$$A^2_{n,\pi}(\mathbb{C}) = \mathcal{F}^n_2(\mathbb{C}). \tag{31}$$

5.2 *Displaced Fock States*

This section summarizes work to be developed further in [7]. Now, we can define a set of coherent states $\mid z >_n$ for each Landau level n. This can be done by displacing via the representation β_w the vector $\mid 0 >_n$ of $\mathcal{F}^n_2(\mathbb{C})$ with the following wavefunction

$$\langle z \mid 0 \rangle_n = e_{n,n}(z).$$

Precisely,

$$|w >_n = \beta_w |0 >_n$$

and the wavefunction is given by

$$\langle z \mid w \rangle_n = \beta_w e_{n,n}(z).$$

We will call this the *true polyanalytic representation* of the Landau level coherent states. Now, observing that

$$e^{\pi|z|^2}\partial_z\left[e^{-\pi|z|^2}F(z)\right] = \partial_z F(z) - \pi\overline{z}F(z),$$

we conclude from the unitarity of the true polyanalytic Bargmann transform that the operator

$$(\partial_z - \pi\overline{z}) : \mathcal{F}_2^n(\mathbb{C}) \to \mathcal{F}_2^{n+1}(\mathbb{C})$$

is unitary and that

$$e_{n,n}(z) = (\partial_z - \pi\overline{z})^n\, e_n(z).$$

Now, combining this identity with the intertwining property (17) gives

$$\begin{aligned}
\beta_w e_{n,n}(z) &= \beta_w \mathcal{B}^{n+1}(h_n)(z)\\
&= \mathcal{B}^{n+1}(\pi_w h_n)(z)\\
&= (\partial_z - \pi\overline{z})^n\, \mathcal{B}^n(\pi_w h_n)(z)\\
&= (\partial_z - \pi\overline{z})^n\, \beta_w e_n(z).
\end{aligned}$$

Since the operator $(\partial_z - \pi\overline{z})^n$ is unitary $\mathcal{F}_2(\mathbb{C}) \to \mathcal{F}_2^{n+1}(\mathbb{C})$ we have the following equivalent representation of the Landau level coherent states in the *analytic* Fock space

$$\langle z \mid w\rangle_n = \beta_w e_n(z).$$

A similar representation has been obtained by Wünsche [78], using quite different methods. These coherent states are called displaced Fock states, since they are obtained by displacing by a Fock shift an already excited Fock state. A natural question concerns the completeness properties of these coherent states. More precisely, given a lattice Λ what are the complete discrete subsystems $\{|\lambda >_n, \lambda \in \Lambda\}$ of this system of coherent states? In order to do so, we go back to the section about "Fock frames" where the completeness and basis properties of the above have been shown to be equivalent to those of Gabor frames with Hermite functions and to the results about sampling in the true polyanalytic spaces. Using the true polyanalytic transform, the results about Gabor frames with Hermite function translate to sampling in true polyanalytic Fock spaces as follows.

Proposition 2. *The lattice $\alpha(\mathbb{Z} + i\mathbb{Z})$ is a set of sampling for $\mathcal{F}_2^n(\mathbb{C})$ if and only $\mathcal{G}(h_n, \alpha(\mathbb{Z}+i\mathbb{Z}))$ is a Gabor frame.*

Thus, we conclude that, in particular, if $\alpha^2 < \frac{1}{n+1}$, the subsystems of states constituted by the lattice $\alpha(\mathbb{Z}+i\mathbb{Z})$ are complete in the Landau levels. Now, take $B=1$ and observe that

$$\widetilde{\Delta_z} = (-\partial_z + \overline{z})\,(\partial_{\overline{z}})\,.$$

This suggests us to consider the operators

$$\begin{aligned} \mathfrak{a}^+ &= -\partial_z + \overline{z} \\ \mathfrak{a}^- &= \partial_{\overline{z}}, \end{aligned}$$

which are formally adjoint to each other and satisfy the commutation relations for the quantum mechanic creation and annihilation operators. Vasilevski [76, Theorem 2.9] proved that the operators

$$\begin{aligned} &\alpha_{l-k}\left(\mathfrak{a}^+\right)^{l-k} |_{\mathcal{F}_2^k(\mathbb{C})} : \mathcal{F}_2^k(\mathbb{C}) \to \mathcal{F}_2^l(\mathbb{C}) \\ &\alpha_{l-k}(\mathfrak{a}^-)^{l-k} |_{\mathcal{F}_2^k(\mathbb{C})} : \mathcal{F}_2^l(\mathbb{C}) \to \mathcal{F}_2^k(\mathbb{C}), \end{aligned}$$

with $\alpha_{l-k} = \sqrt{(k-1)!(l-1)!}$are Hilbert spaces isomorphisms (and one is the inverse of the other). Given our identification (31) we conclude that *the operators a^+ and a^- are, respectively, the raising and lowering operators between two different Landau levels.* For other quantum physics applications of Gabor transforms which include as special cases polyanalytic Fock spaces see [19]. In [78], Wünsche derived the following representation for the displaced Fock states $|z,n>$:

$$|z,n> = \frac{(-1)^n}{\sqrt{n!}}\,(-\partial_z + \overline{z})^n\,|z>\,. \tag{32}$$

In view of our remarks in this section, one realizes that (32) is essentially the map $T : \mathcal{F}_2(\mathbb{C}) \to \mathcal{F}_2^{n+1}(\mathbb{C})$ such that

$$T : F(z) \to e^{\pi|z|^2}\,(\partial_z)^n \left[e^{-\pi|z|^2} F(z)\right].$$

and the displaced Fock states are also true polyanalytic Fock spaces. We can now use Gröchenig and Lyubarskii result to show that if $\alpha^2 < \frac{1}{n+1}$ then the subsystem of these coherent states constituted by the square lattice on the plane is overcomplete. From Ramathan and Steeger general result [66], we know that if $\alpha^2 > 1$ they are not. This can be seen as analogues of Perelomov completeness result [64] in the setting of displaced Fock states.

6 Polyanalytic Ginibre Ensembles

The polyanalytic Ginibre ensemble has been introduced by Haimi and Hendenmalm [47]. It has a physical motivation based on the Landau level interpretation described in the previous section. For the kth Landau level, consider the wave functions of the form

$$\psi_{k,j}(z) = e_{k,j}(z)e^{-\pi|z|^2},$$

where

$$e_{k,j}(z) = \left(\frac{\pi^k}{k!}\right)^{\frac{1}{2}} e^{\pi|z|^2} (\partial_z)^k \left[e^{-\pi|z|^2} e_j(z)\right], \quad j \geq 0.$$

Thus, the wave function of a system consisting of the first n Landau levels with N non-interacting fermions at each Landau level k, with wavefunctions $\psi_{k,j}$ is given by the determinant

$$\Psi_{n,N} = \det[\psi_{k,j}(z_{s,i})]_{nN\times nN}, \quad 1 \leq i,j \leq N, \quad 1 \leq k,s \leq n.$$

This can be rewritten as the probability density of a determinantal point process

$$\Psi_{n,N} = \frac{1}{(nN)!}\det[K_{n,N}(z_{s,i}, z_{k,j})]_{nN\times nN}, \quad 1 \leq i,j \leq N, \quad 1 \leq k,s \leq n,$$

whose correlation kernel is given as

$$K_{n,N}(z,w) = \sum_{k=1}^{n}\sum_{j=1}^{N} \psi_{k,j}(z)\overline{\psi_{k,j}(w)}.$$

To have a glimpse of what we are talking about, we first give a brief description of what is a determinantal point process and then give more details about the polyanalytic Ginibre ensemble.

6.1 Determinantal Point Processes

Consider an infinite dimensional Hilbert space H with continuous functions having their values in $X \subseteq \mathbb{C}$, with a reproducing kernel $K(z,w)$, that is, for every $f \in H$,

$$f(z) = \langle f, K(z,.)\rangle_H \,.$$

Suppose that $\{\varphi_j\}$ is a basis of H. Define the *N-dimensional kernel*

$$K^N(z,w) = \sum_{j=0}^{N} \varphi_j(z)\overline{\varphi_j(w)}. \tag{33}$$

Using N-dimensional kernels one can describe *determinantal point processes* [18] for N points $(z_1,\ldots,z_N) \in X^N$ using their n-point intensities

$$dP^N(z_1,\ldots,z_n) = \frac{1}{n!} \det\left[K^N(z_i,z_j)\right]_{i,j=1}^{n} d\mu(z_1)\ldots d\mu(z_n).$$

We will be more concerned with the 1-point intensity

$$dP^N(z) = K^N(z,z)d\mu(z),$$

which allows us to evaluate the expected number of points to be found in a certain region if they are distributed according to the determinantal point process.

Using the reproducing kernels of the polyanalytic and true polyanalytic Fock spaces, one can define interesting determinantal point processes which are generalizations of the Ginibre ensemble, a determinantal point process in $\mathbb{C}$ related to the reproducing kernel of a Gabor space associated with a Gaussian window h_0:

$$K_{h_0}(z,w) = e^{-i\pi(u\eta - x\xi) - \pi\frac{|z|^2+|w|^2}{2}} e^{\pi\overline{w}z}.$$

The corresponding polynomial kernel (33) is

$$K_{h_0}^N(z,w) = e^{-\pi\frac{|z|^2+|w|^2}{2}} \sum_{j=0}^{N} \frac{(\pi\overline{w}z)^j}{j!}.$$

This is the kernel of the *Ginibre ensemble* [61, Chapter 15].

6.2 *The Polyanalytic Ginibre Ensemble*

In [47], a variant of this setting is used in the investigation of the *polyanalytic Ginibre ensemble.* The authors consider the space with reproducing kernel

$$\mathbf{K}_m^n(z,w) = mL_{n-1}^1(m\,|z-w|^2)e^{mz\overline{w}}$$

and the polynomial space

$$Pol_{m,n,N} = span\{z^j\overline{z}^l : 0 \leq j \leq N-1, 0 \leq l \leq n-1\}.$$

Using the polyanalytic Hermite polynomials $e_{j,l}(z,\overline{z})$ as defined in (9), this is equivalent to writing

$$Pol_{m,n,N} = span\{e_{j,l}(z,\overline{z}) : 0 \leq j \leq N-1, 0 \leq l \leq n-1\}.$$

There is, however, a way of writing down an orthogonal basis in terms of Laguerre polynomials which is more practical for the study of the asymptotics of determinantal point process. In [47] it is shown that, for $n \leq N$, the following functions form an orthogonal basis for $Pol_{m,n,N}$:

$$e^1_{i,r}(z) = \sqrt{\frac{r!}{(r+i)!}} m^{\frac{i+1}{2}} z^i L^i_r(m|z|^2), \quad 0 \leq i \leq N-r-1, \ 0 \leq r \leq n-1,$$

$$e^2_{j,k}(z) = \sqrt{\frac{j!}{(j+k)!}} m^{\frac{k+1}{2}} \overline{z}^k L^k_j(m|z|^2), \quad 0 \leq j \leq n-k-1, \ 1 \leq k \leq n-1.$$

The reproducing kernel of $Pol_{m,n,N}$ can thus be written as

$$\mathbf{K}^n_{m,N}(z,w) = m \sum_{r=0}^{n-1} \sum_{i=0}^{N-r-1} \frac{r!}{(r+i)!} (mz\overline{w})^i L^i_r(m|z|^2) L^i_r(m|w|^2)$$
$$+ m \sum_{j=0}^{n-2} \sum_{k=1}^{N-j-1} \frac{j!}{(j+k)!} (m\overline{z}w)^k L^k_j(m|z|^2) L^k_j(m|w|^2).$$

Remark 2. In his book [65, p. 35], Perelomov points out that this formula, seen as the explicit expression for the matrix elements of the displacement operator had been used by Feynman and Schwinger in a somewhat different form. This makes a case for naming the above polynomials the *Feynman–Schwinger basis* for the polyanalytic Fock space.

Let us illustrate some of the obtained asymptotic results. Denoting the reproducing kernel of $Pol_{m,n,N}$ by $\mathbf{K}^n_{m,N}(z,w)$, it is proved that, if $z, w \in \mathbb{D}$, when $m, N \to \infty$ with $|m-N|$ bounded and $1-|zw| \geq \tau > 0$, then

$$\mathbf{K}^n_{m,N}(z,w) = \mathbf{K}^n_m(z,w) + O(e^{-\frac{1}{2}m\tau^2} e^{m|zw|}).$$

Another result is the blow-up of the 1-point intensity function of the determinantal point process associated with $\mathbf{K}^n_{m,N}(z,w)$. The one point intensity function is $e^{-m|z|^2}\mathbf{K}^n_{m,N}(z,z)$ and its localized version with $z = 1 + m^{-\frac{1}{2}}\xi$ is

$$\mathcal{U}_{m,N,n}(\xi) = \frac{1}{m} e^{-m\left|1+m^{-\frac{1}{2}}\xi\right|^2} \mathbf{K}^n_{m,N}(1+m^{-\frac{1}{2}}\xi, 1+m^{-\frac{1}{2}}\xi)$$

In [47, p. 29] the authors observe that, considered the Hermite polynomials $H_j(t)$ normalized such that

$$e^{tz-\frac{1}{2}z^2}=\sum_{j=0}^{\infty} H_j(t)\frac{z^j}{j!},$$

this can be written as

$$\mathcal{U}_{m,N,n}(\xi) = \sum_{j=0}^{n-1} \frac{1}{r!\sqrt{2\pi}} \int_{-\infty}^{-2\mathrm{Re}\xi} H_j(t)e^{-\frac{1}{2}t^2}dt + O(m^{-\frac{1}{2}+\epsilon})$$

Thus, when $m, N \to \infty$ with $|m - N|$ bounded, $\mathcal{U}_{m,N,n}(\xi)$ is essentially determined by the density

$$\rho(t) = \sum_{j=0}^{n-1} \frac{1}{j!\sqrt{2\pi}} H_j(t)^2 e^{-\frac{1}{2}t^2},$$

which is the one point intensity in the Gaussian Unitary Ensemble. Thus, if we make the polyanaliticity degree increase, $\rho(t)$ approaches the *Wigner semicircle law* (see [61]) and

$$\mathcal{U}_{m,N,n}(\xi) \approx \frac{2n}{\pi} \int_{-1}^{-n^{-\frac{1}{2}}\mathrm{Re}\xi} \sqrt{1-\tau^2}d\tau.$$

Something similar happens to the Berezin measure

$$dB^{(z)}_{m,n,N}(w) = \frac{\left|\mathbf{K}^n_{m,N}(z,w)\right|^2}{\mathbf{K}^n_{m,N}(z,z)} e^{-m|z|^2} dz;$$

while the asymptotics of $dB^{(z)}_{m,n,N}(w)$ are similar to the case $n = 1$, defining the blow-up Berezin density at 1 by $\hat{B}^{(1)}_{m,n,N}(w) = m^{-1}B^{(1)}_{m,n,N}(1 + m^{-1/2}w)$, we have, as $m, N \to +\infty$ with $N = m + O(1)$, the following asymptotics, with uniform control on compact sets:

$$\hat{B}^{(1)}_{m,n,N}(w) = \frac{1}{\pi n} \left| \sum_{j=0}^{n-1} \frac{1}{j!} \int_{-\infty}^{-w} H_j(t+w)H_j(t-\overline{w})e^{-\frac{1}{2}t^2}dt \right|^2 + O(m^{-\frac{1}{2}+\epsilon}).$$

Remark 3. Comparing this setup with Sect. 4.1, one recognizes the parameter m as the strength of the magnetic field B. Therefore, the physical interpretation of the above limit $m, N \to \infty$ consists of increasing the strength of the magnetic field and simultaneously the number of independent states in the system. In the analytic case this has been done for more general weights [11].

Remark 4. A version of the polyanalytic Ginibre ensemble allowing for more general weights in the corresponding Fock space has been recently considered in [48], providing a polyanalytic setting similar to the one considered in [11] for analytic functions.

7 Hyperbolic Analogues: Wavelets and Bergman Spaces

Most of what we have seen about polyanalytic Fock spaces has an analogue in the hyperbolic setting. However, the hyperbolic setting presents several difficulties and the topic is far from being understood. We will outline some of what is known and what one would expect applying time-frequency intuition to this setting. Some facts contained on the material which is currently under investigation in [6] will be included.

7.1 Wavelets and Laguerre Functions

For every $x \in \mathbb{R}$ and $s \in \mathbb{R}^+$, let $z = x + is \in \mathbb{C}^+$ and define

$$\pi_z g(t) = s^{-\frac{1}{2}} g(s^{-1}(t-x)).$$

We will define the wavelet transform on Hardy spaces $H^2(\mathbb{C}^+)$. Since we will always work on the frequency side, we can resort to the Paley–Wiener theorem and work in $L^2(\mathbb{R}^+)$ in a way that we only need to compute real integrals. Fix a function $g \neq 0$. Then the continuous wavelet transform of a function f with respect to a wavelet g is defined, for every $z = x + is$, $x \in \mathbb{R}$, $s > 0$, as

$$W_g f(z) = \langle f, \pi_z g \rangle_{H^2(\mathbb{C}^+)} . \tag{34}$$

We will use *the orthogonal relations for the wavelet transform:*

$$\langle W_{g_1} f_1, W_{g_2} f_2 \rangle_{\mathcal{L}^2_2(\mathbb{C}^+)} = \langle \mathcal{F}g_1, \mathcal{F}g_2 \rangle_{L^2(\mathbb{R}^+, t^{-1})} \langle f_1, f_2 \rangle_{H^2(\mathbb{C}^+)} , \tag{35}$$

valid for all $f_1, f_2 \in H^2(\mathbb{C}^+)$ and for functions $g_1, g_2 \in H^2(\mathbb{C}^+)$ *admissible*, meaning that

$$\|\mathcal{F}g\|^2_{L^2(\mathbb{R}^+, t^{-1})} = c_g^2,$$

where c_g is a constant. For a vector $\mathbf{g} = (g_1, \ldots, g_n)$ such that the Fourier transforms of any two functions g_i and g_j are orthogonal in $L^2(\mathbb{R}^+, t^{-1})$, define π_z pointwisely as

$$\pi_z \mathbf{g} = (\pi_z g_1, \ldots, \pi_z g_n).$$

Let $\mathcal{H} = H^2(\mathbb{C}^+, \mathbb{C}^n)$ be the inner product space whose vector components belong to $H^2(\mathbb{C}^+)$, the standard Hardy space of the upper half-plane, equipped with the natural inner product

$$\langle \mathbf{f}, \mathbf{g} \rangle_{\mathcal{H}} = f_1 g_1 + \ldots + f_n g_n.$$

We say that the vector-valued system $\mathcal{W}(\mathbf{g}, \Lambda)$ is a *wavelet superframe* for $\mathcal{H}$ if there exist constants A and B such that, for every $\mathbf{f} \in \mathcal{H}$,

$$A \, \|\mathbf{f}\|_{\mathcal{H}}^2 \leq \sum_{\lambda \in \Lambda} |\langle \mathbf{f}, \pi_\lambda \mathbf{g} \rangle_{\mathcal{H}}|^2 \leq B \, \|\mathbf{f}\|_{\mathcal{H}}^2 . \tag{36}$$

The orthogonality conditions imposed on the entries of the vector $\mathbf{g}$ allow us to recover the original definitions of superframes [14]. Indeed, it serves the same original motivation for the introduction of superframes in signal analysis: a tool for the *multiplexing of signals.* We consider wavelet superframes with analyzing wavelets $\left(\frac{\Phi_0^\alpha}{c_{\Phi_0^\alpha}}, \ldots, \frac{\Phi_n^\alpha}{c_{\Phi_n^\alpha}} \right)$, where $c_{\Phi_n^\alpha}^2 = \frac{\Gamma(n+\alpha+1)}{n!}$ is the admissibility constant of the vector component Φ_n^α defined via the Fourier transforms as

$$\mathcal{F}\Phi_n^\alpha(t) = t^{\frac{1}{2}} l_n^\alpha(2t), \quad \text{with} \quad l_n^\alpha(t) = t^{\frac{\alpha}{2}} e^{-\frac{t}{2}} \sum_{k=0}^{n} (-1)^k \binom{n+\alpha}{n-k} \frac{t^k}{k!}. \tag{37}$$

The functions Φ_n^α above have been chosen as the substitutes of the Hermite functions used by Gröchenig and Lyubarskii to construct Gabor superframes [45]. As we will see, our choice is well justified by physical, operator theory, and function theory arguments. We start with the remark that the functions Φ_n^α provide an orthogonal basis for all the $g \in H^2\left(\mathbb{C}^+\right)$ satisfying the admissibility condition

$$\|\mathcal{F}g\|_{L^2(\mathbb{R}^+, t^{-1})}^2 < \infty,$$

and we note in passing that, according to the results in [30], such admissible functions constitute a Bergman space.

7.2 Wavelet Frames with Laguerre Functions

One of the fundamental questions about wavelet frames is, given a wavelet g, to characterize the sets of points Λ such that $\mathcal{W}(g, \Lambda)$ is a wavelet frame (and the corresponding problem for the superframes defined above). This problem is much more difficult than the corresponding one for Gabor frames. The only characterization known so far concerns the case $n = 0$ in (37). In this case, the problem can be reduced to the density of sampling in the Bergman spaces, which has been completely understood by Seip in [72]. An important research problem is to understand how Seip's results extend to the whole family $\{\Phi_n^\alpha\}$. The only thing known to the present date is a necessary condition obtained in [4] in terms of a convenient set of points for discretization known as the "hyperbolic lattice" $\Gamma(a,b) = \{a^m bk, a^m\}_{k,m\in\mathbb{Z}}$:

Theorem 7. *If $\mathcal{W}(\Phi_n^{2\alpha-1}, \Gamma(a,b))$ is a wavelet frame for $H^2(\mathbb{C}^+)$, then*

$$b \log a < 2\pi \frac{n+1}{\alpha}.$$

It seems that the above paragraph completely describes the state of the art in the topic, as it relates to special windows. Beyond [72] and [4] nothing seems to be known. For other directions of research on wavelet frames, see [57].

One of the limitations of wavelet theory is the absence of duality theorems like those used in [45] and [1] in order to obtain precise conditions on the lattice generating Gabor superframes with Hermite functions. For this reason, there are no known analogue results for Wavelet superframes with Laguerre functions. Still, a lot can be said. The existence of wavelet frames with windows $\mathbf{\Phi}_n^\alpha$ follows from coorbit theory along the lines of [36], where the case $n = 0$ is explained in detail. Using the time-frequency intuition provided by the results in [45] (see Sect. 3 of this survey), one would expect the following to be true.

Conjecture 2. $\mathcal{W}(\mathbf{\Phi}_n^{2\alpha-1}, \Gamma(a,b))$ is a wavelet superframe for $\mathcal{H}$ if and only if

$$b \log a < \frac{2\pi}{n+\alpha}. \tag{38}$$

The case $n = 0$ follows from the results in [72]. This conjecture is well beyond the existing tools, since the duality between Riesz basis and frames has no known extension to wavelet frames. See the chapter about wavelets in [42]. We have reasonable expectations on the possibility that the connection to polyanalytic Bergman spaces to be described in the next sections may be put in good use for the investigation of this conjecture.

7.3 *The Hyperbolic Landau Levels*

The most notable fact about the wavelet transforms associated with Φ_n^α is that they provide a phase space representation for the higher hyperbolic Landau levels with a constant magnetic field introduced in Physics by Comtet ([22], see also [60]).

We will now briefly sketch this connection. Let H_B denote the Landau Hamiltonian of a charged particle with a uniform magnetic field on $\mathbb{C}^+$, with magnetic length proportional to $|B|$,

$$H_B = s^2 \left(\frac{\partial^2}{\partial x^2} + \frac{\partial^2}{\partial s^2} \right) - 2iBs \frac{\partial}{\partial x}.$$

It is well known that, if $|B| > 1/2$, the spectrum of the operator H_B consists of both a continuous and a *finite* discrete part. The choice of the functions Φ_n^α as analyzing wavelets yields the eigenspaces associated with the discrete part. Denote by $E_{B,m}(\mathbb{C}^+)$ the subspace of $L^2(\mathbb{C}^+)$ defined as

$$E_{B,n} = \{f : L^2(\mathbb{C}^+) : H_B\psi = e_n\psi\},$$

with $e_n = (|B| - n)(|B| - n - 1)$. Mouayn has introduced a system of coherent states [62] for the operator H_B. Once we overcome the differences in the notation and parameters, we can identify Mouayn's coherent states [62] with our wavelets and conclude that

$$E_{B,n} = W_{\Phi_n^{2(B-n)-1}} \left(H^2 \left(\mathbb{C}^+ \right) \right).$$

Thus, wavelet frames with the functions Φ_n^α are discrete subsystems of coherent states attached to the hyperbolic Landau levels. The spectral analysis of the operator H_B is also important in number theory, since the solutions of H_B satisfying an automorphy condition are half-integral weight Maass forms, [31, 68]. Maass forms have become prominent in modern number theory in part thanks to the striking relations to Ramanujan's Mock theta functions (see [21]) As we will see below, the spaces $E_{B,n}$ are, up to a multiplier isomorphism, *true polyanalytic Bergman spaces.*

7.4 Bergman Spaces of Polyanalytic Functions

The Hardy space $H^2(\mathbb{C}^+)$ is constituted by the analytic functions on the upper half-plane such that

$$\sup_{0<s<\infty} \int_{-\infty}^{\infty} |f(z)|^2 \, dx < \infty.$$

Let $\mathcal{L}_\alpha^2 \left(\mathbb{C}^+ \right)$ be the space of square-integrable functions in $\mathbb{C}^+$ with respect to $d\mu_\alpha^+(z) = s^\alpha d\mu^+(z)$, where $d\mu^+(z)$ is the standard area measure in $\mathbb{C}^+$. The *weighted analytic Bergman space*, $A_\alpha(\mathbb{C}^+)$, is constituted by $\mathcal{L}_\alpha^2 \left(\mathbb{C}^+ \right)$ functions analytic in $\mathbb{C}^+$. We will also require the space $\mathbf{A}_\alpha^n(\mathbb{C}^+)$, which is the *polyanalytic Bergman space* consisting of all functions in $\mathcal{L}_\alpha^2 \left(\mathbb{C}^+ \right)$ which can be written in the form

$$F(z) = \sum_{p=0}^{n-1} \overline{z}^p F_p(z), \tag{39}$$

with $F_0(z), \ldots, F_{n-1}(z)$ analytic on $\mathbb{C}^+$. There is also a decomposition in true polyanalytic spaces $\mathcal{A}_\alpha^k(\mathbb{C}^+)$, also due to Vasilevski (see the original paper in [75] and the book [77]):

$$\mathcal{L}_\alpha^2(\mathbb{C}^+) = \bigoplus_{n=0}^{\infty} \mathcal{A}_\alpha^n(\mathbb{C}^+).$$

The true polyanalytic spaces $\mathcal{A}_\alpha^k(\mathbb{C}^+)$ can be defined as

$$\mathcal{A}_\alpha^k(\mathbb{C}^+) = \mathbf{A}_\alpha^k(\mathbb{C}^+) \ominus \mathbf{A}_\alpha^{k-1}(\mathbb{C}^+)$$

such that

$$\mathbf{A}_\alpha^n(\mathbb{C}^+) = \mathcal{A}_\alpha^1(\mathbb{C}^+) \oplus \ldots \oplus \mathcal{A}_\alpha^n(\mathbb{C}^+).$$

The *Bergman transform* of order α is the wavelet transform with a Poisson wavelet times a weight:

$$Ber_\alpha\, f(z) = s^{-\frac{\alpha}{2}-1} W_{\overline{\Phi_0^\alpha}} f(z) = \int_0^\infty t^{\frac{\alpha+1}{2}} \mathcal{F} f(t) e^{izt} dt. \tag{40}$$

It is an isomorphism $Ber_\alpha : H^2(\mathbb{C}^+) \to A_\alpha(\mathbb{C}^+)$ and plays the role of the Bargmann transform in the hyperbolic case. The *true* polyanalytic Bergman transform for $f \in H^2(\mathbb{C}^+)$ is given by the formula

$$Ber_\alpha^{n+1} f = \frac{(2i)^n}{n!} s^{-\alpha} (\partial_z)^n \left[s^{\alpha+n} (Ber_\alpha f)(z) \right]. \tag{41}$$

It is an isomorphism $Ber_\alpha : H^2(\mathbb{C}^+) \to A_\alpha(\mathbb{C}^+)$. This has been proved in [5] for the case $\alpha = 1$ relying on operator decompositions from [75] and [52]. It can be shown for general α by an argument involving special functions. A basis of $\mathrm{A}_\alpha^n(\mathbb{C}^+)$ is provided by the rational functions

$$\Psi_{n,k}^\alpha(z) = s^{-\alpha} \left(\frac{d}{dz} \right)^n \left[s^{\alpha+n} \Psi_k^\alpha(z) \right], \quad z \in \mathbb{C}^+,$$

where Ψ_k^α is the well-known orthogonal basis of $\mathcal{A}_\alpha(\mathbb{C}^+)$ defined as

$$\Psi_k^\alpha(z) = \frac{(2i)^k}{k!} \left(\frac{\Gamma(n+2+\alpha)}{k!\Gamma(2+\alpha)} \right)^{\frac{1}{2}} \left(\frac{z-i}{z+i} \right)^k \left(\frac{1}{z+i} \right)^{\alpha+1}, \qquad z \in \mathbb{C}^+.$$

Then, we have the connection to wavelets provided by the formula

$$Ber_{\alpha}^{n+1} f(z) = s^{-\frac{\alpha}{2}-1} W_{\Phi_n^{\alpha}} f(z).$$

Thus, we can identify the true polyanalytic Bergman spaces as the eigenspaces associated to the hyperbolic Landau levels. A thought experience concerning multiplexing of signals in this context leads to similar constructions to those we found in the Fock/Gabor case. Rephrasing the result about wavelet frames in terms of sampling sequences for polyanalytic Bergman spaces yields the following consequence of Theorem 6:

Corollary 1. *If the sequence* $\{a^m bk + a^m i\}$ *is a sampling sequence for* $\mathrm{A}_{2\alpha-1}^{n+1}(\mathbb{C}^+)$ *then* $b \log a < 2\pi \frac{n+1}{\alpha}$.

An in equivalence to the Conjecture 2, we have the following.

Conjecture 3. The sequence $\{a^m bk + a^m i\}$ is a sampling sequence for $\mathbf{A}_{2\alpha-1}^{n+1}(\mathbb{C}^+)$ if and only if $b \log a < \frac{2\pi}{n+\alpha}$.

Much of the results about polyanalytic Fock spaces are likely to find analogues in the hyperbolic setting, but the methods of proof can be quite different. The topic is currently under investigation.

7.5 An Orthogonal Decomposition of $\mathcal{L}_{\alpha}^{2}(\mathbb{D})$

A standard Cayley transform provides and isomorphism between $\mathcal{L}_{\alpha}^{2}(\mathbb{C}^+)$ and $\mathcal{L}_{\alpha}^{2}(\mathbb{D})$. This motivates the study of certain polyanalytic spaces in the unit disc which, although not having a direct connection to wavelets, have been recently considered for their intrinsic mathematical content [23, 67]. Let $\mathcal{L}_{\alpha}^{2}(\mathbb{D})$ be the space of square-integrable functions in the unit disc, with respect to $dA_{\alpha}(w) = (1-|w|)^{\alpha} dA(w)$, where $dA(w)$ is the standard area measure in the unit disc. Denote by $\mathbf{A}_{\alpha}^{n}(\mathbb{D})$ the *polyanalytic Bergman space* consisting of all functions in $\mathcal{L}_{\alpha}^{2}(\mathbb{D})$ satisfying the higher order Cauchy–Riemann equation. The spaces $\mathbf{A}_{\alpha}^{n}(\mathbb{D})$ can be decomposed in a direct sum of *true polyanalytic Bergman spaces* [50, 67]:

$$\mathbf{A}^{n}(\mathbb{D}) = \mathcal{A}_{\alpha}^{1}(\mathbb{D}) \oplus \ldots \oplus \mathcal{A}_{\alpha}^{n}(\mathbb{D}). \tag{42}$$

The space $\mathcal{A}_{\alpha}^{1}(\mathbb{D}) = A_{\alpha}(\mathbb{D}) = \mathbf{A}_{\alpha}^{1}(\mathbb{D})$ is the Bergman space of analytic functions in the unit disc. The spaces $\mathcal{A}_{\alpha}^{n+1}(\mathbb{D})$ are constituted by functions in $\mathcal{L}_{\alpha}^{2}(\mathbb{D})$ which can be written in the form

$$F(z) = \left(1-|z|^2\right)^{-\alpha} \left(\frac{d}{dz}\right)^{n} \left[\left(1-|z|^2\right)^{\alpha+n} f(z)\right],$$

for some $f(z) \in A_\alpha(\mathbb{D})$. In companion with (42) they provide an orthogonal decomposition for the whole $\mathcal{L}^2_\alpha$ space:

$$\mathcal{L}^2_\alpha(\mathbb{D}) = \bigoplus_{n=0}^{\infty} \mathcal{A}^n_\alpha(\mathbb{D}).$$

The reproducing kernel of $\mathcal{A}^n_\alpha(\mathbb{D})$ is given explicitly as follows [67]:

$$\mathcal{K}^n_\alpha(z,w) = \frac{C_{\alpha,n}}{(1-|z|^2)^\alpha(1-|w|^2)^\alpha} \tag{43}$$

$$\times \left(\frac{d^2}{dzd\overline{w}}\right)^n \left[(1-|z|^2)^{\alpha+n}(1-|w|^2)^{\alpha+n}\frac{n(1-z\overline{w})+\alpha+1}{(1-z\overline{w})^{\alpha+2}}\right], \tag{44}$$

where $C_{\alpha,n} = \frac{\Gamma(\alpha+1)}{\pi\Gamma(n+1+\alpha)n!}$

Remark 5. The interested reader can verify that the correspondence

$$F \to \left(\frac{1}{1-w}\right)^{\alpha+1} F\left(i\frac{w+1}{1-w}\right)$$

does not provide an unitary mapping between the spaces $\mathbf{A}^n_\alpha(\mathbb{C}^+)$ and the spaces $\mathbf{A}^n_\alpha(\mathbb{D})$, and *does not* provide an unitary mapping between the spaces $\mathcal{A}^n_\alpha(\mathbb{D})$ and $\mathcal{A}^n_\alpha(\mathbb{C}^+)$. This means that the "true" spaces are different if we move from the unit circle to the upper half-plane and need to be studied separately.

Acknowledgments The authors wish to thank Radu Frunza for sharing his MATLAB-code and to Franz Luef, José Luis Romero, Karlheinz Gröchenig, Luis V. Pessoa, and Tomasz Hrycak for interesting discussions and comments on early versions of these notes.

L.D. Abreu was supported by CMUC and FCT project PTDC/MAT/114394/2009 through COMPETE/FEDER and by Austria Science Foundation (FWF) projects "'Frames and Harmonic Analysis"' and START-project FLAME.

References

1. L.D. Abreu, Sampling and interpolation in Bargmann-Fock spaces of polyanalytic functions. Appl. Comp. Harm. Anal. **29**, 287–302 (2010)
2. L.D. Abreu, On the structure of Gabor and super Gabor spaces. Monatsh. Math. **161**, 237–253 (2010)
3. L.D. Abreu, K. Gröchenig, Banach Gabor frames with Hermite functions: polyanalytic spaces from the Heisenberg group. Appl. Anal. **91**, 1981–1997 (2012)
4. L.D. Abreu, Wavelet frames with Laguerre functions. C. R. Acad. Sci. Paris Ser. I **349**, 255–258 (2011)

5. L.D. Abreu, Super-wavelets versus poly-Bergman spaces. Int. Eq. Op. Theor. **73**, 177–193 (2012)
6. L.D. Abreu, *Wavelet (super)frames with Laguerre functions*, ongoing work
7. L.D. Abreu, M. de Gosson, *Displaced coherent states and true polyanalytic Fock spaces*, ongoing work
8. L.D. Abreu, N. Faustino, On toeplitz operators and localization operators. Proc. Am. Math. Soc. (to appear)
9. M.L. Agranovsky, Characterization of polyanalytic functions by meromorphic extensions into chains of circles. J. d'Analyse Math. **113**, 305–329 (2011)
10. S.T. Ali, J.P. Antoine, J.P. Gazeau, *Coherent States and Their Generalizations* (Springer, Berlin, 2000)
11. Y. Ameur, H. Hedenmalm, N. Makarov, Berezin transform in polynomial Bergman spaces. Comm. Pure Appl. Math. **63**, 1533–1584 (2010)
12. G. Ascensi, J. Bruna, Model space results for the Gabor and Wavelet transforms. IEEE Trans. Inform. Theory **55**, 2250–2259 (2009)
13. N. Askour, A. Intissar, Z. Mouayn, Espaces de Bargmann généralisés et formules explicites pour leurs noyaux reproduisants. C. R. Acad. Sci. Paris Sér. I Math. **325**, 707–712 (1997)
14. R. Balan, *Multiplexing of signals using superframes*, In SPIE *Wavelets applications*, vol. 4119 *of Signal and Image processing XIII*, pp. 118–129 (2000)
15. M.B. Balk, *Polyanalytic Functions* (Akad. Verlag, Berlin, 1991)
16. H. Begehr, G.N. Hile, A hierarchy of integral operators. Rocky Mountain J. Math. **27**, 669–706 (1997)
17. H. Begehr, Orthogonal decompositions of the function space $L^2(\overline{D}, \mathbb{C})$. J. Reine Angew. Math. **549**, 191–219 (2002)
18. J. Ben Hough, M. Krishnapur, Y. Peres, B. Virág, *Zeros of Gaussian Analytic Functions and Determinantal Point Processes*, University Lecture Series, vol. 51, x+154 (American Mathematical Society, Providence, RI, 2009)
19. A.J. Bracken, P. Watson, The quantum state vector in phase space and Gabor's windowed Fourier transform. J. Phys. A **43**, art. no. 395304 (2010)
20. S. Brekke, K. Seip. Density theorems for sampling and interpolation in the Bargmann-Fock space. III. Math. Scand. **73**, 112–126 (1993)
21. K. Bringmann, K. Ono, Dyson's ranks and Maass forms. Ann. Math. **171**, 419–449 (2010)
22. A. Comtet, On the Landau levels on the hyperbolic plane. Ann. Phys. **173**, 185–209 (1987)
23. Z. Cuckovic, T. Le, Toeplitz operators on Bergman spaces of polyanalytic functions. Bull. Lond. Math. Soc. **44**(5), 961–973 (2012)
24. I. Daubechies, *"Ten Lectures On Wavelets"*, CBMS-NSF Regional conference series in applied mathematics (1992)
25. I. Daubechies, J.R. Klauder, Quantum-mechanical path integrals with Wiener measure for all polynomial Hamiltonians. II. J. Math. Phys. **26**(9), 2239–2256 (1985)
26. M. de Gosson, Spectral properties of a class of generalized Landau operators. Comm. Partial Differ. Equat. **33**(10–12), 2096–2104 (2008)
27. M. de Gosson, F. Luef, Spectral and regularity properties of a pseudo-differential calculus related to Landau quantization. J. Pseudo-Differ. Oper. Appl. **1**(1), 3–34 (2010)
28. M. Dörfler, J.L. Romero, Frames adapted to a phase-space cover, preprint arXiv:1207.5383 (2012)
29. P. Duren, A. Schuster, "*Bergman Spaces*", Mathematical Surveys and Monographs, vol. 100 (American Mathematical Society, Providence, RI, 2004)
30. P. Duren, E.A. Gallardo-Gutiérrez, A. Montes-Rodríguez, A Paley-Wiener theorem for Bergman spaces with application to invariant subspaces. Bull. Lond. Math. Soc. **39**, 459–466 (2007)
31. W. Duke, Hyperbolic distribution problems and half-integral weight Maass forms. Invent. Math. **92**, 73–90 (1988)

32. H.G. Feichtinger, Modulation spaces on locally compact abelian groups. In: *Proceedings of "International Conference on Wavelets and Applications" 2002*, pp. 99–140, Chennai, India, 2003. Updated version of a technical report, University of Vienna, 1983
33. H.G. Feichtinger, On a new Segal algebra. Monatsh. Math. **92**(4), 269–289 (1981)
34. H.G. Feichtinger, K. Gröchenig, Banach spaces related to integrable group representations and their atomic decompositions, I. J. Funct. Anal. **86**(2), 307–340 (1989)
35. H.G. Feichtinger, K. Gröchenig, A unified approach to atomic decompositions via integrable group representations. In: *Proc. Function Spaces and Applications, Conf. Lund*, 1986. Lect. Notes Math., vol. 1302 (Springer, New York, 1988), p. 5273
36. H.G. Feichtinger, M. Pap, *Connection between the coorbit theory and the theory of Bergman spaces*, this volume (2013)
37. R.P. Feynman, An operator calculus having applications in quantum electrodynamics. Phys. Rev. **84**, 108–128 (1951)
38. H. Führ, Simultaneous estimates for vector-valued Gabor frames of Hermite functions. Adv. Comput. Math. **29**(4), 357–373 (2008)
39. I. Gertner, G.A. Geri, Image representation using Hermite functions. Biol. Cybernetics **71**(2), 147–151 (1994)
40. J. Ginibre, Statistical ensembles of complex, quaternion, and real matrices. J. Math. Phys. **6**, 440 (1965)
41. K. Gröchenig, Describing functions: Atomic decompositions versus frames. Monatsh. Math. **112**, 1–42 (1991)
42. K. Gröchenig, "*Foundations of Time-Frequency Analysis*" (Birkhäuser, Boston, 2001)
43. K. Gröchenig, Localization of frames, Banach frames, and the invertibility of the frame operator. J. Fourier Anal. Appl. **10**, 105–132 (2004)
44. K. Gröchenig, Y. Lyubarskii, Gabor frames with Hermite functions. C. R. Acad. Sci. Paris Ser. I **344**, 157–162 (2007)
45. K. Gröchenig, Y. Lyubarskii, Gabor (super)frames with Hermite functions. Math. Ann. **345**, 267–286 (2009)
46. K. Gröchenig, J. Stoeckler, Gabor frames and totally positive functions. Duke Math. J. **162**(6), 1003–1031 (2013)
47. A. Haimi, H. Hedenmalm, The polyanalytic Ginibre ensembles. Preprint arXiv:1106.2975 (2012)
48. A. Haimi, H. Hedenmalm, Asymptotic expansions of polyanalytic Bergman kernels. Preprint arXiv:1303.0720 (2013)
49. D. Han, D.R. Larson, Frames, bases and group representations. Mem. Am. Math. Soc. **147**, 697 (2000)
50. C. Heil, History and evolution of the Density Theorem for Gabor frames. J. Fourier Anal. Appl. **13**, 113–166 (2007)
51. H. Hedenmalm, B. Korenblum, K. Zhu, *The Theory of Bergman Spaces* (Springer, New York, 2000). ISBN 978-0-387-98791-0
52. O. Hutnik, A note on wavelet subspaces. Monatsh. Math. **160**, 59–72 (2010)
53. O. Hutník, M. Hutníková, An alternative description of Gabor spaces and Gabor-Toeplitz operators. Rep. Math. Phys. **66**(2), 237–250 (2010)
54. A. Jaffe, F. Quinn, Theoretical mathematics: Toward a cultural synthesis of mathematics and theoretical physics. Bull. Am. Math. Soc. (N.S.) **29**, 1–13 (1993)
55. G.V. Kolossov, Sur les problêms d'élasticité a deux dimensions. C. R. Acad. Sci. **146**, 522–525 (1908)
56. A.D. Koshelev, On kernel functions for the Hilbert space of polyanalytic functions in the disk. Dokl. Akad. Nauk. SSSR [Soviet Math. Dokl.] **232**, 277–279 (1977)
57. G. Kuttyniok, *Affine Density in Wavelet Analysis*, Lecture Notes in Mathematics, vol. 1914 (Springer, Berlin, 2007)
58. Y. Lyubarskii, P.G. Nes, Gabor frames with rational density. Appl. Comp. Harm. Anal. **34**, 488–494 (2013)

59. Y. Lyubarskii, Frames in the Bargmann space of entire functions, *Entire and subharmonic functions*, Adv. Soviet Math., vol. 11 (Amer. Math. Soc., Providence, RI, 1992), pp. 167–180
60. T. Mine, Y. Nomura, Landau levels on the hyperbolic plane in the presence of Aharonov–Bohm fields. J. Funct. Anal. **263**, 1701–1743 (2012)
61. M.L. Mehta, *Random Matrices*, 3rd edn. (Academic Press, New York, 1991)
62. Z. Mouayn, Characterization of hyperbolic Landau states by coherent state transforms. J. Phys. A Math. Gen. **36**, 8071–8076 (2003)
63. N.I. Muskhelishvili, *Some Basic Problems of Mathematical Elasticity Theory* (in Russian) (Nauka, Moscow, 1968)
64. A.M. Perelomov, On the completeness of a system of coherent states. Theor. Math. Phys. **6**, 156–164 (1971)
65. A.M. Perelomov, *Generalized Coherent States and Their Applications* (Springer, New York, 1986). Texts Monographs Phys. **6**, 156–164 (1971)
66. J. Ramanathan, T. Steger, Incompleteness of sparse coherent states. Appl. Comput. Harmon. Anal. **2**, 148–153 (1995)
67. A.K. Ramazanov, On the structure of spaces of polyanalytic functions. (Russian) Mat. Zametki **72**(5), 750–764 (2002); translation in Math. Notes **72**(5–6), 692–704 (2002)
68. W. Roelcke, Der eigenwertproblem der automorphen Formen in der hyperbolischen Ebene, I. Math. Ann. **167**, 292–337 (1966)
69. A. Ron, Z. Shen, Weyl-Heisenberg frames and Riesz bases in $L^2(\mathbb{R}^d)$. Duke Math. J. **89**, 237–282 (1997)
70. K. Seip, Reproducing formulas and double orthogonality in Bargmann and Bergman spaces. SIAM J. Math. Anal. **22**(3), 856–876 (1991)
71. K. Seip, R. Wallstén, Density Theorems for sampling and interpolation in the Bargmann-Fock space II. J. Reine Angew. Math. **429**, 107–113 (1992)
72. K. Seip, Beurling type density theorems in the unit disc. Invent. Math. **113**, 21–39 (1993)
73. D. Slepian, Some comments on Fourier analysis, uncertainty and modelling. SIAM Rev. **25**, 379–393 (1983)
74. S. Thangavelu, *Lectures on Hermite and Laguerre Expansions*, Math. Notes, vol. 42 (Princeton University Press, New Jersey, 1993)
75. N.L. Vasilevski, On the structure of Bergman and poly-Bergman spaces. Integr. Equat. Oper. Theory **33**, 471–488 (1999)
76. N.L. Vasilevski, *Poly-Fock Spaces*, Differential operators and related topics, vol. I (Odessa, 1997), pp. 371–386, Oper. Theory Adv. Appl., vol. 117 (Birkhäuser, Basel, 2000)
77. N.L. Vasilevski, *Commutative Algebras of Toeplitz Operators On the Bergman Space*, Oper. Theory Adv. Appl., vol. 185 (Birkhäuser, Basel, 2008)
78. A. Wünsche, *Displaced Fock states and their connection to quasiprobabilities*. Quantum Opt. **3**, 359–383 (1991)
79. K. Zhu, *Analysis on Fock Spaces*. Graduate Texts in Mathematics, vol. 263 (Springer, New York, 2012)

Classical and Stochastic Löwner–Kufarev Equations

Filippo Bracci, Manuel D. Contreras, Santiago Díaz-Madrigal, and Alexander Vasil'ev

Abstract In this paper we present a historical and scientific account of the development of the theory of the Löwner–Kufarev classical and stochastic equations spanning the 90-year period from the seminal paper by K. Löwner in 1923 to recent generalizations and stochastic versions and their relations to conformal field theory.

Keywords Brownian motion • Conformal mapping • Evolution family • Integrable system • Kufarev • Löwner • Pommerenke • Schramm • Subordination chain

Mathematics Subject Classification (2000). Primary 01A70 · 30C35; Secondary 17B68 · 70H06 · 81R10

Filippo Bracci (✉)
Dipartimento Di Matematica, Università di Roma "Tor Vergata", Via Della Ricerca Scientifica 1, 00133 Roma, Italy
e-mail: fbracci@mat.uniroma2.it

M.D. Contreras · S. Díaz-Madrigal
Departamento de Matemática Aplicada II, Escuela Técnica Superior de Ingeniería, Universidad de Sevilla, Camino de los Descubrimientos, s/n 41092 Sevilla, Spain
e-mail: contreras@us.es; madrigal@us.es

A. Vasil'ev
Department of Mathematics, University of Bergen, P.O. Box 7803, Bergen N-5020, Norway
e-mail: alexander.vasiliev@math.uib.no

A. Vasil'ev (ed.), *Harmonic and Complex Analysis and its Applications*, Trends in Mathematics, DOI 10.1007/978-3-319-01806-5_2,

1 Introduction

The Löwner theory went through several periods of its development. It was born in 1923 in the seminal paper by Löwner [165], its *formation* was completed in 1965 in the paper by Pommerenke [189] and was finally formulated thoroughly in his monograph [190] unifying Löwner's ideas of semigroups and evolution equations and Kufarev's contribution [144, 146] of the t-parameter differentiability (in the Carathéodory kernel) of a family of conformal maps of simply connected domains $\Omega(t)$ onto the canonical domain $\mathbb{D}$, the unit disk in particular, and of PDE for subordination Löwner chains of general type. It is worth mentioning that the 20-year period between papers [165] and [144] was illuminated by Goluzin's impact [90] on further applications of Löwner's parametric method to extremal problems for univalent functions, as well as works by Fekete and Szegö [75], Peschl [182], Robinson [215], Komatu [135], Bazilevich [27, 28] related to this period.

The next period became *applications* of the *parametric method* to solving concrete problems for conformal maps culminating at the de Branges proof [46] of the celebrating Bieberbach conjecture [37] in 1984. During approximately 16 years after this proof the interest to univalent functions had been somehow decaying. However, the *modern period* has been marked by burst of interest to stochastic (Schramm)–Löwner evolution (SLE) which has implied an elegant description of several 2D conformally invariant statistical physical systems at criticality by means of a solution to the Cauchy problem for the Löwner equation with a random driving term given by 1D Brownian motion, see [157, 230]. At the same time, several connections with mathematical physics were discovered recently, in particular, relations with a singular representation of the Virasoro algebra using Löwner martingales in [24, 81, 82, 130] and with a Hamiltonian formulation and a construction of the KP integrable hierarchies in [169–171].

2 Löwner

Karel Löwner was born on May 29, 1893 in Lány, Bohemia. He also used the German spelling Karl of his first name until his immigration to the U.S.A. in 1939 after Nazis occupied Prague, when he changed it to Charles Loewner as a new start in a new country. Although coming from a Jewish family and speaking Czech at home, all of his education was in German, following his father's wish. After finishing a German Gimnasium in Prague in 1912 he entered the Charles University of Prague and received his Ph.D. from this university in 1917 under the supervision of Georg Pick.

After four and a half years in German Technical University in Prague where the mathematical environment was not stimulating enough, Löwner took up a position at the University of Berlin (now Humboldt Universität Berlin) with great enthusiasm where he started to work surrounded by Schur, Brauer, Hopf, von Neumann, Szegö,

K. Löwner

and other famous names. Following a brief lectureship at Cologne in 1928 Löwner returned to the Charles University of Prague where he hold a chair in Mathematics until the occupation of Czechoslovakia in 1939. When the Nazis occupied Prague, he was put in jail. Luckily, after paying the "emigration tax" twice over he was allowed to leave the country with his family and moved, in 1939, to the U.S.A. After von Neumann arranged a position for him at the Louisville University, Löwner was able to start from the bottom of his career. Following further appointments at the Brown (1944) and Syracuse (1946) Universities, he moved in 1951 to his favorite place, California, where he worked at the Stanford University until his death. His former Ph.D. student in Prague, Lipman Bers, testifies [180] that Löwner was a man whom everybody liked, a man at peace with himself, a man who was incapable of malice. He was a great teacher, Lipman Bers, Adriano Garcia, Gerald Goodman, Carl Fitzgerald, Roger Horn, Ernst Lammel, Charles Titus were his students among 26 in total. Löwner died as a Professor Emeritus in Stanford on January 8, 1968, after a brief illness. For more about Löwner's biography, see [2, 180].

G. Pick

Löwner's work covers wide areas of complex analysis and differential geometry and displays his deep understanding of Lie theory and his passion for semigroup theory. Perhaps it is worth to mention here two of his discoveries. First was his introduction of infinitesimal methods for univalent mappings [165], where he defined basic notions of semigroups of conformal maps of $\mathbb{D}$ and their evolution families leading to the sharp estimate $|a_3| \leq 3$ in the Bieberbach conjecture $|a_n| \leq n$ in the class $\mathbf{S}$ of normalized univalent maps $f(z) = z + a_2z^2 + \dots$ of the unit disk $\mathbb{D}$.

It is wellknown that the Bieberbach conjecture was finally proven by de Branges in 1984 [45, 46]. In his proof, de Branges introduced ideas related to Löwner's but he used only a distant relative of Löwner's equation in connection with his main apparatus, his own rather sophisticated theory of composition operators. However, elaborating on de Branges' ideas, FitzGerald and Pommerenke [76] discovered how to avoid altogether the composition operators and rewrote the proof in classical terms, applying the *bona fide* Löwner equation and providing in this way a direct and powerful application of Löwner's classical method.

The second paper [166], we want to emphasize on, is dedicated to properties of n-monotonic matrix functions, which turned to be of importance for electrical engineering and for quantum physics. Löwner was also interested in the problems of fluid mechanics, see, e.g., [167], while working at Brown University 1944–1946 on a war-related program. Starting with some unusual applications of the theory of univalent functions to the flow of incompressible fluids, he later applied his methods to the difficult problems of the compressible case. This led him naturally to the study of partial differential equations in which he obtained significant differential inequalities and theorems regarding general conservation laws.

3 Kufarev

The counterpart subordination chains and the Löwner–Kufarev PDE for them were introduced by Pavel Parfen'evich Kufarev (Tomsk, March 18, 1909–Tomsk, July 17, 1968). Kufarev entered the Tomsk State University in 1927, first, the Department of Chemistry, and then, the Department of Mathematics, which he successfully finished in 1931. After a year experience in industry in Leningrad he returned to Tomsk in 1932 and all his academic life was connected with this university, the first university in Siberia. Kufarev started to work at the Department of Mathematics led by professor Lev Aleksandrovich Vishnevskiĭ (1887–1937). At the same time, two prominent Western mathematicians came to Tomsk.

One was Stefan Bergman (1895–1977). Being of Jewish origin he was forced from his position in Berlin in 1933 due to the "Restoration of the civil service," an anti-Semitic Hitler's law. Bergman came to Tomsk in 1934 and was working there until 1936, then at Tbilisi in Georgia in 1937 and had to leave Soviet Union under Stalin's oppression towards foreign scientists. For 2 years he was working at the Institute Henri Poincaré in Paris, and then, left France for the USA in 1939 because of German invasion, he finally worked at Stanford from 1952 together with Löwner.

P. P. Kufarev

S. Bergman

The history of the other one is more tragic. Fritz Noether (1884–1941), the youngest son of Max Noether (1844–1921) and the younger brother of Emmy Noether (1882–1935), came to Tomsk in the same year as Bergman, because of the same reason, and remained there until 1937, when Vishnevskiĭ and "his group" were accused of espionage. Noether and Vishnevskiĭ were arrested by NKVD (later KGB) and Noether was transported to Orel concentration camp where he was jailed until 1941. Nazis approached Orel in 1941 and many prisoners were executed on September 10, 1941 (Stalin's order from September 8, 1941). Fritz Noether was among them following the official Soviet version of 1988, cf. [62].

Fitz's wife Regine returned to Germany from Soviet Union in 1935 under psychological depression and soon died. The same year his famous sister Emmy died in Pennsylvania after a feeble operation. His two sons were deported from Soviet Union. Fortunately, they were given refuge in Sweden as a first step. However, many years later Evgeniy Berkovich and Boris M. Schein published a correspondence [35], also [236], in which prof. Boris Shein referred to prof. Saveliĭ

F. Noether

Vladimirovich Falkovich (1911–1982, Saratov), who met Fritz Noether in Moscow metro in the late fall of 1941. Noether and Falkovich knew each other and Noether talked his story on the arrest and tortures in Tomsk NKVD. In particular, NKVD agents confiscated many of his things and books. Noether said that he finally was released from Orel Central and went to Lubyanka (NKVD/KGB headquarters) for some traveling documents to visit his family. Then the train stopped and their conversation was interrupted. Since he did not come to Tomsk (his son's Gottfried Noether testimony), he most probably was arrested again. This ruins the official Soviet version of Noether's death and indicates that the story is not completed yet.[1] One of the possible reasons is that several Jewish prisoners (first of all, originating from Poland, e.g., Henrik Erlich and Victor Alter; and from Germany) were released following Stalin's plans of creating an Anti-Hitler Jewish Committee, which further was not realized. Noether could be among them being rather known mathematician himself and having support from Einstein and Weyl.

Kufarev was also accused of espionage being a member of Vishnevskiĭ's "terrorist group" together with Boris A. Fuchs, but they were not arrested in contrast to Fritz Noether, see [38].

Bergman and Noether supported Kufarev's thesis defense in 1936. Then Kufarev was awarded the Doctor of Sciences degree (analogue of German habilitation) in 1943 and remained the unique mathematician in Siberia with this degree until 1957 when the *Akademgorodok* was founded and Mikhail A. Lavrentiev invited many first-class mathematicians to Novosibirsk. Kufarev became a full professor of the Tomsk State University in 1944, served as a dean of the Faculty of Mechanics and Mathematics 1952–1955, and remained a professor of the university until his death in 1968 after a hard illness.

[1] I have also some personal interest to this story because me, Falkovich and Shein worked in different periods at the same Saratov State University. *A. Vasil'ev*

The main interests of Kufarev were in the theory of univalent functions and applications to fluid mechanics, in particular, in Hele-Shaw flows, see an overview in [253]. His main results on the parametric method were published in two papers [144, 146], where he considered subordination chains of general type and wrote corresponding PDE for mapping functions, but he returned to this method all the time together with his students, combining variational and parametric methods, creating the parametric method for multiply connected domains, half-plane version of the parametric method, etc.

4 Pommerenke and Unification of the Parametric Method

Christian Pommerenke (born December 17, 1933 in Copenhagen, Professor Emeritus at the Technische Universität Berlin) became that person who unified Löwner and Kufarev's ideas and thoroughly combined analytic properties of the ordering of the images of univalent mappings of the unit disk with evolutionary aspects of semigroups of conformal maps. He seems to have been the first one to use the expression "Löwner chain" for describing the family of "increasing" univalent mappings in Löwner's theory.

Ch. Pommerenke

Recall that we denote the unit disk by $\mathbb{D} = \{\zeta : |\zeta| < 1\}$ and the class of normalized univalent maps $f:\mathbb{D} \to \mathbb{C}$, $f(z) = z + a_2z^2 + \dots$ by **S**. A t-parameter family $\Omega(t)$ of simply connected hyperbolic univalent domains forms a *Löwner subordination chain* in the complex plane $\mathbb{C}$, for $0 \leq t < \tau$ (where τ may be ∞), if $\Omega(t) \subseteq \Omega(s)$, whenever $t < s$, and the family is continuous in the Carathéodory sense. We suppose that the origin is a point of $\Omega(0)$.

A Löwner subordination chain $\Omega(t)$ is described by a t-dependent family of conformal maps $z = f(\zeta, t)$ from $\mathbb{D}$ onto $\Omega(t)$, normalized by $f(\zeta, t) = a_1(t)\zeta + a_2(t)\zeta^2 + \dots$, $a_1(t) > 0$, $\dot{a}_1(t) > 0$. Pommerenke [189, 190] described governing evolution equations in partial and ordinary derivatives, known now as the Löwner–Kufarev equations.

One can normalize the growth of evolution of a subordination chain by the conformal radius of $\Omega(t)$ with respect to the origin setting $a_1(t) = e^t$.

We say that the function p is from the Carathéodory class if it is analytic in $\mathbb{D}$, normalized as $p(\zeta) = 1 + p_1\zeta + p_2\zeta^2 + \dots, \quad \zeta \in \mathbb{D}$, and such that $\operatorname{Re} p(\zeta) > 0$ in $\mathbb{D}$. Given a Löwner subordination chain of domains $\Omega(t)$ defined for $t \in [0, \tau)$, there exists a function $p(\zeta, t)$, measurable in $t \in [0, \tau)$ for any fixed $z \in \mathbb{D}$, and from the Carathéodory class for almost all $t \in [0, \tau)$, such that the conformal mapping $f : \mathbb{D} \to \Omega(t)$ solves the equation

$$\frac{\partial f(\zeta, t)}{\partial t} = \zeta \frac{\partial f(\zeta, t)}{\partial \zeta} p(\zeta, t), \tag{1}$$

for $\zeta \in \mathbb{D}$ and for almost all $t \in [0, \tau)$. Equation (1) is called the Löwner–Kufarev equation due to two seminal papers: by Löwner [165] who considered the case when

$$p(\zeta, t) = \frac{e^{iu(t)} + \zeta}{e^{iu(t)} - \zeta}, \tag{2}$$

where $u(t)$ is a continuous function regarding $t \in [0, \tau)$, and by Kufarev [144] who proved differentiability of f in t for all ζ in the case of general p from the Carathéodory class.

Let us consider a reverse process. We are given an initial domain $\Omega(0) \equiv \Omega_0$ (and therefore, the initial mapping $f(\zeta, 0) \equiv f_0(\zeta)$), and an analytic function $p(\zeta, t)$ of positive real part normalized by $p(\zeta, t) = 1 + p_1\zeta + \dots$. Let us solve (1) and ask ourselves, whether the solution $f(\zeta, t)$ defines a subordination chain of simply connected univalent domains $f(\mathbb{D}, t)$. The initial condition $f(\zeta, 0) = f_0(\zeta)$ is not given on the characteristics of the partial differential equation (1); hence, the solution exists and is unique but not necessarily univalent. Assuming s as a parameter along the characteristics we have

$$\frac{dt}{ds} = 1, \quad \frac{d\zeta}{ds} = -\zeta p(\zeta, t), \quad \frac{df}{ds} = 0,$$

with the initial conditions $t(0) = 0$, $\zeta(0) = z$, $f(\zeta, 0) = f_0(\zeta)$, where z is in $\mathbb{D}$. Obviously, we can assume $t = s$. Observe that the domain of ζ is the entire unit disk. However, the solutions to the second equation of the characteristic system range within the unit disk but do not fill it. Therefore, introducing another letter w (in order to distinguish the function $w(z, t)$ from the variable ζ) we arrive at the Cauchy problem for the Löwner–Kufarev equation in ordinary derivatives

$$\frac{dw}{dt} = -wp(w, t), \tag{3}$$

for a function $\zeta = w(z, t)$ with the initial condition $w(z, 0) = z$. Equation (3) is a nontrivial characteristic equation for (1). Unfortunately, this approach requires the extension of $f_0(w^{-1}(\zeta, t))$ into the whole $\mathbb{D}$ (here w^{-1} means the inverse function in ζ) because the solution to (1) is the function $f(\zeta, t)$ given as $f_0(w^{-1}(\zeta, t))$, where $\zeta = w(z, s)$ is a solution of the initial value problem for the characteristic equation (3) that maps $\mathbb{D}$ into $\mathbb{D}$. Therefore, the solution of the initial value problem for (1) may be nonunivalent.

On the other hand, solutions to (3) are holomorphic univalent functions $w(z,t) = e^{-t}(z + a_2(t)z^2 + \ldots)$ in the unit disk that map $\mathbb{D}$ into itself. Every function f from the class **S** can be represented by the limit

$$f(z) = \lim_{t\to\infty} e^t w(z,t), \tag{4}$$

where $w(z,t)$ is a solution to (3) with some function $p(z,t)$ of positive real part for almost all $t \geq 0$ (see [190, pp. 159–163]). Each function $p(z,t)$ generates a unique function from the class **S**. The reciprocal statement is not true. In general, a function $f \in \mathbf{S}$ can be obtained using different functions $p(\cdot, t)$.

Now we are ready to formulate the condition of univalence of the solution to (1) in terms of the limiting function (4), which can be obtained by combination of known results of [190].

Theorem 1 ([190, 207]). *Given a function $p(\zeta, t)$ of positive real part normalized by $p(\zeta, t) = 1 + p_1\zeta + \ldots$, the solution to (1) is unique, analytic, and univalent with respect to ζ for almost all $t \geq 0$, if and only if, the initial condition $f_0(\zeta)$ is taken in the form* (4), *where the function $w(\zeta, t)$ is the solution to* (3) *with the same driving function p.*

Concluding this section we remark that the Löwner and Löwner–Kufarev equations are described in several monographs [4, 14, 57, 72, 93, 108, 122, 190, 218].

5 Half-plane Version

In 1946, Kufarev [145, Introduction] first mentioned an evolution equation in the upper half-plane $\mathbb{H}$ analogous to the one introduced by Löwner in the unit disk and was first studied by Popova [196] in 1954. In 1968, Kufarev et al. [151] introduced a combination of Goluzin–Schiffer's variational and parametric methods for this equation for the class of univalent functions in the upper half-plane, which is known to be related to physical problems in hydrodynamics. They showed its application to the extremal problem of finding the range of $\{\operatorname{Re} e^{i\alpha} f(z), \operatorname{Im} f(z)\}$, $\operatorname{Im} z > 0$. Moreover, during the second half of the past century, the Soviet school intensively studied Kufarev's equations for $\mathbb{H}$. We ought to cite here at least contributions by Aleksandrov [14], Aleksandrov and Sobolev [15], Goryainov and Ba [101, 104]. However, this work was mostly unknown to many Western mathematicians, in particular, because some of it appeared in journals not easily accessible from the outside of the Soviet Union. In fact, some of Kufarev's papers were not even reviewed by Mathematical Reviews. Anyhow, we refer the reader to [7], which contains a complete bibliography of his papers.

In order to introduce Kufarev's equation properly, let us fix some notation. Let γ be a Jordan arc in the upper half-plane $\mathbb{H}$ with starting point $\gamma(0) = 0$. Then there exists a unique conformal map $g_t : \mathbb{H} \setminus \gamma[0,t] \to \mathbb{H}$ with the normalization

$$g_t(z) = z + \frac{c(t)}{z} + O\left(\frac{1}{z^2}\right), \quad z \sim \infty.$$

After a reparametrization of the curve γ, one can assume that $c(t) = 2t$. Under this normalization, one can show that g_t satisfies the following differential equation:

$$\frac{dg_t(z)}{dt} = \frac{2}{g_t(z) - \xi(t)}, \qquad g_0(z) = z. \tag{5}$$

The equation is valid up to a time $T_z \in (0, +\infty]$ that can be characterized as the first time t such that $g_t(z) \in \mathbb{R}$ and where h is a continuous real-valued function. Conversely, given a continuous function $h\colon [0, +\infty) \to \mathbb{R}$, one can consider the following initial value problem for each $z \in \mathbb{H}$:

$$\frac{dw}{dt} = \frac{2}{w - \xi(t)}, \qquad w(0) = z\,. \tag{6}$$

Let $t \mapsto w^z(t)$ denote the unique solution to this Cauchy problem and let $g_t(z) := w^z(t)$. Then g_t maps holomorphically a (not necessarily slit) subdomain of the upper half-plane $\mathbb{H}$ onto $\mathbb{H}$. Equation (6) is nowadays known as the *chordal Löwner differential equation* with the function h as the driving term. The name is due to the fact that the curve $\gamma[0,t]$ evolves in time as t tends to infinity into a sort of chord joining two boundary points. This kind of construction can be used to model evolutionary aspects of decreasing families of domains in the complex plane. Equation (3) with the function p given by (2) in this context is called the *radial Löwner equation*, because in the slit case, the tip of slit tends to the origin in the unit disk.

Quite often it is presented the half-plane version considering the inverse of the functions g_t (see, i.e., [151]). Namely, the conformal mappings $f_t = g_t^{-1}$ from $\mathbb{H}$ onto $\mathbb{H} \setminus \gamma\big([0,t)\big)$ satisfy the PDE

$$\frac{\partial f_t(z)}{\partial t} = -f_t'(z)\frac{2}{z - \xi(t)}. \tag{7}$$

We remark that using the Cayley transform $T(z) = \frac{z+1}{1-z}$ we obtain that the chordal Löwner equation in the unit disk takes the form

$$\frac{\partial h_t(z)}{\partial t} = -h_t'(z)(1-z)^2 p(z,t), \tag{8}$$

where $\operatorname{Re} p(z,t) \geq 0$ for all $t \geq 0$ and $z \in \mathbb{D}$. From a geometric point of view, the difference between this family of parametric functions and those described by the

Löwner–Kufarev equation (1) is clear: the ranges of the solutions of (8) decrease with t while in the former equation (1) increase. This duality "decreasing" versus "increasing" has recently been analyzed in [54] and, roughly speaking, we can say that the "decreasing" setting can be deduced from the "increasing" one.

The stochastic version of (6) with a random entry λ will be discussed in Sect. 13.

Analogues of the Löwner–Kufarev methods appeared also in the theory of planar quasiconformal maps but only few concrete problems were solved using them, see [233] and [87].

G. M. Goluzin

6 Applications to Extremal Problems: Optimal Control

After Löwner himself, one of the first who applied Löwner's method to extremal problems in the theory of univalent functions in 1936 was Gennadiĭ Mikhailovich Goluzin (1906–1952, St. Petersburg–Leningrad, Russia) [90, 91] and Ernst Peschl (1906, Passau–1986, Eitorf, Germany) [182]. Goluzin was a founder of Leningrad school in geometric function theory, a student of Vladimir I. Smirnov. He obtained in an elegant way several new and sharp estimates. The most important of them is the sharp estimate in the rotation theorem. Namely, if $f \in \mathbf{S}$, then

$$|\arg f'(z)| \leq \begin{cases} 4 \arcsin |z| & \text{if } 0 < |z| \leq \frac{1}{\sqrt{2}}, \\ \pi + \log \frac{|z|^2}{1-|z|^2} & \text{if } \frac{1}{\sqrt{2}} < |z| < 1. \end{cases}$$

Goluzin himself proved [91] sharpness only for the case $0 < |z| \leq \frac{1}{\sqrt{2}}$, Bazilevich [27] completed the proof for $\frac{1}{\sqrt{2}} < |z| < 1$ later during the same year.

E. Peschl (right) with H. Weyl

Ernst Peschl was a student of Constantin Carathéodory and obtained his doctorate at Munich in 1931. He spent 2 years 1931–1933 in Jena working with Robert König, then moved to Münster, Bonn, Braunschweig and finally returned to the Rheinische Friedrich-Wilhelms University in Bonn where he worked until his retirement in 1974. In contrast to other heroes of our story Peschl remained in Germany under Nazi and even was a member of the Union of National Socialist Teachers 1936–1938 (he was thrown out of the Union of National Socialist Teachers since he had not paid his membership fees), however, it was only in order to continue his academic career. Being fluent in French, Peschl served as a military interpreter during the II-nd World War until 1943. Peschl [182] applied Löwner's method to prove the following statement. If $f \in \mathbf{S}$ and $f(z) = z + a_2z^2 + \dots$, then $2\Phi\left(\mathrm{Re}\,\frac{a_2}{2}\right) - 1 \le \mathrm{Re}\,(a_3 - a_2^2) \le 1$, where

$$\Phi(x) = \frac{x^2}{\varphi^2(x)}(2\varphi(x) - 1),$$

where $\varphi(x)$ is a unique solution to the equation $x + \varphi e^{1-\varphi} = 0$. The result is sharp. The paper is much deeper than only this inequality and became the first serious treatment of the problem of description of the coefficient body $(a_2, a_3, \dots, a_n)$ in the class S, which was later treated in a nice monograph by Schaeffer and Spencer [226]. Let us mention that the above inequality was repeated by Goryainov [97] in 1980 who also presented all possible extremal functions.

Related result is a nice matter of the paper by Fekete and Szegö [75], see also [72, p. 104] who ingeniously applied Löwner's method in order to disprove a conjecture $|c_{2n+1}| \le 1$ by Littlewood and Paley [164] on coefficients of odd univalent functions $\mathbf{S}^{(2)} \subset \mathbf{S}$ defined by the expansion $h(z) = z + c_3z^3 + c_5z^5 + \dots$.

Theorem 2 ([75]). *Suppose that* $f \in \mathbf{S}$ *and* $0 < \alpha < 1$. *Then*

$$|a_3 - \alpha a_2^2| \leq 1 + 2^{-2\alpha/(1-\alpha)}.$$

This bound is sharp for all $0 < \alpha < 1$.

The choice $\alpha = \frac{1}{4}$ and a simple recalculation of coefficients of functions from $\mathbf{S}^{(2)}$ versus $\mathbf{S}$

$$c_3 = \frac{1}{2}a_2, \quad c_5 = \frac{1}{2}\big(a_3 - \frac{1}{4}a_2\big),$$

lead to the corollary $|c_5| \leq \frac{1}{2} + e^{-2/3} = 1.013\ldots$. Since the result is sharp, there exist odd univalent functions with coefficients bigger than 1. Using similar method Robertson [213] proved that $|c_3|^2 + |c_5|^2 \leq 2$, which is a particular case of his conjecture $\sum_{k=1}^{n} |c_{2k-1}|^2 \leq n$ for the class $\mathbf{S}^{(2)}$ with $c_1 = 1$.

Among other papers on coefficient estimates, let us distinguish the first proof of the Bieberbach conjecture for $n = 4$ by Garabedian and Schiffer [85] in 1955 where Löwner's method was also used. Walter Hayman wrote in his review on this paper that the method "not only carries Bieberbach's conjecture one step further from the point where Löwner placed it over 30 years ago, but also gives a hope, in principle at least, to prove the conjecture for the next one or two coefficients by an increase in labour, rather than an essentially new method." Finally, the complete proof of the Bieberbach conjecture by de Branges [45, 46] in 1984 used the parametric method. It is worth mentioning that the coefficient problem for the inverse functions is much simpler and was solved by Löwner in the same 1923 paper.

Theorem 3 ([165]). *Suppose that* $f \in \mathbf{S}$ *and that*

$$z = \phi(w) = f^{-1}(w) = w + b_2 w^2 + \ldots$$

is the inverse function. Then

$$|b_n| \leq \frac{1 \cdot 3 \cdot 5 \cdots (2n-1)}{(n+1)!} 2^n,$$

with the equality for the function $f(z) = \frac{z}{(1+z)^2}$.

Let us just mention that most of the elementary estimates of functionals in the class $\mathbf{S}$, such as $|f(z)|$, $|\arg \frac{f(z)}{z}|$, $|f'(z)|$, can be obtained by Löwner's method, see, e.g., [14, 72, 122, 190]. In particular, the sharp estimate

$$\left| \arg \frac{z f'(z)}{f(z)} \right| \leq \log \frac{1+|z|}{1-|z|}$$

implies that for every $0 < r \leq \tanh(\pi/4)$ the disk $|z| < r$ is mapped by $f \in \mathbf{S}$ onto a starlike domain with respect to the origin. The constant $\tanh(\pi/4)$ is sharp and is called the radius of starlikeness. Krzyż [143] used the Löwner method to obtain the radius of close-to-convexity 0.80... in the class **S**.

In 1950–1952, Kufarev and his student Fales [148–150] solved a Lavrentiev problem on the weighted product of the conformal radii of two non-overlapping domains in $\mathbb{D}$ using his parametric method. Namely, let z_1 and z_2 be two points in $\mathbb{D}$ and let Γ be a Jordan curve splitting $\mathbb{D}$ into two domains B_1 and B_2 such that $z_1 \in B_1$ and $z_2 \in B_2$. Let f_1 and f_2 be two conformal maps of $\mathbb{D}$ onto B_1 and B_2, respectively, $f_k(0) = z_k$, $f_k'(0) > 0$, $k = 1, 2$. The problem is to find Γ which maximizes the functional $J(\Gamma; z_1, z_2) = |f_1'(0)|^\alpha |f_2'(0)|^\beta$, where $\alpha, \beta \geq 0$. For $\alpha = \beta$ the maximum is attained if Γ is the non-Euclidean line in $\mathbb{D}$ which bisects orthogonally the non-Euclidean segment which connects the two points z_1 and z_2. This was obtained by Lavrentiev [154] himself by different approach. For $\alpha \neq \beta$ the answer is much more complicated.

After Kufarev, the Tomsk school in geometric function theory was led by Igor Aleksandrovich Aleksandrov (born May 11, 1932, Novosibirsk, Russia) who completed his Ph.D. in the Tomsk State University in 1958 under supervision by Kufarev. Together with his students, he developed the parametric method combining it with the Goluzin–Schiffer variational method in 1960 and solved several important extremal problems. In 1963 he defended the Doctor of Sciences degree. This time specialists in univalent functions turned from estimation of functionals on the class **S** to evaluation of ranges of systems of functionals. One of the natural systems is $I(f) = \{\log \frac{f(z)}{z}, \log \frac{zf'(z)}{f(z)}\}$ for a chosen branch of log, such that $I(id) = (0, 0, 0, 0)$. When f runs over **S**, the range $\mathfrak{B} = \{I(f) \colon f \in \mathbf{S}\}$ of $I(f)$ fills a closed bounded set in $\mathbb{R}^4$. The boundary of this set was completely described by Popov [194], Gutlyanskiĭ [115], Goryainov and Gutlyanskiĭ [96] (for the subclass $\mathbf{S}_M$ of bounded functions $|f(z)| \leq M$ from **S**). Goryainov [98–100] moreover obtained the uniqueness of the extremal functions for the boundary of

I. A. Aleksandrov

$\mathfrak{B}$ and their form. Goryainov, based on Löwner–Kufarev equations and on results by Pommerenke [190] and Gutlyanskiĭ [115] on representation of functions from **S** by (4), created a method of determination of all boundary functions for this and other systems of functionals, see also [97]. It was time when a part of mathematicians led by prof. Georgiĭ Dmitrievich Suvorov moved from Tomsk to Donetsk (Ukraine) to a newly established (1966) Institute of Applied Mathematics of the Academy of Sciences of the Ukraine. Vladimir Ya. Gutlyanskiĭ was a student of Aleksandrov, Georgiĭ D. Suvorov of Kufarev, and Victor V. Goryainov of Gutlyanskiĭ. There were several related works on studying $\mathfrak{B}$, its projections, generalizations to other similar results for the classes of univalent functions by Chernetskiĭ, Astakhov, and other members of this active group. Let us mention earlier works by Aleksandrov and Kopanev [9] who determined the domain of variability $\log f'(z)$ for functions in **S** for each fixed z by Löwner–Kufarev method. Earlier Arthur Grad [226, pp. 263–291] obtained this domain by variational method. The same authors [8] found the range of the system of functionals $\{\log |\frac{f(z)}{z}|, \log |f'(z)|\}$ in the class **S**.

That time several powerful methods for solution of extremal problems competed. Most known are the variational method of Schiffer and Goluzin [226, 227], area principle [159], method of extremal metrics [153]. The problem is hard first of all because the univalence condition for the class **S** makes it a nonlinear manifold and all variational methods must be very special. In this situation, a considerable progress was achieved when a general optimization principle [192] appeared in 1964, now known as the Pontryagin maximum principle (PMP). It turned out that the Löwner and Löwner–Kufarev equations being viewed as evolution equations give controllable systems of differential equations where the driving term or control function is provided by the Carathéodory function p. Perhaps it was Löwner himself who first noticed all advantages of combination of his parametric method with PMP. His last student Gerald S. Goodman [94, 95] first explicitly stated this combination in 1966. However, he did not return to this topic, and later (and independently) Aleksandrov and Popov [10] gave a real start to investigations in this direction. Further applications of this combination were performed in [11] finding the range of the system of functionals $\{\text{Re}\,(e^{i\alpha}a_2), \text{Re}\,(a_3 - a_2^2)\}$ in the class **S**, $\alpha \in [0, \pi/2]$. We observe that during this period Vladimir Ivanovich Popov was a real driving force of this process, see, e.g., [195]. However, after some first success [12, 13] with already known problems or their modifications, some insuperable difficulties did not allow to continue with a significant progress although the ideas were rather clear. Real breakthrough happened in 1984 when Dmitri Prokhorov [199] started a series of papers in which he solved several new problems by applying PMP together with the Löwner–Kufarev equation (3). Let us describe a typical application of PMP in the Löwner–Kufarev theory. As an example we consider one of the most difficult problems in the theory of univalent functions, namely, description of the boundary of the coefficient body in the class **S**. A beautiful book by Schaeffer and Spenser [226] can be our starting point. If V_n denotes the nth coefficient body, $V_n = V_n(f) = \{a_2, a_3, \dots, a_n\}$, $f = z + a_2z^2 + a_3z^3 + \dots$ in the class **S**, then the first nontrivial body V_3 was completely described in [226]. The authors also give some qualitative description of general V_n. Several qualitative results on ∂V_n were obtained in the monograph [39]. In particular, Bobenko developed a method

D. V. Prokhorov

of second variation more general than that by Duren and Schiffer [71] and proved that ∂V_n is smooth except sets of smaller dimension.

Now let $f(z,t) = e^t w(z,t)$, where $w(z,t)$ is a solution to the Löwner equation (3) with the function p given by (2). We introduce the matrix $A(t)$ and the vector $a(t)$ as

$$A(t) = \begin{pmatrix} 0 & 0 & \dots & 0 & 0 \\ a_1(t) & 0 & \dots & 0 & 0 \\ a_2(t) & a_1(t) & \dots & 0 & 0 \\ \vdots & \vdots & \ddots & \vdots & \vdots \\ a_{n-1}(t) & a_{n-2}(t) & \dots & a_1(t) & 0 \end{pmatrix}, \quad a(t) = \begin{pmatrix} a_1(t) \\ a_2(t) \\ a_3(t) \\ \vdots \\ a_n(t) \end{pmatrix},$$

where $a_1(t) \equiv 1$ and $a_2(t), a_3(t), \dots$ are the coefficients of the function $f(z,t)$. Substituting $f(z,t)$ in (3) we obtain a controllable system

$$\frac{da(t)}{dt} = -2 \sum_{k=1}^{n-1} e^{-k(t+iu(t))} A^k(t) a(t), \tag{9}$$

with the initial condition $a^T(0) = (1, 0, 0, \dots, 0)$. The coefficient body V_n is a reachable set for the controllable system (9).

The results concerning the structure and properties of V_n include

(i) V_n is homeomorphic to a $(2n-2)$-dimensional ball and its boundary ∂V_n is homeomorphic to a $(2n-3)$-dimensional sphere;
(ii) every point $x \in \partial V_n$ corresponds to exactly one function $f \in \mathbf{S}$ which will be called a *boundary function* for V_n;
(iii) with the exception for a set of smaller dimension, at every point $x \in \partial V_n$ there exists a normal vector satisfying the Lipschitz condition;

(iv) there exists a connected open set X_1 on ∂V_n, such that the boundary ∂V_n is an analytic hypersurface at every point of X_1. The points of ∂V_n corresponding to the functions that give the extremum to a linear functional belong to the closure of X_1.

It is worth noticing that all boundary functions have a similar structure. They map the unit disk $\mathbb{D}$ onto the complex plane $\mathbb{C}$ minus piecewise analytic Jordan arcs forming a tree with a root at infinity and having at most $n-1$ tips. This assertion underlines the importance of multi-slit maps in the coefficient problem for univalent functions.

Solutions $a(t)$ to (9) for different control function $u(t)$ (piecewise continuous, in general) represent all points of ∂V_n as $t \to \infty$. The trajectories $a(t)$, $0 \le t < \infty$, fill V_n so that every point of V_n belongs to a certain trajectory $a(t)$. The endpoints of these trajectories can be interior or else boundary points of V_n. This way, we set V_n as the closure of the reachable set for the control system (9).

According to property (ii) of V_n, every point $x \in \partial V_n$ is attained by exactly one trajectory $a(t)$ which is determined by a choice of the piecewise continuous control function $u(t)$. The function $f \in \mathbf{S}$ corresponding to x is a multi-slit map of $\mathbb{D}$. If the boundary tree of f has only one tip, then there is a unique continuous control function $u(t)$ in $t \in [0, \infty)$ that corresponds to f. Otherwise one obtains multi-slit maps for piecewise continuous control u.

In order to reach a boundary point $x \in \partial V_n$ corresponding to a one-slit map, the trajectory $a(t)$ has to obey extremal properties, i.e., to be an *optimal trajectory*. The continuous control function $u(t)$ must be optimal, and hence it satisfies a necessary condition of optimality. PMP is a powerful tool to be used that provides a joint interpretation of two classical necessary variational conditions: the Euler equations and the Weierstrass inequalities (see, e.g., [192]).

To realize the maximum principle we consider an *adjoint vector* (or momenta)

$$\psi(t) = \begin{pmatrix} \psi_1(t) \\ \cdot \\ \cdot \\ \cdot \\ \psi_n(t) \end{pmatrix},$$

with the complex-valued coordinates $\psi_1, \dots, \psi_n$, and the *Hamiltonian function*

$$H(a, \psi, u) = \operatorname{Re}\left[\left(-2\sum_{k=1}^{n-1} e^{-k(t+iu(t))} A^k(t) a(t)\right)^T \bar{\psi}\right],$$

where $\bar{\psi}$ means the vector with complex conjugate coordinates. To come to the Hamiltonian formulation for the coefficient system we require that $\bar{\psi}$ satisfies the adjoint system of differential equations

$$\frac{d\bar{\psi}}{dt} = -\frac{\partial H}{\partial a}, \quad 0 \le t < \infty. \tag{10}$$

Taking into account (9) we rewrite (10) as

$$\frac{d\bar{\psi}}{dt} = 2\sum_{k=1}^{n-1} \left(e^{-k(t+iu(t))}(k+1)(A^T)^k\right)\bar{\psi}, \quad \psi(0) = \xi. \tag{11}$$

The maximum principle states that any optimal control function $u^*(t)$ possesses a maximizing property for the Hamiltonian function along the corresponding trajectory, i.e.,

$$\max_u H(a^*(t), \psi^*(t), u) = H(a^*(t), \bar{\psi}^*(t), u^*), \quad t \ge 0, \tag{12}$$

where a^* and ψ^* are solutions to the system (9), (11) with $u = u^*(t)$.

The maximum principle (12) yields that

$$\left.\frac{\partial H(a^*(t), \psi^*(t), u)}{\partial u}\right|_{u=u^*(t)} = 0. \tag{13}$$

Evidently, (9), (10), and (13) imply that

$$\frac{d\, H(a^*(t), \psi^*(t), u^*(t))}{dt} = 0, \tag{14}$$

for an optimal control function $u^*(t)$.

If the boundary function f gives a point at the boundary hypersurface ∂V_n, then its rotation is also a boundary function. The rotation operation $f(z) \to e^{-i\alpha} f(e^{i\alpha}z)$ gives a curve on ∂V_n and establishes a certain symmetry of the boundary hypersurface and a change of variables $u \to u + \alpha$, $\psi \to e^{i\alpha}\psi$. This allows us to normalize the adjoint vector as $\operatorname{Im}\psi_n = 0$ and $\psi_n = \pm 1$, which corresponds to the projection of V_n onto the hypersurface $\operatorname{Im} a_n = 0$ of dimension $2n - 3$. By abuse of notation, we continue to write V_n for this projection. Further study reduces to investigation of critical points of the equation $\frac{\partial}{\partial u} H(a(t), \psi(t), u) = 0$, and search and comparison of local extrema. In the case when two local maxima coincide, the Gamkrelidze theory [84] of sliding regimes comes into play. Then one considers the generalized Löwner equation in which the function p takes the form

$$p(z,t) = \sum_{k=1}^{n} \lambda_k \frac{e^{iu_k(t)} + z}{e^{iu_k(t)} - z}, \quad \sum_{k=1}^{n} \lambda_k = 1, \tag{15}$$

where $u_k(t)$ is a continuous functions regarding $t \in [0, \infty)$, and $\lambda_k \ge 0$ are constant.

Theorem 4 ([200, 202]). *Let $f \in \mathbf{S}$ give a nonsingular boundary point of the set V_n and let f map the unit disk onto the plane with piecewise-analytic slits having m finite tips. Then there exist m real-valued functions $u_1, \dots, u_m$ continuous on* $[0, \infty)$ *and positive numbers $\lambda_1, \dots, \lambda_m$, $\sum_{k=1}^{m} \lambda_k = 1$, such that a solution $w = w(z, t)$ to the Cauchy problem for the generalized Löwner differential equation (3,15) represents f according to the formula $f(z) = \lim_{t\to\infty} f(z, t)$. This representation is unique.*

The boundary ∂V_n is then parametrized by the initial conditions ξ in (11). If (11) allows to choose only one optimal control u^*, then the boundary point is given by a boundary function that maps the unit disk onto the plane with piecewise-analytic slit having a unique finite tips. If the Hamiltonian function has some m equal local maximums, then this case is called the sliding regime, and it is provided by $m - 1$ equations with respect to ξ, which determine the equality of the values of the Hamiltonian function at m critical points. The constants λ_k represent additional controls of the problem.

Therefore, the sliding regime of the optimal control problem with m maximum points of the Hamiltonian function is realized when $\xi \in \mathfrak{M}_m$, where the manifold $\mathfrak{M}_m$ of dimension $2n - 3 - m$ is defined by $m - 1$ equations for the values of the Hamiltonian function at m critical points for $t = 0$. It is shown in [202] that the sliding regime with m maximum points of function H at $t = 0$ preserves this property for $t > 0$. The number of maximum points may only decrease and only because of the joining of some of them at certain instants t.

Theorem 5 ([200, 202]). *The boundary hypersurface ∂V_n, $n \geq 2$, is a union of the sets $\Omega_1, \dots, \Omega_{n-1}$, every pair of which does not have mutual interior points. Each set Ω_m, $1 \geq m \geq n - 1$, corresponds to the manifold $\mathfrak{M}_m$, $\mathfrak{M}_1 = \mathbb{R}^{2n-4}$, so that the parametric representation*

$$\Omega_m = \left\{ a(\infty, \xi, \lambda) \colon \xi \in \mathfrak{M}_m;\ \xi_n = \pm 1;\ \lambda_1, \dots, \lambda_m \geq 0;\ \sum_{k=1}^{m} \lambda_k = 1 \right\}$$

holds, where $a(\infty, \xi, \lambda)$ is the manifold coordinate of the system of bicharacteristics a, ψ for the Hamiltonian system (9), (11) with continuous branches of the optimal controls given by (13).

This theorem can be used in investigations of topologic, metric, smooth, and analytic properties of the boundary hypersurface ∂V_n. In particular, the qualitative results in [39, 226] can be obtained as corollaries. Equally, this method work for the subclass $\mathbf{S}^M$ of bounded functions $|f(z)| \leq M$ from $\mathbf{S}$. In this case, one stops at the moment $t = \log M$.

Let us here mention a slightly different approach proposed by Friedland and Schiffer [79, 80] in 1976–1977 to the problem of description of V_n. Instead of the ordinary Löwner differential equation (3), they considered the Löwner partial differential equation (1) with the function p in the form (2). Their main conclusion states that if the initial condition for (1) is a boundary function for V_n, then the solutions to (1) form a one-parameter curves on ∂V_n (for special choice of u). The conclusions on the special character of the Koebe functions reflect the angular property of the corresponding points at the boundary hypersurface.

Besides these general results on the coefficient body several concrete problems and conjectures were solved by this method. In particular,several projections of V_3 were described in [250]. The range of the system of set of functionals $\{f(r_1), f(r_2)\}$, $0 < r_1 < r_2 < 1$, was given in [252] in the subclass $\mathbf{S}_R$ of functions from $\mathbf{S}$ with real coefficients. Prokhorov [201] described V_4 in the class $\mathbf{S}_R^M = \mathbf{S}_R \cap \mathbf{S}^M$ generalizing and improving Tammi's results [246, 247]. The range of the system of functionals $\{f(r), a_2, a_3\}$ was described in [251] in the class $\mathbf{S}_R^M$.

We would like to mention here some conjectures which were confirmed or disproved by the above described method. First let us introduce the so-called Pick function that plays in the class $\mathbf{S}^M$ a role analogous to the Koebe function $k_\theta(z) = \frac{z}{(1-e^{i\theta}z)^2}$ in the class $\mathbf{S}$. The Pick function $p_\theta(z)$ satisfies the equation

$$\frac{M^2 p_\theta}{(M - p_\theta)^2} = k_\theta(z), \quad M > 1, \quad z \in \mathbb{D}.$$

The coefficients of $p_0(z) = z + p_2(M)z^2 + \ldots$ we denote by $p_n(M)$. The Jakubowski conjecture says that the estimate $|a_n| \leq p_n(M)$ holds for even n in the class $\mathbf{S}^M$ for sufficiently large M. This conjecture was proved in [203]. As for odd coefficients of univalent functions, it is easy to verify that the necessary maximality conditions for the Pick functions fail for large M, see [203].

Bombieri [40] in 1967 posed the problem to find

$$\sigma_{mn} := \liminf_{f \to K, f \in S} \frac{n - \mathbb{R}a_n}{m - \mathbb{R}a_m}, \quad m, n \geq 2,$$

$f \to K$ locally uniformly in U. We call σ_{mn} the Bombieri numbers. He conjectured that $\sigma_{mn} = B_{mn}$, where

$$B_{mn} = \min_{\theta \in [0, 2\pi)} \frac{n \sin\theta - \sin(n\theta)}{m \sin\theta - \sin(m\theta)}.$$

and proved that $\sigma_{mn} \leq B_{mn}$ for $m = 3$ and n odd. It is noteworthy that Bshouty and Hengartner [47] proved Bombieri's conjecture for functions from S having real coefficients in their Taylor's expansion. Continuing this contribution by Bshouty and Hengartner, the conjecture for the whole class S has been disproved by Greiner and Roth [112] for $n = 2$, $m = 3$, $f \in S$. Actually, they have got the sharp Bombieri number $\sigma_{32} = (e - 1)/4e < 1/4 = B_{32}$.

It is easily seen that $\sigma_{43} = B_{43} = \sigma_{23} = B_{23} = 0$. Applying Löwner's parametric representation for univalent functions and the optimal control method we found [208] the exact Bombieri numbers $\sigma_{42}, \sigma_{24}, \sigma_{34}$ and their numerical approximations $\sigma_{42} \approx 0.050057\ldots$, $\sigma_{24} \approx 0.969556\ldots$, and $\sigma_{34} \approx 0.791557\ldots$ (the Bombieri conjecture for these permutations of m, n suggests $B_{42} = 0.1$, $B_{24} = 1$, $B_{34} = 0.828427\ldots$). Of course, our method permits us to reprove the result of [112] about σ_{32}.

Our next target is the fourth coefficient a_4 of a function from $\mathbf{S}^M$. The sharp estimate $|a_2| \leq 2(1 - 1/M) = p_2(M)$ in the class S^M is rather trivial and has been obtained by Pick [186] in 1917. The next coefficient a_3 was estimated independently by Schaeffer and Spencer [225] in 1945 and Tammi [244] in 1953. The Pick function does not give the maximum to $|a_3|$ and the estimate is much more difficult. Schiffer and Tammi [228] in 1965 found that $|a_4| \leq p_4(M)$ for any $f \in S^M$ with $M > 300$. This result was repeated by Tammi [246, p. 210] in a weaker form ($M > 700$) and there it was conjectured that this constant could be decreased until 11. The case of function with real coefficients is simpler: the Pick function gives the maximum to $|a_4|$ for $M \geq 11$ and this constant is sharp (see [245], [247, p. 163]). By our suggested method we showed [208] that the Pick function locally maximizes $|a_4|$ on S^M if $M > M_0 = 22.9569\ldots$ and does not for $1 < M < M_0$. This disproves Tammi's conjecture.

Among other results obtained by PMP we mention here the Charzyński–Tammi conjecture on functions from $\mathbf{S}^M$ close to identity (or M close to 1). The suggested extremal function analogous to the Koebe and Pick functions is $n-1$-symmetric function

$$Q_{n-1}^M(z) = \left(p_0^{M^{n-1}}(z^{n-1})\right)^{1/(n-1)}.$$

The conjecture proved by Schiffer and Tammi [228] and Siewierski [237] states that the coefficient of the function $Q_{n-1}^M(z)$ gives the extremum to the coefficient a_n of a function from $\mathbf{S}^M$, namely

$$|a_n| \leq \frac{2}{n-1}\left(1 - \frac{1}{M^{n-1}}\right), \quad n \geq 2.$$

Prokhorov [204] found an asymptotic estimate in the above problem in terms of $\log^2 M$ as M is close to 1. For $M \to \infty$, Prokhorov and Nikulin [206] obtained also asymptotic estimates in the coefficient problem for the class $\mathbf{S}^M$ with. In particular,

$$|a_n| \leq n - \frac{n(n^2-1)}{3}\frac{1}{M} + o\left(\frac{1}{M}\right), \quad M \to \infty.$$

More about coefficient problems solved by PMP and Löwner–Kufarev theory, see [205].

7 One-Slit Maps

It is interesting that in spite of many known properties of the Löwner equations several geometric questions remained unclear until recently.

Let us return to the original Löwner equation (3) with the driving function p given by (2)

$$\frac{dw}{dt} = -w\frac{e^{iu(t)} + w}{e^{iu(t)} - w}, \quad w(z,0) \equiv z, \tag{16}$$

Solutions $w(z,t)$ to (16) map $\mathbb{D}$ onto $\Omega(t) \subset \mathbb{D}$. If $\Omega(t) = \mathbb{D} \setminus \gamma(t)$, where $\gamma(t)$ is a Jordan curve in $\mathbb{D}$ except one of its endpoints, then the driving term $u(t)$ is uniquely defined and we call the corresponding map w a *slit map*. However, from 1947 [147] it is known that solutions to (16) with continuous $u(t)$ may give non-slit maps, in particular, $\Omega(t)$ can be a family of hyperbolically convex digons in $\mathbb{D}$.

Marshall and Rohde [173] addressed the following question: *Under which condition on the driving term $u(t)$ the solution to (16) is a slit map?* Their result states that if $u(t)$ is Lip(1/2) (Hölder continuous with exponent 1/2), and if for a certain constant $C_{\mathbb{D}} > 0$, the norm $\|u\|_{1/2}$ is bounded $\|u\|_{1/2} < C_{\mathbb{D}}$, then the solution w is a slit map, and moreover, the Jordan arc $\gamma(t)$ is s quasislit (a quasiconformal image of an interval within a Stolz angle). As they also proved, a converse statement without the norm restriction holds. The absence of the norm restriction in the latter result is essential. On the one hand, Kufarev's example [147] contains $\|u\|_{1/2} = 3\sqrt{2}$, which means that $C_{\mathbb{D}} \leq 3\sqrt{2}$. On the other hand, Kager et al. [129] constructed exact slit solutions to the half-plane version of the Löwner equation with arbitrary norms of the driving term.

Reminding the half-plane version of the Löwner equation let $\mathbb{H} = \{z : \operatorname{Im} z > 0\}$, $\mathbb{R} = \partial\mathbb{H}$. The functions $h(z,t)$, normalized near infinity by $h(z,t) = z - 2t/z + b_{-2}(t)/z^2 + \dots$, solving the equation

$$\frac{dh}{dt} = \frac{-2}{h - \lambda(t)}, \quad h(z,0) \equiv z, \tag{17}$$

where $\lambda(t)$ is a real-valued continuous driving term, map $\mathbb{H}$ onto a subdomain of $\mathbb{H}$. The difference in the sign between (17) and (6) is because in (6) the equation is for the inverse mapping. In some papers, e.g., [129, 162], the authors work with (16), (17) changing $(-)$ to $(+)$ in their right-hand sides, and with the mappings of slit domains onto $\mathbb{D}$ or $\mathbb{H}$. However, the results remain the same for both versions.

The question about the slit mappings and the behavior of the driving term $\lambda(t)$ in the case of the half-plane $\mathbb{H}$ was addressed by Lind [162]. The techniques used by Marshall and Rohde carry over to prove a similar result in the case of (17), see [173, p. 765]. Let us denote by $C_{\mathbb{H}}$ the corresponding bound for the norm $\|\lambda\|_{1/2}$. The main result by Lind is the sharp bound, namely $C_{\mathbb{H}} = 4$.

Marshall and Rohde [173] remarked that there exist many examples of driving terms $u(t)$ which are not Lip(1/2), but which generate slit solutions with simple arcs $\gamma(t)$. In particular, if $\gamma(t)$ is tangent to $\mathbb{T}$, then $u(t)$ is never Lip(1/2).

Our result [209] states that if $\gamma(t)$ is a circular arc tangent to $\mathbb{R}$, then the driving term $\lambda(t)$ ∈Lip(1/3). Besides, we prove that $C_{\mathbb{D}} = C_{\mathbb{H}} = 4$ and consider properties of singular solutions to the one-slit Löwner equation. Moreover, examples of non-slit solutions filling the whole spectrum $[4, \infty)$ were given in [127, 163].

The authors analyzed in [163] Löwner traces $\gamma(t)$ driven by $\lambda(t)$ asymptotic to $k\sqrt{1-t}$. They proved a form of stability of the self-intersection for such $\lambda(t)$. Being slightly rephrased it reads as follows.

Theorem 6 ([163]). *Let the driving term* $\lambda\colon [0,1] \to \mathbb{R}$ *be sufficiently regular with the above asymptotic of* $\lambda(t)$

$$\lim_{t\to 1} \frac{|\lambda(1)-\lambda(t)|}{\sqrt{1-t}} = k > 4.$$

Then $\gamma(1-0)$ *exists, is real, and* γ *intersects* $\mathbb{R}$ *at the same angle as the trace for* $\lambda = k\sqrt{1-t}$.

Namely,

$$\lim_{t\to 1} \arg(\gamma(t)-\gamma(1)) = \pi \frac{1-\sqrt{1-16/k^2}}{1+\sqrt{1-16/k^2}}.$$

The method of proof of the above theorem also applies to the case $|k| < 4$. In this case the trace γ driven by λ is a Jordan arc, and γ is asymptotically similar to the logarithmic spiral at $\gamma(1) \in \mathbb{H}$.

Another our result [127] states that an analytic orthogonal slit requires the 1/2 Lipschitz vanishing norm, exactly as in Kadanoff's et al. examples [129] with a line-slit and a circular slit. In this case the conformal radius approaching the origin is of order Lip 1/2 (compare with Earle and Epstein [73]).

8 Univalence Criteria

Several important univalence criteria can be obtained by means of the Löwner–Kufarev differential equation. For example, a function $f(z) = z + \dots$ analytic in $\mathbb{D}$ is spirallike of type $\alpha \in (-\pi/2, \pi/2)$ (and therefore, univalent) if and only if

$$\operatorname{Re}\left(e^{i\alpha} z \frac{f'(z)}{f(z)}\right) > 0, \quad \text{in } \mathbb{D},$$

see [214, 242] and [190, p. 172]. Spirallikeness means that a function $f(z)$ is analytic, univalent, and if $w \in f(\mathbb{D})$, $\tau \geq 0$, then $we^{-e^{-i\alpha}\tau} \in f(\mathbb{D})$. If $\alpha = 0$, then we obtain the usual class of starlike functions S^*.

Next criterion is obtained by integration of the Löwner–Kufarev equation (3) with a special choice of the driving function p from the Carathéodory class. Let us choose

$$p(z,t) = \frac{1}{h(z) + th_0(z)}, \quad h_0(z) = i\beta + \alpha\frac{zg'(z)}{g(z)}, \quad g \in S^*.$$

Integrating (3) as a Bernoulli-type equation, and letting $t \to \infty$ leads to the limiting function

$$f(z) = \left(\frac{\alpha + i\beta}{1 + i\alpha}\int_0^z h(z)z^{i\beta-1}g^{\alpha}(z)\,dz\right)^{1/(\alpha+i\beta)} = z + \dots,$$

$\alpha > 0$, $\beta \in \mathbb{R}$, which is the Bazilevich class $B_{\alpha,\beta}$ introduced in [29, 30]. Prokhorov [198] and Sheil-Small [235] proved that the class $B_{\alpha,\beta}$ is characterized by the condition

$$\int_{\theta_1}^{\theta_2} \operatorname{Re} F(re^{i\theta})\,d\theta > -\pi, \qquad 0 < r < 1, \quad 0 < \theta_2 - \theta_1 < 2\pi,$$

$$F(z) = 1 + z\frac{f''(z)}{f'(z)} + (\alpha - 1 + i\beta)z\frac{f'(z)}{f(z)},$$

under the additional assumption that $f(z)f'(z)/z \neq 0$ on $\mathbb{T} = \partial\mathbb{D}$. Moreover, the boundary of $f(\mathbb{D})$ is accessible from outside by the curves $w = a(1 + bt)^{1/(\alpha+i\beta)}$.

In 1972, Becker [31] assumed that the driving function p in (3) satisfied the inequality

$$\left|\frac{p(z,t) - 1}{p(z,t) + 1}\right| \leq k < 1, \quad z \in \mathbb{D}, \quad 0 \leq t < \infty.$$

Then the solutions in the form (4) have k-quasiconformal extension to the Riemann sphere $\hat{\mathbb{C}}$. This allowed him to show that the inequality

$$(1 - |z|^2)\left|z\frac{f''(z)}{f'(z)}\right| \leq k < 1$$

in $\mathbb{D}$ for an analytic function $f(z) = z + \dots$ implies that f is univalent and has a k-quasiconformal extension to $\hat{\mathbb{C}}$. This improves a result of Duren et al. [68]. Later in 1984 Becker and Pommerenke [32] established the criteria

$$(1-|z|^2)|zf''(z)/f'(z)| \le 1,\ f'(0) \ne 0\ (z \in \mathbb{D})$$
$$2\,\mathrm{Re}\, z|f''(z)/f'(z)| \le 1\ (z \in \mathbb{H})$$
$$(|z|^2-1)|zf''(z)/f'(z)| \le 1\ (z \in \mathbb{D}^-),$$

where $\mathbb{H}$ is the right half-plane and $\mathbb{D}^-$ is the exterior of $\mathbb{D}$. In all inequalities the constant 1 is the best possible. The first criterium implies that the boundary of $f(\mathbb{D})$ is Jordan whereas the second and the third do not necessary imply this. Various univalence conditions were obtained later, see [5] for more complete list of them.

9 Semigroups

Looking at the classical radial Löwner–Kufarev equation (3) and the classical chordal Löwner–Kufarev equation (6), one notices that there is a similitude between the two. Indeed, we can write both equations in the form

$$\frac{dz(t)}{dt} = G(z,t),$$

with

$$G(z,t) = (\tau - z)(1 - \overline{\tau} z)p(z,t),$$

where $\tau = 0, 1$ and $\mathrm{Re}\, p(z,t) \ge 0$ for all $z \in \mathbb{D}$ and $t \ge 0$.

The reason for the previous formula is not at all by chance, but it reflects a very important feature of "Herglotz vector fields" (see Sect. 10 for the definition). In order to give a rough idea of what we are aiming, consider the case $\tau = 0$ (the radial case). Fix $t_0 \in [0, +\infty)$. Consider the holomorphic vector field $G(z) := G(z, t_0)$. Let $g(z) := |z|^2$. Then,

$$dg_z(G(z)) = 2\mathrm{Re}\,\langle G(z), z\rangle = -|z|^2 \mathrm{Re}\, p(z,t_0) \le 0, \quad \forall z \in \mathbb{D}. \tag{18}$$

This Lyapunov-type inequality has a deep geometrical meaning. Indeed, (18) tells that G points toward the center of the level sets of g, which are concentric circles centered at 0. For each $z_0 \in \mathbb{D}$, consider then the Cauchy problem

$$\begin{cases} \frac{dw(t)}{dt} = G(w(t)), \\ w(0) = z_0 \end{cases} \tag{19}$$

and let $w^{z_0} : [0, \delta) \to \mathbb{D}$ be the maximal solution (such a solution can propagate also in the "past," but we just consider the "future" time). Since G points inward with respect to all circles centered at 0, the flow $t \mapsto w^{z_0}(t)$ cannot escape from the circle of radius $g(z_0)$. Therefore, the flow is defined for all future times, namely, $\delta = +\infty$. This holds for all $z_0 \in \mathbb{D}$.

Hence, the vector field G has the feature to be $\mathbb{R}^+$-semicomplete, that is, the maximal solution of the initial value problem (19) is defined in the interval $[0, +\infty)$. To understand how one can unify both radial and chordal Löwner theory, we dedicate this section to such vector fields and their flows.

A family of holomorphic self-maps of the unit disk (ϕ_t) is a *(one-parameter) semigroup (of holomorphic functions)* if $\phi: (\mathbb{R}^+, +) \to \mathsf{Hol}(\mathbb{D}, \mathbb{D})$ is a continuous homomorphism between the semigroup of nonnegative real numbers and the semigroup of holomorphic self-maps of the disk with respect to composition, endowed with the topology of uniform convergence on compact sets. In other words:

- $\phi_0 = \mathrm{id}_{\mathbb{D}}$;
- $\phi_{t+s} = \phi_s \circ \phi_t$ for all $s, t \geq 0$;
- ϕ_t converges to ϕ_{t_0} uniformly on compact sets as t goes to t_0.

It can be shown that if (ϕ_t) is a semigroup, then ϕ_t is univalent for all $t \geq 0$.

Semigroups of holomorphic maps are a classical subject of study, both as (local/global) flows of continuous dynamical systems and from the point of view of "fractional iteration," the problem of embedding the discrete set of iterates generated by a single self-map into a one-parameter family (a problem that is still open even in the disk). It is difficult to exactly date the birth of this notion but it seems that the first paper dealing with semigroups of holomorphic maps and their asymptotic behavior is due to Tricomi in 1917 [248]. Semigroups of holomorphic maps also appear in connection with the theory of Galton–Watson processes (branching processes) started in the 1940s by Kolmogorov and Dmitriev [118]. An extensive recent survey [106] gives a complete overview on details. Furthermore, they are an important tool in the theory of strongly continuous semigroups of operators between spaces of analytic functions (see, for example, [239]).

A very important contribution to the theory of semigroups of holomorphic self-maps of the unit disk is due to Berkson and Porta [36]. They proved that:

Theorem 7 ([36]). *A semigroup of holomorphic self-maps of the unit disk (ϕ_t) is in fact real-analytic in the variable t, and is the solution of the Cauchy problem*

$$\frac{\partial \phi_t(z)}{\partial t} = G(\phi_t(z)), \quad \phi_0(z) = z\,, \tag{20}$$

where the map G, the infinitesimal generator of the semigroup, has the form

$$G(z) = (z - \tau)(\overline{\tau} z - 1)\, p(z) \tag{21}$$

for some $\tau \in \overline{\mathbb{D}}$ and a holomorphic function $p: \mathbb{D} \to \mathbb{C}$ with $\operatorname{Re} p \geq 0$.

Conversely, any vector field of the form (21) *is semicomplete and if, for $z \in \mathbb{D}$, we take w^z the solution of the initial value problem*

$$\frac{dw}{dt} = G(w), \quad w(0) = z,$$

then $\phi_t(z) := w^z(t)$ *is a semigroup of analytic functions.*

Expression (21) of the infinitesimal generator is known as its *Berkson–Porta decomposition*. Other characterizations of vector fields which are infinitesimal generators of semigroups can be seen in [41] and references therein.

The dynamics of the semigroup (ϕ_t) are governed by the analytical properties of the infinitesimal generator G. For instance, all the functions of the semigroup have a common fixed point at τ (in the sense of non-tangential limit if τ belongs to the boundary of the unit disk) and asymptotically tends to τ, which can thus be considered a sink point of the dynamical system generated by G.

When $\tau = 0$, it is clear that (21) is a particular case of (3), because the infinitesimal generator G is of the form $-wp(w)$, where p belongs to the Carathéodory class. As a consequence, when the semigroup has a fixed point in the unit disk (which, up to a conjugation by an automorphism of the disk, amounts to taking $\tau = 0$), *once differentiability in t is proved* Berkson–Porta's theorem can be easily deduced from Löwner's theory. However, when the semigroup has no common fixed points in the interior of the unit disk, Berkson–Porta's result is really a new advance in the theory.

We have already remarked that semigroups give rise to evolution families; they also provide examples of Löwner chains. Indeed, Heins [123] and Siskasis [238] have independently proved that if (ϕ_t) is a semigroup of holomorphic self-maps of the unit disk, then there exists a (unique, when suitably normalized) holomorphic function $h\colon \mathbb{D} \to \mathbb{C}$, the *Königs function* of the semigroup, such that $h(\phi_t(z)) = m_t(h(z))$ for all $t \geq 0$, where m_t is an affine map (in other words, the semigroup is semiconjugated to a semigroup of affine maps). Then it is easy to see that the maps $f_t(z) := m_t^{-1}(h(z))$, for $t \geq 0$, form a Löwner chain (in the sense explained in the next section).

The theory of semigroups of holomorphic self-maps has been extensively studied and generalized: to Riemann surfaces (in particular, Heins [123] has shown that Riemann surfaces with non-Abelian fundamental group admit no nontrivial semigroup of holomorphic self-maps); to several complex variables; and to infinitely dimensional complex Banach spaces, by Baker, Cowen, Elin, Goryainov, Poggi-Corradini, Pommerenke, Reich, Shoikhet, Siskakis, Vesentini, and many others. We refer to [41] and the books [1] and [212] for references and more information on the subject.

10 General Theory: Herglotz Vector Fields, Evolution Families, and Löwner Chains

Although the chordal and radial versions of the Löwner theory share common ideas and structure, on their own they can only be regarded as parallel but independent theories. The approach of [70, 101–105] does show that there can be (and actually are) much more independent variants of Löwner evolution bearing similar structures, but does not solve the problem of constructing a unified theory covering all the cases.

Recently a new unifying approach has been suggested by Gumenyuk and the three first authors [20, 42, 43, 53]. In the previous section we saw that the vector fields which appear in radial and chordal Löwner equations have the property to be infinitesimal generators for all fixed times. We will exploit such a fact to define a general family of Herglotz vector fields. Thus, relying partially on the theory of one-parametric semigroups, which can be regarded as the autonomous version of Löwner theory, we can build a new general theory.

Definition 1 ([42]). Let $d \in [1, +\infty]$. A *weak holomorphic vector field of order d* in the unit disk $\mathbb{D}$ is a function $G : \mathbb{D} \times [0, +\infty) \to \mathbb{C}$ with the following properties:

WHVF1. for all $z \in \mathbb{D}$, the function $[0, +\infty) \ni t \mapsto G(z, t)$ is measurable,
WHVF2. for all $t \in [0, +\infty)$, the function $\mathbb{D} \ni z \mapsto G(z, t)$ is holomorphic,
WHVF3. for any compact set $K \subset \mathbb{D}$ and any $T > 0$ there exists a nonnegative function $k_{K,T} \in L^d([0, T], \mathbb{R})$ such that

$$|G(z, t)| \leq k_{K,T}(t)$$

for all $z \in K$ and for almost every $t \in [0, T]$.

We say that G is a *(generalized) Herglotz vector field* of order d if, in addition to conditions WHVF1–WHVF3 above, for almost every $t \in [0, +\infty)$ the holomorphic function $G(\cdot, t)$ is an infinitesimal generator of a one-parametric semigroup in $\mathsf{Hol}(\mathbb{D}, \mathbb{D})$.

Herglotz vector fields in the unit disk can be decomposed by means of Herglotz functions (and this the reason for the name). We begin with the following definition:

Definition 2. Let $d \in [1, +\infty]$. A *Herglotz function of order d* is a function $p : \mathbb{D} \times [0, +\infty) \mapsto \mathbb{C}$ with the following properties:

1. For all $z \in \mathbb{D}$, the function $[0, +\infty) \ni t \mapsto p(z, t) \in \mathbb{C}$ belongs to $L^d_{loc}([0, +\infty), \mathbb{C})$;
2. For all $t \in [0, +\infty)$, the function $\mathbb{D} \ni z \mapsto p(z, t) \in \mathbb{C}$ is holomorphic;
3. For all $z \in \mathbb{D}$ and for all $t \in [0, +\infty)$, we have $\operatorname{Re} p(z, t) \geq 0$.

Then we have the following result which, using the Berkson–Porta formula, gives a general form of the classical Herglotz vector fields:

Theorem 8 ([42]). *Let $\tau : [0, +\infty) \to \overline{\mathbb{D}}$ be a measurable function and let $p : \mathbb{D} \times [0, +\infty) \to \mathbb{C}$ be a Herglotz function of order $d \in [1, +\infty)$. Then the map $G_{\tau,p} : \mathbb{D} \times [0, +\infty) \to \mathbb{C}$ given by*

$$G_{\tau,p}(z, t) = (z - \tau(t))(\overline{\tau(t)}z - 1)p(z, t),$$

for all $z \in \mathbb{D}$ *and for all* $t \in [0, +\infty)$, *is a Herglotz vector field of order* d *on the unit disk.*

Conversely, if $G : \mathbb{D} \times [0, +\infty) \to \mathbb{C}$ *is a Herglotz vector field of order* $d \in [1, +\infty)$ *on the unit disk, then there exist a measurable function* $\tau : [0, +\infty) \to \overline{\mathbb{D}}$ *and a Herglotz function* $p : \mathbb{D} \times [0, +\infty) \to \mathbb{C}$ *of order* d *such that* $G(z, t) = G_{\tau,p}(z, t)$ *for almost every* $t \in [0, +\infty)$ *and all* $z \in \mathbb{D}$.

Moreover, if $\tilde{\tau} : [0, +\infty) \to \overline{\mathbb{D}}$ *is another measurable function and* $\tilde{p} : \mathbb{D} \times [0, +\infty) \to \mathbb{C}$ *is another Herglotz function of order* d *such that* $G = G_{\tilde{\tau},\tilde{p}}$ *for almost every* $t \in [0, +\infty)$, *then* $p(z, t) = \tilde{p}(z, t)$ *for almost every* $t \in [0, +\infty)$ *and all* $z \in \mathbb{D}$ *and* $\tau(t) = \tilde{\tau}(t)$ *for almost all* $t \in [0, +\infty)$ *such that* $G(\cdot, t) \not\equiv 0$.

Remark 1. The generalized Löwner–Kufarev equation

$$\frac{dw}{dt} = G(w, t), \quad t \geq s, \quad w(s) = z, \tag{22}$$

resembles the radial Löwner–Kufarev ODE when $\tau \equiv 0$ and $p(0, t) \equiv 1$. Furthermore, with the help of the Cayley map between $\mathbb{D}$ and $\mathbb{H}$, the chordal Löwner equation appears to be the special case of (22) with $\tau \equiv 1$.

We also give a generalization of the concept of evolution families in *the whole semigroup* $\mathsf{Hol}(\mathbb{D}, \mathbb{D})$ as follows:

Definition 3. ([42]) A family $(\varphi_{s,t})_{t \geq s \geq 0}$ of holomorphic self-maps of the unit disk is an *evolution family of order* d with $d \in [1, +\infty]$ if

EF1. $\varphi_{s,s} = \mathrm{Id}_{\mathbb{D}}$ for all $s \geq 0$,
EF2. $\varphi_{s,t} = \varphi_{u,t} \circ \varphi_{s,u}$ whenever $0 \leq s \leq u \leq t < +\infty$,
EF3. for every $z \in \mathbb{D}$ and every $T > 0$ there exists a nonnegative function $k_{z,T} \in L^d([0, T], \mathbb{R})$ such that

$$|\varphi_{s,u}(z) - \varphi_{s,t}(z)| \leq \int_u^t k_{z,T}(\xi) d\xi$$

whenever $0 \leq s \leq u \leq t \leq T$.

Condition EF3 is to guarantee that any evolution family can be obtained via solutions of an ODE which resembles both the radial and chordal Löwner–Kufarev equations. The vector fields that drive this generalized Löwner–Kufarev ODE are referred to as *Herglotz vector fields.*

Remark 2. Definition 3 does not require elements of an evolution family to be univalent. However, this condition is satisfied. Indeed, by Theorem 9, any evolution family $(\varphi_{s,t})$ can be obtained via solutions to the generalized Löwner–Kufarev ODE. Hence the univalence of $\varphi_{s,t}$'s follows from the uniqueness of solutions to this ODE. For an essentially different direct proof, see [43, Proposition 3].

Remark 3. Different notions of evolution families considered previously in the literature can be reduced to special cases of L^d-evolution families defined above.

The Schwarz lemma and distortion estimates imply that solutions of the classical Löwner–Kufarev equation (3) are evolution families of order ∞. Also, it can be proved that for all semigroups of analytic maps (ϕ_t), the biparametric family $(\varphi_{s,t}) := (\phi_{t-s})$ is also an evolution family of order ∞.

Equation (22) establishes a 1-to-1 correspondence between evolution families of order d and Herglotz vector fields of the same order. Namely, the following theorem takes place.

Theorem 9 ([42, Theorem 1.1]). *For any evolution family $(\varphi_{s,t})$ of order $d \in [1, +\infty]$ there exists an (essentially) unique Herglotz vector field $G(z,t)$ of order d such that for every $z \in \mathbb{D}$ and every $s \geq 0$ the function $[s, +\infty) \ni t \mapsto w_{z,s}(t) := \varphi_{s,t}(z)$ solves the initial value problem (22).*

Conversely, given any Herglotz vector field $G(z,t)$ of order $d \in [1, +\infty]$, for every $z \in \mathbb{D}$ and every $s \geq 0$ there exists a unique solution $[s, +\infty) \ni t \mapsto w_{z,s}(t)$ to the initial value problem (22). The formula $\varphi_{s,t}(z) := w_{z,s}(t)$ for all $s \geq 0$, all $t \geq s$, and all $z \in \mathbb{D}$, defines an evolution family $(\varphi_{s,t})$ of order d.

Here by *essential uniqueness* we mean that two Herglotz vector fields $G_1(z,t)$ and $G_2(z,t)$ corresponding to the same evolution family must coincide for a.e. $t \geq 0$.

The strong relationship between semigroups and evolution families on the one side and Herglotz vector fields and infinitesimal generators on the other side is very much reflected by the so-called product formula in convex domains of Reich and Shoikhet [212]. Such a formula can be rephrased as follows: let $G(z,t)$ be a Herglotz vector field. For almost all $t \geq 0$, the holomorphic vector field $\mathbb{D} \ni z \mapsto G(z,t)$ is an infinitesimal generator. Let (ϕ_r^t) be the associated semigroups of holomorphic self-maps of $\mathbb{D}$. Let $(\varphi_{s,t})$ be the evolution family associated with $G(z,t)$. Then, uniformly on compacta of $\mathbb{D}$ it holds

$$\phi_t^r = \lim_{m\to\infty} \varphi_{t,t+\frac{r}{m}}^{\circ m} = \lim_{m\to\infty} \underbrace{(\varphi_{t,t+\frac{r}{m}} \circ \ldots \circ \varphi_{t,t+\frac{r}{m}})}_{m}.$$

Using such a formula for the case of the unit disk $\mathbb{D}$, in [44] it has been proved the following result which gives a description of semigroups-type evolution families:

Theorem 10. *Let $G(z,t)$ be a Herglotz vector field of order d in $\mathbb{D}$ and let $(\varphi_{s,t})$ be its associated evolution family. The following are equivalent:*

1. *there exists a function $g \in L^d_{loc}([0, +\infty), \mathbb{C})$ and an infinitesimal generator H such that $G(z,t) = g(t)H(z)$ for all $z \in \mathbb{D}$ and almost all $t \geq 0$,*
2. *$\varphi_{s,t} \circ \varphi_{u,v} = \varphi_{u,v} \circ \varphi_{s,t}$ for all $0 \leq s \leq t$ and $0 \leq u \leq v$.*

In order to end up the picture started with the classical Löwner theory, we should put in the frame also the Löwner chains. The general notion of a Löwner chain has been given[2] in [53].

Definition 4 ([53]). A family $(f_t)_{t\geq 0}$ of holomorphic maps of $\mathbb{D}$ is called a *Löwner chain of order* d with $d \in [1, +\infty]$ if it satisfies the following conditions:

LC1. each function $f_t : \mathbb{D} \to \mathbb{C}$ is univalent,
LC2. $f_s(\mathbb{D}) \subset f_t(\mathbb{D})$ whenever $0 \leq s < t < +\infty$,
LC3. for any compact set $K \subset \mathbb{D}$ and any $T > 0$ there exists a nonnegative function $k_{K,T} \in L^d([0, T], \mathbb{R})$ such that

$$|f_s(z) - f_t(z)| \leq \int_s^t k_{K,T}(\xi) d\xi$$

whenever $z \in K$ and $0 \leq s \leq t \leq T$.

This definition of (generalized) Löwner chains matches the abstract notion of evolution family introduced in [42]. In particular the following statement holds.

Theorem 11 ([53, Theorem 1.3]). *For any Löwner chain* (f_t) *of order* $d \in [1, +\infty]$, *if we define*

$$\varphi_{s,t} := f_t^{-1} \circ f_s \quad \textit{whenever } 0 \leq s \leq t,$$

then $(\varphi_{s,t})$ *is an evolution family of the same order* d. *Conversely, for any evolution family* $(\varphi_{s,t})$ *of order* $d \in [1, +\infty]$, *there exists a Löwner chain* (f_t) *of the same order* d *such that*

$$f_t \circ \varphi_{s,t} = f_s \quad \textit{whenever } 0 \leq s \leq t.$$

In the situation of this theorem we say that the Löwner chain (f_t) and the evolution family $(\varphi_{s,t})$ are *associated with* each other. It was proved in [53] that given an evolution family $(\varphi_{s,t})$, an associated Löwner chain (f_t) is unique up to conformal maps of $\cup_{t\geq 0} f_t(\mathbb{D})$. Thus there are essentially one two different types of Löwner chains: those such that $\cup_{t\geq 0} f_t(\mathbb{D}) = \mathbb{C}$ and those such that $\cup_{t\geq 0} f_t(\mathbb{D})$ is a simply connected domain different from $\mathbb{C}$ (see [53] for a characterization in terms of the evolution family associated with).

Thus in the framework of the approach described above the essence of Löwner theory is represented by the essentially 1-to-1 correspondence among Löwner chains, evolution families, and Herglotz vector fields.

---

[2]See also [20] for an extension of this notion to complex manifolds and with a complete different approach even in the unit disk. The construction of a Löwner chain associated with a given evolution family proposed there differs essentially from the one we used in [53, Theorems 1.3 and 1.6].

Once the previous correspondences are established, given a Löwner chain (f_t) of order d, the general Löwner–Kufarev PDE

$$\frac{\partial f_t(z)}{\partial t} = -G(z,t)\frac{\partial f_t(z)}{\partial z}. \tag{23}$$

follows by differentiating the structural equation

$$\frac{dz}{dt} = G(z,t), \qquad z(0) = z. \tag{24}$$

Conversely, given a Herglotz vector field $G(z,t)$ of order d, one can build the associated Löwner chain (of the same order d), solving (23) by means of the associated evolution family.

The Berkson–Porta decomposition of a Herglotz vector field $G(z,t)$ also gives information on the dynamics of the associated evolution family. For instance, when $\tau(t) \equiv \tau \in \mathbb{D}$, the point τ is a (common) fixed point of $(\varphi_{s,t})$ for all $0 \leq s \leq t < +\infty$. Moreover, it can be proved that, in such a case, there exists a unique locally absolutely continuous function $\lambda\colon [0,+\infty) \to \mathbb{C}$ with $\lambda' \in L^d_{loc}([0,+\infty),\mathbb{C})$, $\lambda(0) = 0$ and $\operatorname{Re}\lambda(t) \geq \operatorname{Re}\lambda(s) \geq 0$ for all $0 \leq s \leq t < +\infty$ such that for all $s \leq t$

$$\varphi'_{s,t}(\tau) = \exp(\lambda(s) - \lambda(t)).$$

A similar characterization holds when $\tau(t) \equiv \tau \in \partial\mathbb{D}$ (see [41]).

Let us summarize this section. In the next scheme we show the main three notions of Löwner theory we are dealing with in this paper and the relationship between them:

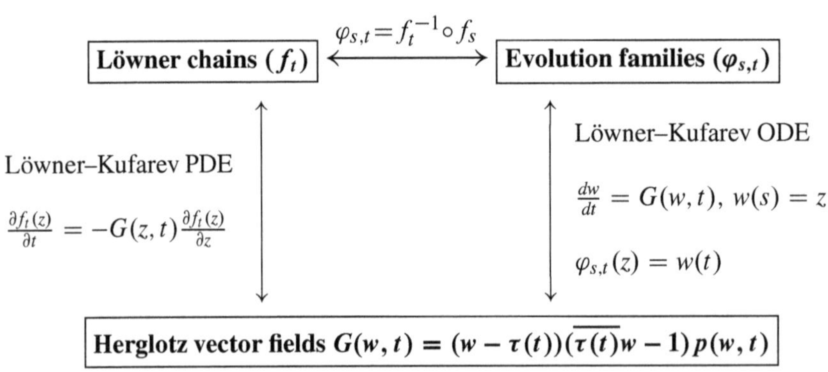

where p is a Herglotz function.

11 Extensions to Multiply Connected Domains

Y. Komatu
(from Oberwolfach Photo Collection)

Yûsaku Komatu (1914–2004), in 1942 [135], was the first to generalize Löwner's parametric representation to univalent holomorphic functions defined in a circular annulus and with images in the exterior of a disc. Later, Goluzin [92] gave a much simpler way to establish Komatu's results. With the same techniques, Li [161] considered a slightly different case, when the image of the annulus is the complex plane with two slits (ending at infinity and at the origin, respectively). See also [116, 136, 158]. The monograph [14] contains a self-contained detailed account on the Parametric Representation in the multiply connected setting.

Another way of adapting Löwner's method to multiply connected domains was developed by Kufarev and Kuvaev [152]. They obtained a differential equation satisfied by automorphic functions realizing conformal covering mappings of the unit disc onto multiply connected domains with a gradually erased slit. This differential approach has also been followed by Tsai [249]. Roughly speaking, these results can be considered as a version for multiply connected domains of the slit-radial Löwner equation. In a similar way, Bauer and Friedrich have developed a slit-chordal theory for multiply connected domains. Moreover, they have even dealt with stochastic versions of both the radial and the chordal cases. In this framework the situation is more subtle than in the simply connected case, because moduli spaces enter the picture [25, 26].

Following the guide of the new and general approach to Löwner theory in the unit disk as described in Sect. 9, the second and the third author of this survey jointly with Pavel Gumenyuk have developed a global approach to Löwner theory for double connected domains which give a uniform framework to previous works of Komatu, Goluzin, Li en Pir, and Lebedev. More interestingly, this abstract theory shows some phenomena not considered before and also poses many new questions.

Besides the similarities between the general approach for simple and double connected domains, there are a number of significative differences both in concepts and in results. For instance, in order to develop an interesting substantial theory for the double connected case, instead of a static reference domain D, one has to consider a family of expanding annuli $D_t = \mathbb{A}_{r(t)} := \{z : r(t) < |z| < 1\}$, where $r : [0, +\infty) \to [0, 1)$ is non-increasing and continuous. The first who noticed this fact, in a very special case, was already Komatu [135]. Such a family (D_t), it is usually called a canonical domain system of order d, whenever $-\pi/\log(r)$ belongs to $AC^d([0, +\infty), \mathbb{R})$.

This really non-optional dynamic context forces to modify the definitions of (again) the three basic elements of the general theory: semicomplete vector fields, evolution families, and Löwner chains. Nevertheless, there is still a (essentially) one-to-one correspondence between these three notions.

As in the unit disk, (weak) vector fields are introduced in this picture bearing in mind Carathéodory's theory of ODEs.

Definition 5. Let $d \in [1, +\infty]$. A function $G : \mathcal{D} \to \mathbb{C}$ is said to be a *weak holomorphic vector field* of order d in the domain

$$\mathcal{D} := \{(z, t) : t \geq 0, z \in D_t\},$$

if it satisfies the following three conditions:

WHVF1. For each $z \in \mathbb{C}$ the function $G(z, \cdot)$ is measurable in $E_z := \{t : (z, t) \in \mathcal{D}\}$.

WHVF2. For each $t \in E$ the function $G(\cdot, t)$ is holomorphic in D_t.

WHVF3. For each compact set $K \subset \mathcal{D}$ there exists a nonnegative function

$$k_K \in L^d\big(pr_{\mathbb{R}}(K), \mathbb{R}\big), \quad pr_{\mathbb{R}}(K) := \{t \in E : \exists z \in \mathbb{C} \quad (z, t) \in K\},$$

such that

$$|G(z, t)| \leq k_K(t), \quad \text{for all } (z, t) \in K.$$

Given a weak holomorphic vector field G in $\mathcal{D}$ and an initial condition $(z, s) \in \mathcal{D}$, it is possible to consider the initial value problem,

$$\dot{w} = G(w, t), \quad w(s) = z. \tag{25}$$

A solution to this problem is any continuous function $w : J \to \mathbb{C}$ such that $J \subset E$ is an interval, $s \in J$, $(w(t), t) \in \mathcal{D}$ for all $t \in J$ and

$$w(t) = z + \int_s^t G(w(\xi), \xi)\, d\xi, \quad t \in J. \tag{26}$$

When these kinds of problems have solutions well defined globally to the right for any initial condition, the vector field G is called semicomplete.

Putting together the main properties of the flows generated by semicomplete weak vector fields, we arrive to the concept of evolution families for the doubly connected setting.

Definition 6. A family $(\varphi_{s,t})_{0\leq s\leq t<+\infty}$ of holomorphic mappings $\varphi_{s,t} : D_s \to D_t$ is said to be an *evolution family of order d over* (D_t) (in short, an L^d*-evolution family*) if the following conditions are satisfied:

EF1. $\varphi_{s,s} = \mathrm{id}_{D_s}$,
EF2. $\varphi_{s,t} = \varphi_{u,t} \circ \varphi_{s,u}$ whenever $0 \leq s \leq u \leq t < +\infty$,
EF3. for any closed interval $I := [S, T] \subset [0, +\infty)$ and any $z \in D_S$ there exists a nonnegative function $k_{z,I} \in L^d([S, T], \mathbb{R})$ such that

$$|\varphi_{s,u}(z) - \varphi_{s,t}(z)| \leq \int_u^t k_{z,I}(\xi) d\xi$$

whenever $S \leq s \leq u \leq t \leq T$.

As expected, there is a one-to-one correspondence between evolution families over canonical domain systems and semicomplete weak holomorphic vector fields, analogous to the correspondence between evolution families and Herglotz vector fields in the unit disk.

Theorem 12 ([55, Theorem 5.1]). *The following two assertions hold:*

(A) For any L^d-evolution family $(\varphi_{s,t})$ over the canonical domain system (D_t) there exists an essentially unique semicomplete weak holomorphic vector field $G : \mathscr{D} \to \mathbb{C}$ of order d and a null-set $N \subset [0, +\infty)$ such that for all $s \geq 0$ the following statements hold:

(i) the mapping $[s, +\infty) \ni t \mapsto \varphi_{s,t} \in \mathrm{Hol}(D_s, \mathbb{C})$ is differentiable for all $t \in [s, +\infty) \setminus N$;
(ii) $d\varphi_{s,t}/dt = G(\cdot, t) \circ \varphi_{s,t}$ for all $t \in [s, +\infty) \setminus N$.

(B) For any semicomplete weak holomorphic vector field $G : \mathscr{D} \to \mathbb{C}$ of order d the formula $\varphi_{s,t}(z) := w_s^*(z, t)$, $t \geq s \geq 0$, $z \in D_s$, where $w_s^*(z, \cdot)$ is the unique non-extendable solution to the initial value problem

$$\dot{w} = G(w, t), \quad w(s) = z, \tag{27}$$

defines an L^d-evolution family over the canonical domain system (D_t).

With the corresponding perspective, there is also a true version of the non-autonomous Berkson–Porta description of Herglotz vector fields. Now, the role played by functions associated with Scharwz kernel is fulfilled by a natural class of functions associated with the Villat kernel (see [55, Theorem 5.6]).

Löwner chains in the double connected setting are introduced in a similar way as it was done in the case of the unit disc.

Definition 7. A family $(f_t)_{t\geq 0}$ of holomorphic functions $f_t : D_t \to \mathbb{C}$ is called a *Loewner chain* of order d over (D_t) if it satisfies the following conditions:

LC1. each function $f_t : D_t \to \mathbb{C}$ is univalent,
LC2. $f_s(D_s) \subset f_t(D_t)$ whenever $0 \leq s < t < +\infty$,
LC3. for any compact interval $I := [S, T] \subset [0, +\infty)$ and any compact set $K \subset D_S$ there exists a nonnegative function $k_{K,I} \in L^d([S, T], \mathbb{R})$ such that

$$|f_s(z) - f_t(z)| \leq \int_s^t k_{K,I}(\xi)d\xi$$

for all $z \in K$ and all (s, t) such that $S \leq s \leq t \leq T$.

The following theorem shows that every Löwner chain generates an evolution family of the same order.

Theorem 13 ([56, Theorem 1.9]). *Let (f_t) be a Löwner chain of order d over a canonical domain system (D_t) of order d. If we define*

$$\varphi_{s,t} := f_t^{-1} \circ f_s, \qquad 0 \leq s \leq t < \infty, \tag{28}$$

then $(\varphi_{s,t})$ is an evolution family of order d over (D_t).

An interesting consequence of this result is that any Löwner chain over a canonical system of annuli satisfies a PDE driven by a semicomplete weak holomorphic vector field. Moreover, the concrete formulation of this PDE clearly resembles the celebrated Löwner–Kufarev PDE appearing in the simple connected case.

The corresponding converse of the above theorem is more subtle than the one shown in the simple connected setting. Indeed, for any evolution family $(\varphi_{s,t})$ there exists a (essentially unique) Löwner chain (f_t) of the same order such that (28) holds but the selection of this fundamental chain is affected by the different conformal types of the domains D_t as well as the behavior of their elements with the index $I(\gamma)$ (with respect to zero) of certain closed curves $\gamma \subset \mathbb{D}_t$.

Theorem 14 ([56, Theorem 1.10]). *Let $(\varphi_{s,t})$ be an evolution family of order $d \in [1, +\infty]$ over the canonical domain system $D_t := \mathbb{A}_{r(t)}$ with $r(t) > 0$ (a non-degenerate system). Let $r_\infty := \lim_{t\to+\infty} r(t)$. Then there exists a Löwner chain (f_t) of order d over (D_t) such that*

1. *$f_s = f_t \circ \varphi_{s,t}$ for all $0 \leq s \leq t < +\infty$, i.e. (f_t) is associated with $(\varphi_{s,t})$;*
2. *$I(f_t \circ \gamma) = I(\gamma)$ for any closed curve $\gamma \subset D_t$ and any $t \geq 0$;*
3. *If $0 < r_\infty < 1$, then $\cup_{t\in[0,+\infty)} f_t(D_t) = \mathbb{A}_{r_\infty}$;*
4. *If $r_\infty = 0$, then $\cup_{t\in[0,+\infty)} f_t(D_t)$ is either $\mathbb{D}^*$, $\mathbb{C} \setminus \overline{\mathbb{D}}$, or $\mathbb{C}^*$.*

If (g_t) is another Löwner chain over (D_t) associated with $(\varphi_{s,t})$, then there is a biholomorphism $F : \cup_{t\in[0,+\infty)} g_t(D_t) \to \cup_{t\in[0,+\infty)} f_t(D_t)$ such that $f_t = F \circ g_t$ for all $t \geq 0$.

In general, a Löwner chain associated with a given evolution family is not unique. We call a Löwner chain (f_t) to be *standard* if it satisfies conditions (2)–(4) from Theorem 14. It follows from this theorem that the standard Löwner chain (f_t) associated with a given evolution family is defined uniquely up to a rotation (and scaling if $\cup_{t\in[0,+\infty)} f_t(D_t) = \mathbb{C}^*$).

Furthermore, combining Theorems 13 and 14 one can easily conclude that for any Löwner chain (g_t) of order d over a canonical domain system (D_t) there exists a biholomorphism $F : \cup_{t\in[0,+\infty)} g_t(D_t) \to L[(g_t)]$, where $L[(g_t)]$ is either $\mathbb{D}^*$, $\mathbb{C} \setminus \overline{\mathbb{D}}$, $\mathbb{C}^*$, or $\mathbb{A}_\rho$ for some $\rho > 0$, and such that the formula $f_t = F \circ g_t$, $t \geq 0$, defines a standard Löwner chain of order d over the canonical domain system (D_t). Hence, the conformal type of any Löwner chain $\cup_{t\in[0,+\infty)} g_t(D_t)$ can be identified with $\mathbb{D}^*$, $\mathbb{C} \setminus \overline{\mathbb{D}}$, $\mathbb{C}^*$, or $\mathbb{A}_\rho$ for some $\rho > 0$. Moreover, it is natural (well defined) to say that the (conformal) type of an evolution family $(\varphi_{s,t})$ is the conformal type of any Löwner chain associated with it.

The following statement characterizes the conformal type of a Löwner chain via dynamical properties of two associated evolution families over (D_t). One of them is the usual one $(\varphi_{s,t})$ and the other one is defined as follows: for each $s \geq 0$ and $t \geq s$,

$$\tilde{\varphi}_{s,t}(z) := r(t)/\varphi_{s,t}(r(s)/z).$$

It is worth mentioning that at least one of the families $(\varphi_{0,t})$ and $(\tilde{\varphi}_{0,t})$ converges to 0 as $t \to +\infty$ provided $r_\infty = \lim_{t\to+\infty} r(t) = 0$.

Theorem 15 ([56, Theorem 1.13]). *Let $\big((D_t), (\varphi_{s,t})\big)$ be a non-degenerate evolution family and denote as before $r_\infty := \lim_{t\to+\infty} r(t)$. In the above notation, the following statements hold:*

(i) the conformal type of the evolution family $(\varphi_{s,t})$ is $\mathbb{A}_\rho$ for some $\rho > 0$ if and only if $r_\infty > 0$;
(ii) the conformal type of evolution family $(\varphi_{s,t})$ is $\mathbb{D}^*$ if and only if $r_\infty = 0$ and $\varphi_{0,t}$ does not converge to 0 as $t \to +\infty$;
(iii) the conformal type of evolution family $(\varphi_{s,t})$ is $\mathbb{C} \setminus \overline{\mathbb{D}}$ if and only if $r_\infty = 0$ and $\tilde{\varphi}_{0,t}$ does not converge to 0 as $t \to +\infty$;
(iv) the conformal type of evolution family $(\varphi_{s,t})$ is $\mathbb{C}^*$ if and only if $r_\infty = 0$ and both $\varphi_{0,t} \to 0$ and $\tilde{\varphi}_{0,t} \to 0$ as $t \to +\infty$.

12 Integrability

In this section we plan to reveal relations between contour dynamics tuned by the Löwner–Kufarev equations and the Liouville (infinite dimensional) integrability, which was exploited actively since establishment of the Korteweg–de Vries equation as an equation for spectral stability in the Sturm–Liouville problem, and construction of integrable hierarchies in a series of papers by Gardner, Green, Kruskal, Miura, Zabusky, et al., see, e.g., [86, 262], and exact integrability results by Zakharov and Faddeev [263]. In this section we mostly follow our results in [170, 171].

Recently, it has become clear that one-parameter expanding evolution families of simply connected domains in the complex plane in some special models has been governed by infinite systems of evolution parameters, conservation laws. This phenomenon reveals a bridge between a nonlinear evolution of complex shapes emerged in physical problems, dissipative in most of the cases, and exactly solvable models. One of such processes is the Laplacian growth, in which the harmonic (Richardson's) moments are conserved under the evolution, see, e.g., [114, 174]. The infinite number of evolution parameters reflects the infinite number of degrees of freedom of the system and clearly suggests to apply field theory methods as a natural tool of study. The Virasoro algebra provides a structural background in most of field theories, and it is not surprising that it appears in soliton-like problems, e.g., KP, KdV, or Toda hierarchies, see [74, 88].

Another group of models, in which the evolution is governed by an infinite number of parameters, can be observed in controllable dynamical systems, where the infinite number of degrees of freedom follows from the infinite number of driving terms. Surprisingly, the same algebraic structural background appears again for this group. We develop this viewpoint here.

One of the general approaches to the homotopic evolution of shapes starting from a canonical shape, the unit disk in our case, is given by the Löwner–Kufarev theory. A shape evolution is described by a time-dependent conformal parametric map from the canonical domain onto the domain bounded by the shape at any fixed instant. In fact, these one-parameter conformal maps satisfy the Löwner–Kufarev differential equation (3), or an infinite dimensional controllable system, for which the infinite number of conservation laws is given by the *Virasoro generators* in their covariant form.

Recently, Friedrich and Werner [82], and independently Bauer and Bernard [24], found relations between SLE (stochastic or Schramm–Löwner evolution) and the highest weight representation of the Virasoro algebra. Moreover, Friedrich developed the Grassmannian approach to relate SLE to the highest weight representation of the Virasoro algebra in [81].

All the above results encourage us to conclude that the *Virasoro algebra* is a common algebraic structural basis for these and possibly other types of contour dynamics and we present the development in this direction here. At the same time, the infinite number of conservation laws suggests a relation with exactly solvable models.

The geometry underlying classical integrable systems is reflected in Sato's and Segal–Wilson's constructions of the infinite dimensional *Grassmannian* Gr. Based on the idea that the evolution of shapes in the plane is related to an evolution in a general universal space, the Segal–Wilson Grassmannian in our case, we provide an embedding of the Löwner–Kufarev evolution into a fiber bundle with the cotangent bundle over $\mathscr{F}_0$ as a base space, and with the smooth Grassmannian Gr_∞ as a typical fiber. Here $\mathscr{F}_0 \subset \mathbf{S}$ denotes the space of all conformal embeddings f of the unit disk into $\mathbb{C}$ normalized by $f(z) = z\left(1 + \sum_{n=1}^{\infty} c_n z^n\right)$ smooth on the boundary S^1, and under the *smooth Grassmannian* Gr_∞ we understand a dense subspace Gr_∞ of Gr defined further on.

So our plan is as follows. We will

- Consider homotopy in the space of shapes starting from the unit disk: $\mathbb{D} \to \Omega(t)$ given by the Löwner–Kufarev equation;
- Give its Hamiltonian formulation;
- Find conservation laws;
- Explore their algebraic structure;
- Embed Löwner–Kufarev trajectories into a moduli space (Grassmannian);
- Construct τ-function, Baker–Akhiezer function, and finally KP hierarchy.

Finally we present a class of solutions to KP hierarchy, which are preserving their form along the Löwner–Kufarev trajectories.

Let us start with two useful lemmas [170, 171].

Lemma 1. *Let the function $w(z,t)$ be a solution to the Cauchy problem* (3). *If the driving function $p(\cdot,t)$, being from the Carathéodory class for almost all $t \geq 0$, is C^∞ smooth in the closure $\hat{\mathbb{D}}$ of the unit disk $\mathbb{D}$ and summable with respect to t, then the boundaries of the domains $\Omega(t) = w(\mathbb{D},t) \subset \mathbb{D}$ are smooth for all t and $w(\cdot,t)$ extended to S^1 is injective on S^1.*

Lemma 2. *With the above notations let $f(z) \in \mathscr{F}_0$. Then there exists a function $p(\cdot,t)$ from the Carathéodory class for almost all $t \geq 0$, and C^∞ smooth in $\hat{\mathbb{D}}$, such that $f(z) = \lim_{t\to\infty} f(z,t)$ is the final point of the Löwner–Kufarev trajectory with the driving term $p(z,t)$.*

12.1 Witt and Virasoro Algebras

The complex *Witt algebra* is the Lie algebra of holomorphic vector fields defined on $\mathbb{C}^* = \mathbb{C} \setminus \{0\}$ acting by derivation over the ring of Laurent polynomials $\mathbb{C}[z, z^{-1}]$. It is spanned by the basis $L_n = z^{n+1}\frac{\partial}{\partial z}$, $n \in \mathbb{Z}$. The Lie bracket of two basis vector fields is given by the commutator $[L_n, L_m] = (m-n)L_{n+m}$. Its central extension is the complex *Virasoro algebra* $\mathfrak{vir}_{\mathbb{C}}$ with the central element c commuting with all L_n, $[L_n, c] = 0$, and with the Virasoro commutation relation

$$[L_n, L_m] = (m-n)L_{n+m} + \frac{c}{12}n(n^2-1)\delta_{n,-m}, \quad n, m \in \mathbb{Z},$$

where $c \in \mathbb{C}$ is the central charge denoted by the same character. These algebras play important role in conformal field theory. In order to construct their representations one can use an analytic realization.

12.2 Segal–Wilson Grassmannian

Sato's (universial) Grassmannian appeared first in 1982 in [222] as an infinite dimensional generalization of the classical finite dimensional Grassmannian manifolds and they are described as "the topological closure of the inductive limit of" a finite dimensional Grassmanian as the dimensions of the ambient vector space and its subspaces tend to infinity.

M. Sato

It turned out to be a very important infinite dimensional manifold being related to the representation theory of loop groups, integrable hierarchies, micrological analysis, conformal and quantum field theories, the second quantization of fermions, and to many other topics [59,177,224,259]. In the Segal and Wilson approach [224] the infinite dimensional Grassmannian $\mathrm{Gr}(H)$ is taken over the separable Hilbert space H. The first systematic description of the infinite dimensional Grassmannian can be found in [197].

We present here a general definition of the infinite dimensional smooth Grassmannian $\mathrm{Gr}_\infty(H)$. As a separable Hilbert space we take the space $L^2(S^1)$ and consider its dense subspace $H = C^\infty_{\|\cdot\|_2}(S^1)$ of smooth complex-valued functions defined on the unit circle endowed with $L^2(S^1)$ inner product $\langle f, g\rangle = \frac{1}{2\pi}\int\limits_{S^1} f\bar{g}\,dw$, $f, g \in H$. The orthonormal basis of H is $\{z^k\}_{k\in\mathbb{Z}} = \{e^{ik\theta}\}_{k\in\mathbb{Z}}$, $e^{i\theta} \in S^1$.

Let us split all integers $\mathbb{Z}$ into two sets $\mathbb{Z}^+ = \{0, 1, 2, 3, \ldots\}$ and $\mathbb{Z}^- = \{\ldots, -3, -2, -1\}$, and let us define a polarization by

$$H_+ = \mathrm{span}_H\{z^k,\ k \in \mathbb{Z}^+\}, \qquad H_- = \mathrm{span}_H\{z^k,\ k \in \mathbb{Z}^-\}.$$

Here and further span is taken in the appropriate space indicated as a subscription. The Grassmanian is thought of as the set of closed linear subspaces W of H, which are commensurable with H_+ in the sense that they have finite codimension in both H_+ and W. This can be defined by means of the descriptions of the orthogonal projections of the subspace $W \subset H$ to H_+ and H_-.

Definition 8. The infinite dimensional smooth Grassmannian $\mathrm{Gr}_\infty(H)$ over the space H is the set of subspaces W of H, such that

1. the orthogonal projection $pr_+ \colon W \to H_+$ is a Fredholm operator,
2. the orthogonal projection $pr_- \colon W \to H_-$ is a compact operator.

The requirement that pr_+ is Fredholm means that the kernel and cokernel of pr_+ are finite dimensional. More information about Fredholm operators the reader can find in [64]. It was proved in [197], that $\mathrm{Gr}_\infty(H)$ is a dense submanifold in a Hilbert manifold modeled over the space $\mathscr{L}_2(H_+, H_-)$ of Hilbert–Schmidt operators from H_+ to H_-, that itself has the structure of a Hilbert space, see [210]. Any $W \in \mathrm{Gr}_\infty(H)$ can be thought of as a graph W_T of a Hilbert–Schmidt operator $T \colon W \to W^\perp$, and points of a neighborhood U_W of $W \in \mathrm{Gr}_\infty(H)$ are in one-to-one correspondence with operators from $\mathscr{L}_2(W, W^\perp)$.

G. Segal

Let us denote by $\mathfrak{S}$ the set of all collections $\mathbb{S} \subset \mathbb{Z}$ of integers such that $\mathbb{S} \setminus \mathbb{Z}^+$ and $\mathbb{Z}^+ \setminus \mathbb{S}$ are finite. Thus, any sequence $\mathbb{S}$ of integers is bounded from below and contains all positive numbers starting from some number. It is clear that the sets $H_{\mathbb{S}} = \mathrm{span}_H\{z^k,\ k \in \mathbb{S}\}$ are elements of the Grassmanian $\mathrm{Gr}_\infty(H)$ and they are usually called *special points*. The collection of neighborhoods $\{U_{\mathbb{S}}\}_{\mathbb{S}\in\mathfrak{S}}$,

$$U_{\mathbb{S}} = \{W \mid \text{there is an orthogonal projection } \pi \colon W \to H_{\mathbb{S}} \text{ that is an isomorphism}\}$$

forms an open cover of $\mathrm{Gr}_\infty(H)$. The virtual cardinality of $\mathbb{S}$ defines the *virtual dimension* (v.d.) of $H_\mathbb{S}$, namely:

$$\mathrm{virtcard}(\mathbb{S}) = \mathrm{virtdim}(H_\mathbb{S}) = \dim(\mathbb{N} \setminus \mathbb{S}) - \dim(\mathbb{S} \setminus \mathbb{N}) = \mathrm{ind}(pr_+). \qquad (29)$$

The expression $\mathrm{ind}(pr_+) = \dim\ker(pr_+) - \dim\mathrm{coker}(pr_-)$ is called the index of the Fredholm operator pr_+. According to their virtual dimensions the points of $\mathrm{Gr}_\infty(H)$ belong to different components of connectivity. The Grassmannian is the disjoint union of connected components parametrized by their virtual dimensions.

12.3 Hamiltonian Formalism

Let the driving term $p(z,t)$ in the Löwner–Kufarev ODE (3) be from the Carathéodory class for almost all $t \geq 0$, C^∞-smooth in $\hat{\mathbb{D}}$, and summable with respect to t as in Lemma 1. Then the domains $\Omega(t) = f(\mathbb{D},t) = e^t w(\mathbb{D},t)$ have smooth boundaries $\partial\Omega(t)$ and the function f is injective on S^1, i.e.; $f \in \mathscr{F}_0$. So the Löwner–Kufarev equation can be extended to the closed unit disk $\hat{\mathbb{D}} = \mathbb{D} \cup S^1$.

Let us consider the sections ψ of $T^*\mathscr{F}_0 \otimes \mathbb{C}$, that are from the class $C^\infty_{\|\cdot\|_2}$ of smooth complex-valued functions $S^1 \to \mathbb{C}$ endowed with L^2 norm,

$$\psi(z) = \sum_{k\in\mathbb{Z}} \psi_k z^{k-1}, \quad |z| = 1.$$

We also introduce the space of observables on $T^*\mathscr{F}_0 \otimes \mathbb{C}$, given by integral functionals

$$\mathscr{R}(f,\bar{\psi},t) = \frac{1}{2\pi}\int_{z\in S^1} r(f(z),\bar{\psi}(z),t)\frac{dz}{iz},$$

where the function $r(\xi,\eta,t)$ is smooth in variables ξ,η and measurable in t.

We define a special observable, the time-dependent pseudo-Hamiltonian $\mathscr{H}$, by

$$\mathscr{H}(f,\bar{\psi},p,t) = \frac{1}{2\pi}\int_{z\in S^1} \bar{z}^2 f(z,t)(1 - p(e^{-t}f(z,t),t))\bar{\psi}(z,t)\frac{dz}{iz}, \qquad (30)$$

with the driving function (control) $p(z,t)$ satisfying the above properties. The Poisson structure on the space of observables is given by the canonical brackets

$$\{\mathscr{R}_1,\mathscr{R}_2\} = 2\pi\int_{z\in S^1} z^2\left(\frac{\delta\mathscr{R}_1}{\delta f}\frac{\delta\mathscr{R}_2}{\delta\bar{\psi}} - \frac{\delta\mathscr{R}_1}{\delta\bar{\psi}}\frac{\delta\mathscr{R}_2}{\delta f}\right)\frac{dz}{iz},$$

where $\frac{\delta}{\delta f}$ and $\frac{\delta}{\delta\bar{\psi}}$ are the variational derivatives, $\frac{\delta}{\delta f}\mathscr{R} = \frac{1}{2\pi}\frac{\partial}{\partial f}r$, $\frac{\delta}{\delta\bar{\psi}}\mathscr{R} = \frac{1}{2\pi}\frac{\partial}{\partial\bar{\psi}}r$.

Representing the coefficients c_n and $\bar{\psi}_m$ of f and $\bar{\psi}$ as integral functionals

$$c_n = \frac{1}{2\pi}\int_{z\in S^1} \bar{z}^{n+1} f(z,t)\frac{dz}{iz}, \quad \bar{\psi}_m = \frac{1}{2\pi}\int_{z\in S^1} z^{m-1}\bar{\psi}(z,t)\frac{dz}{iz},$$

$n \in \mathbb{N}$, $m \in \mathbb{Z}$, we obtain $\{c_n, \bar{\psi}_m\} = \delta_{n,m}$, $\{c_n, c_k\} = 0$, and $\{\bar{\psi}_l, \bar{\psi}_m\} = 0$, where $n, k \in \mathbb{N}$, $l, m \in \mathbb{Z}$.

The infinite dimensional Hamiltonian system is written as

$$\frac{dc_k}{dt} = \{c_k, \mathscr{H}\}, \quad c_k(0) = 0, \tag{31}$$

$$\frac{d\bar{\psi}_k}{dt} = \{\bar{\psi}_k, \mathscr{H}\}, \quad \psi_k(0) = \xi_k, \tag{32}$$

where $k \in \mathbb{Z}$ and $c_0 = c_{-1} = c_{-2} = \cdots = 0$, or equivalently, multiplying by corresponding powers of z and summing up,

$$\frac{df(z,t)}{dt} = f(1 - p(e^{-t}f,t)) = 2\pi\frac{\delta\mathscr{H}}{\delta\bar{\psi}}z^2 = \{f, \mathscr{H}\}, \quad f(z,0) \equiv z, \tag{33}$$

$$\frac{d\bar{\psi}}{dt} = -(1 - p(e^{-t}f,t) - e^{-t}fp'(e^{-t}f,t))\bar{\psi} = -2\pi\frac{\delta\mathscr{H}}{\delta f}z^2 = \{\bar{\psi}, \mathscr{H}\}, \tag{34}$$

where $\psi(z,0) = \xi(z) = \sum_{k\in\mathbb{Z}} \xi_k z^{k-1}$ and $z \in S^1$. So the phase coordinates $(f, \bar{\psi})$ play the role of the canonical Hamiltonian pair and the coefficients ξ_k are free parameters. Observe that the Eq. (33) is the Löwner–Kufarev equation (3) for the function $f = e^t w$.

Let us set up the *generating function* $\mathscr{G}(z) = \sum_{k\in\mathbb{Z}} \mathscr{G}_k z^{k-1}$, such that

$$\bar{\mathscr{G}}(z) := f'(z,t)\bar{\psi}(z,t).$$

Consider the "nonpositive" $(\bar{\mathscr{G}}(z))_{\le 0}$ and "positive" $(\bar{\mathscr{G}}(z))_{>0}$ parts of the Laurent series for $\bar{\mathscr{G}}(z)$:

$$\begin{aligned}(\bar{\mathscr{G}}(z))_{\le 0} &= (\bar{\psi}_1 + 2c_1\bar{\psi}_2 + 3c_2\bar{\psi}_3 + \ldots) + (\bar{\psi}_2 + 2c_1\bar{\psi}_3 + \ldots)z^{-1} + \ldots \\ &= \sum_{k=0}^{\infty} \bar{\mathscr{G}}_{k+1} z^{-k}.\end{aligned}$$

$$\begin{aligned}(\bar{\mathscr{G}}(z))_{>0} &= (\bar{\psi}_0 + 2c_1\bar{\psi}_1 + 3c_2\bar{\psi}_2 + \ldots)z + (\bar{\psi}_{-1} + 2c_1\bar{\psi}_0 + 3c_2\bar{\psi}_1 \ldots)z^2 + \ldots \\ &= \sum_{k=1}^{\infty} \bar{\mathscr{G}}_{-k+1} z^k.\end{aligned}$$

Proposition 1. *Let the driving term $p(z,t)$ in the Löwner–Kufarev ODE be from the Carathéodory class for almost all $t \geq 0$, C^∞-smooth in $\hat{\mathbb{D}}$, and summable with respect to t. The functions $\mathscr{G}(z)$, $(\mathscr{G}(z))_{<0}$, $(\mathscr{G}(z))_{\geq 0}$, and all coefficients $\mathscr{G}_n$ are time-independent for all $z \in S^1$.*

Proof. It is sufficient to check the equality $\dot{\bar{\mathscr{G}}} = \{\bar{\mathscr{G}}, \mathscr{H}\} = 0$ for the function $\mathscr{G}$, and then, the same holds for the coefficients of the Laurent series for $\mathscr{G}$. □

Proposition 2. *The conjugates $\bar{\mathscr{G}_k}$, $k = 1, 2, \ldots$, to the coefficients of the generating function satisfy the Witt commutation relation $\{\bar{\mathscr{G}_m}, \bar{\mathscr{G}_n}\} = (n-m)\bar{\mathscr{G}}_{n+m}$ for $n, m \geq 1$, with respect to our Poisson structure.*

The isomorphism $\iota : \bar{\psi}_k \to \partial_k = \frac{\partial}{\partial c_k}$, $k > 0$, is a Lie algebra isomorphism

$$(T_f^{*(0,1)} \mathscr{F}_0, \{\ ,\ \}) \to (T_f^{(1,0)} \mathscr{F}_0, [\ ,\ ]).$$

It makes a correspondence between the conjugates $\bar{\mathscr{G}_n}$ of the coefficients $\mathscr{G}_n$ of $(\mathscr{G}(z))_{\geq 0}$ at the point $(f, \bar{\psi})$ and the Kirillov vectors $L_n[f] = \partial_n + \sum_{k=1}^{\infty} (k+1) c_k \partial_{n+k}$, $n \in \mathbb{N}$, see [133]. Both satisfy the Witt commutation relations

$$[L_n, L_m] = (m-n) L_{n+m}.$$

12.4 Curves in Grassmannian

Let us recall that the underlying space for the universal smooth Grassmannian $\mathrm{Gr}_\infty(H)$ is $H = C^\infty_{\|\cdot\|_2}(S^1)$ with the canonical L^2 inner product of functions defined on the unit circle. Its natural polarization

$$H_+ = \mathrm{span}_H\{1, z, z^2, z^3, \ldots\}, \qquad H_- = \mathrm{span}_H\{z^{-1}, z^{-2}, \ldots\},$$

was introduced before. The pseudo-Hamiltonian $\mathscr{H}(f, \bar{\psi}, t)$ is defined for an arbitrary $\psi \in L^2(S^1)$, but we consider only smooth solutions of the Hamiltonian system, therefore, $\psi \in H$. We identify this space with the dense subspace of $T_f^* \mathscr{F}_0 \otimes \mathbb{C}$, $f \in \mathscr{F}_0$. The generating function $\mathscr{G}$ defines a linear map $\bar{\mathscr{G}}$ from the dense subspace of $T_f^* \mathscr{F}_0 \otimes \mathbb{C}$ to H, which being written in a block matrix form becomes

$$\begin{pmatrix} \bar{\mathscr{G}}_{>0} \\ \bar{\mathscr{G}}_{\leq 0} \end{pmatrix} = \begin{pmatrix} C_{1,1} & C_{1,2} \\ 0 & C_{1,1} \end{pmatrix} \begin{pmatrix} \bar{\psi}_{>0} \\ \bar{\psi}_{\leq 0} \end{pmatrix}, \tag{35}$$

where

$$\begin{pmatrix} C_{1,1} & C_{1,2} \\ 0 & C_{1,1} \end{pmatrix} = \left(\begin{array}{cccccc|ccccc} \ddots & \ddots & \ddots & \ddots & \ddots & & \ddots & \ddots & \ddots & \ddots & \ddots \\ \cdots & 0 & \mathbf{1} & 2c_1 & 3c_2 & & 4c_3 & 5c_4 & 6c_5 & 7c_6 & \cdots \\ \cdots & 0 & 0 & \mathbf{1} & 2c_1 & & 3c_2 & 4c_3 & 5c_4 & 6c_5 & \cdots \\ \cdots & 0 & 0 & 0 & \mathbf{1} & & 2c_1 & 3c_2 & 4c_3 & 5c_4 & \cdots \\ \hline \cdots & 0 & 0 & 0 & 0 & & \mathbf{1} & 2c_1 & 3c_2 & 4c_3 & \cdots \\ \cdots & 0 & 0 & 0 & 0 & & 0 & \mathbf{1} & 2c_1 & 3c_2 & \cdots \\ \cdots & 0 & 0 & 0 & 0 & & 0 & 0 & \mathbf{1} & 2c_1 & \cdots \\ \ddots & \ddots & \ddots & \ddots & \ddots & & \ddots & \ddots & \ddots & \ddots & \ddots \end{array}\right).$$

Proposition 3. *The operator $C_{1,1}\colon H_+ \to H_+$ is invertible.*

The generating function also defines a map $\mathscr{G}\colon T^*\mathscr{F}_0 \otimes \mathbb{C} \to H$ by

$$T^*\mathscr{F}_0 \otimes \mathbb{C} \ni (f(z), \psi(z)) \mapsto \mathscr{G} = \bar{f}'(z)\psi(z) \in H.$$

Observe that any solution $\left(f(z,t), \bar{\psi}(z,t)\right)$ of the Hamiltonian system is mapped into a single point of the space H, since all $\mathscr{G}_k$, $k \in \mathbb{Z}$ are time-independent by Proposition 1.

Consider a bundle $\pi\colon \mathscr{B} \to T^*\mathscr{F}_0 \otimes \mathbb{C}$ with a typical fiber isomorphic to $\mathrm{Gr}_\infty(H)$. We are aimed at construction of a curve $\Gamma\colon [0,T] \to \mathscr{B}$ that is traced by the solutions to the Hamiltonian system, or in other words, by the Löwner–Kufarev evolution. The curve Γ will have the form

$$\Gamma(t) = \left(f(z,t), \psi(z,t), W_{T_n}(t)\right)$$

in the local trivialization. Here W_{T_n} is the graph of a finite rank operator $T_n\colon H_+ \to H_-$, such that W_{T_n} belongs to the connected component of U_{H_+} of virtual dimension 0. In other words, we build a hierarchy of finite rank operators $T_n\colon H_+ \to H_-$, $n \in \mathbb{Z}^+$, whose graphs in the neighborhood U_{H_+} of the point $H_+ \in \mathrm{Gr}_\infty(H)$ are

$$T_n((\mathscr{G}(z))_{>0}) = T_n(\mathscr{G}_1, \mathscr{G}_2, \dots, \mathscr{G}_k, \dots) = \begin{cases} G_0(\mathscr{G}_1, \mathscr{G}_2, \dots, \mathscr{G}_k, \dots) \\ G_{-1}(\mathscr{G}_1, \mathscr{G}_2, \dots, \mathscr{G}_k, \dots) \\ \dots \\ G_{-n+1}(\mathscr{G}_1, \mathscr{G}_2, \dots, \mathscr{G}_k, \dots), \end{cases}$$

with $G_0 z^{-1} + G_{-1} z^{-2} + \ldots + G_{-n+1} z^{-n} \in H_-$. Let us denote by $G_k = \mathscr{G}_k, k \in \mathbb{N}$. The elements $G_0, G_{-1}, G_{-2}, \ldots$ are constructed so that all $\{\bar{G}_k\}_{k=-n+1}^{\infty}$ satisfy the truncated Witt commutation relations

$$\{\bar{G}_k, \bar{G}_l\}_n = \begin{cases} (l-k)\bar{G}_{k+l}, & \text{for } k+l \geq -n+1, \\ 0, & \text{otherwise,} \end{cases}$$

and are related to Kirilov's vector fields [133] under the isomorphism ι. The projective limit as $n \leftarrow \infty$ recovers the whole Witt algebra and the Witt commutation relations. Then the operators T_n such that their conjugates are $\bar{T}_n = (\tilde{B}^{(n)} + C_{2,1}^{(n)}) \circ C_{1,1}^{-1}$, are operators from H_+ to H_- of finite rank and their graphs $W_{T_n} = (\mathrm{id} + T_n)(H_+)$ are elements of the component of virtual dimension 0 in $\mathrm{Gr}_\infty(H)$. We can construct a basis $\{e_0, e_1, e_2, \ldots\}$ in W_{T_n} as a set of Laurent polynomials defined by means of operators T_n and $\bar{C}_{1,1}$ as a mapping

$$\{\psi_1, \psi_2, \ldots\} \overset{\bar{C}_{1,1}}{\longrightarrow} \{G_1, G_2, \ldots\} \overset{\mathrm{id}+T_n}{\longrightarrow} \{G_{-n+1}, G_{-n+2}, \ldots, G_0, G_1, G_2, \ldots\},$$

of the canonical basis $\{1, 0, 0, \ldots\}$, $\{0, 1, 0, \ldots\}$, $\{0, 0, 1, \ldots\}$,...

Let us formulate the result as the following main statement.

Proposition 4. *The operator T_n defines a graph $W_{T_n} = \mathrm{span}\{e_0, e_1, e_2, \ldots\}$ in the Grassmannian Gr_∞ of virtual dimension 0. Given any*

$$\psi = \sum_{k=0}^{\infty} \psi_{k+1} z^k \in H_+ \subset H,$$

the function

$$G(z) = \sum_{k=-n}^{\infty} G_{k+1} z^k = \sum_{k=0}^{\infty} \psi_{k+1} e_k,$$

is an element of W_{T_n}.

Proposition 5. *In the autonomous case of the Cauchy problem* (3)*, when the function $p(z,t)$ does not depend on t, the pseudo-Hamiltonian $\mathscr{H}$ plays the role of time-dependent energy and $\mathscr{H}(t) = \bar{G}_0(t) + const$, where $\bar{G}_0\big|_{t=0} = 0$. The constant is defined as $\sum_{n=1}^{\infty} p_k \bar{\psi}_k(0)$.*

Remark 4. The Virasoro generator L_0 plays the role of the energy functional in CFT. In view of the isomorphism ι, the observable $\bar{G}_0 = \iota^{-1}(L_0)$ plays an analogous role.

Thus, we constructed a countable family of curves $\Gamma_n \colon [0, T] \to \mathscr{B}$ in the trivial bundle $\mathscr{B} = T^*\mathscr{F}_0 \otimes \mathbb{C} \times \mathrm{Gr}_\infty(H)$, such that the curve Γ_n admits the form

$\Gamma_n(t) = \big(f(z,t), \psi(z,t), W_{T_n}(t)\big)$, for $t \in [0,T]$ in the local trivialization. Here $\big(f(z,t), \bar{\psi}(z,t)\big)$ is the solution of the Hamiltonian system (31–32). Each operator $T_n(t)\colon H_+ \to H_-$ that maps $\mathcal{G}_{>0}$ to

$$\big(G_0(t), G_{-1}(t), \dots, G_{-n+1}(t)\big)$$

defined for any $t \in [0,T]$, $n = 1,2,\dots$, is of finite rank and its graph $W_{T_n}(t)$ is a point in $\mathrm{Gr}_\infty(H)$ for any t. The graphs W_{T_n} belong to the connected component of the virtual dimension 0 for every time $t \in [0,T]$ and for fixed n. The coordinates $(G_{-n+1},\dots,G_{-2},G_{-1},G_0,G_1,G_2,\dots)$ of a point in the graph W_{T_n} considered as a function of ψ are isomorphic to the Kirilov vector fields

$$(L_{-n+1},\dots,L_{-2},L_{-1},L_0,L_1,L_1,L_2,\dots)$$

under the isomorphism ι.

12.5 τ-Function

Remind that any function g holomorphic in the unit disc, nonvanishing on the boundary and normalized by $g(0) = 1$ defines the multiplication operator $g\varphi$, $\varphi(z) = \sum_{k\in\mathbb{Z}} \varphi_k z^k$, that can be written in the matrix form

$$\begin{pmatrix} a & b \\ 0 & d \end{pmatrix} \begin{pmatrix} \varphi_{\geq 0} \\ \varphi_{<0} \end{pmatrix}. \tag{36}$$

All these upper triangular matrices form a subgroup GL^+_{res} of the group of automorphisms GL_{res} of the Grassmannian $\mathrm{Gr}_\infty(H)$.

With any function g and any graph W_{T_n} constructed in the previous section (which is transverse to H_-) we can relate the τ-function $\tau_{W_{T_n}}(g)$ by the following formula

$$\tau_{W_{T_n}}(g) = \det(1 + a^{-1} b T_n),$$

where a, b are the blocks in the multiplication operator generated by g^{-1}. If we write the function g in the form $g(z) = \exp(\sum_{n=1}^{\infty} t_n z^n) = 1 + \sum_{k=1}^{\infty} S_k(\mathbf{t}) z^k$, where the coefficients $S_k(\mathbf{t})$ are the homogeneous elementary Schur polynomials, then the coefficients $\mathbf{t} = (t_1, t_2, \dots)$ are called generalized times. For any fixed W_{T_n} we get an orbit in $\mathrm{Gr}_\infty(H)$ of curves Γ_n constructed in the previous section under the action of the elements of the subgroup GL^+_{res} defined by the function g. On the other hand, the τ-function defines a section in the determinant bundle over $\mathrm{Gr}_\infty(H)$ for any fixed $f \in \mathcal{F}_0$ at each point of the curve Γ_n.

12.6 Baker–Akhiezer Function, KP Flows, and KP Equation

Let us consider the component Gr^0 of the Grassmannian Gr_∞ of virtual dimension 0, and let g be a holomorphic function in $\mathbb{D}$ considered as an element of GL^+_{res} analogously to the previous section. Then g is an upper triangular matrix with 1s on the principal diagonal. Observe that $g(0) = 1$ and g does not vanish on S^1. Given a point $W \in \mathrm{Gr}^0$ let us define a subset $\Gamma^+ \subset GL^+_{\mathrm{res}}$ as

$$\Gamma^+ = \{g \in GL^+_{\mathrm{res}} : g^{-1}W \text{ is transverse to } H_-\}.$$

Then there exists [224] a unique function $\Psi_W[g](z)$ defined on S^1, such that for each $g \in \Gamma^+$, the function $\Psi_W[g]$ is in W, and it admits the form

$$\Psi_W[g](z) = g(z)\left(1 + \sum_{k=1}^{\infty} \omega_k(g, W)\frac{1}{z^k}\right).$$

The coefficients $\omega_k = \omega_k(g, W)$ depend both on $g \in \Gamma^+$ and on $W \in \mathrm{Gr}^0$, besides they are holomorphic on Γ^+ and extend to meromorphic functions on GL^+_{res}. The function $\Psi_W[g](z)$ is called the *Baker–Akhiezer function* of W.

Henry Frederick Baker (1866–1956) was British mathematician known for his contribution in algebraic geometry, PDE, and Lie theory. Naum Ilyich Akhiezer (1901–1980) was a Soviet mathematician known for his contributions in approximation theory and the theory of differential and integral operators, mathematical physics and history of mathematics. His brother Alexander was known theoretical physicist.

The Baker–Akhiezer function plays a crucial role in the definition of the KP (Kadomtsev–Petviashvili) hierarchy which we will define later. We are going to construct the Baker–Akhiezer function explicitly in our case.

Let $W = W_{T_n}$ be a point of Gr^0 defined in Proposition 4. Take a function $g(z) = 1 + a_1 z + a_2 z^2 + \cdots \in \Gamma^+$, and let us write the corresponding bi-infinite series for the Baker–Akhiezer function $\Psi_W[g](z)$ explicitly as

$$\begin{aligned}
\Psi_W[g](z) = \sum_{k\in\mathbb{Z}} \mathscr{W}_k z^k &= (1 + a_1 z + a_2 z^2 + \ldots)\left(1 + \frac{\omega_1}{z} + \frac{\omega_2}{z^2} + \ldots\right) \\
&= \ldots + (a_2 + a_3\omega_1 + a_4\omega_2 + a_5\omega_3 + \ldots)z^2 \\
&\quad + (a_1 + a_2\omega_1 + a_3\omega_2 + a_4\omega_3 + \ldots)z \\
&\quad + (1 + a_1\omega_1 + a_2\omega_2 + a_3\omega_3 + \ldots) \\
&\quad + (\omega_1 + a_1\omega_2 + a_2\omega_3 + \ldots)\frac{1}{z} \\
&\quad + (\omega_2 + a_1\omega_3 + a_2\omega_4 + \ldots)\frac{1}{z^2} + \ldots \\
&\quad \ldots + (\omega_k + a_1\omega_{k+1} + a_2\omega_{k+2} + \ldots)\frac{1}{z^k} + \ldots
\end{aligned}$$

The Baker–Akhiezer function for g and W_{T_n} must be of the form

$$\Psi_{W_{T_n}}[g](z) = g(z)\left(1 + \sum_{k=1}^{n} \omega_k(g)\frac{1}{z^k}\right) = \sum_{k=-n}^{\infty} \mathscr{W}_k z^k.$$

For a fixed $n \in \mathbb{N}$ we truncate this bi-infinite series by putting $\omega_k = 0$ for all $k > n$. In order to satisfy the definition of W_{T_n}, and determine the coefficients $\omega_1, \omega_2, \dots, \omega_n$, we must check that there exists a vector $\{\psi_1, \psi_2, \dots\}$, such that $\Psi_{W_{T_n}}[g](z) = \sum_{k=0}^{\infty} e_k \psi_{k+1}$. First we express ψ_k as linear functions $\psi_k = \psi_k(\omega_1, \omega_2, \dots, \omega_n)$ by

$$(\psi_1, \psi_2, \psi_3, \dots) = \bar{C}_{1,1}^{-1}\Big(\mathscr{W}_0(\omega_1, \omega_2, \dots, \omega_n), \mathscr{W}_1(\omega_1, \omega_2, \dots, \omega_n), \dots\Big). \quad (37)$$

Using Wronski formula we can write

$$\begin{aligned}
\psi_1 &= \mathscr{W}_0 - 2\bar{c}_1\mathscr{W}_1 - (3\bar{c}_2 - 4\bar{c}_1^2)\mathscr{W}_2 - (4\bar{c}_3 - 12\bar{c}_2\bar{c}_1 + 8\bar{c}_1^3)\mathscr{W}_3 + \dots,\\
\psi_2 &= \mathscr{W}_1 - 2\bar{c}_1\mathscr{W}_2 - (3\bar{c}_2 - 4\bar{c}_1^2)\mathscr{W}_3 - (4\bar{c}_3 - 12\bar{c}_2\bar{c}_1 + 8\bar{c}_1^3)\mathscr{W}_4 + \dots,\\
\psi_3 &= \mathscr{W}_2 - 2\bar{c}_1\mathscr{W}_3 - (3\bar{c}_2 - 4\bar{c}_1^2)\mathscr{W}_4 - (4\bar{c}_3 - 12\bar{c}_2\bar{c}_1 + 8\bar{c}_1^3)\mathscr{W}_5 + \dots,\\
&\dots\dots\dots
\end{aligned}$$

Next we define $\omega_1, \omega_2, \dots, \omega_n$ as functions of g and W_{T_n}, or in other words, as functions of $a_k, \bar{c}_k$ by solving linear equations

$$\begin{aligned}
\omega_1 &= \bar{c}_1\psi_1 + 2\bar{c}_2\psi_2 + \dots k\bar{c}_k\psi_k + \dots,\\
\omega_2 &= \sum_{k=1}^{\infty}\Big((k+2)\bar{c}_{k+1} - 2\bar{c}_1\bar{c}_k\Big)\psi_k,\\
&\dots\dots\dots
\end{aligned}$$

where ψ_k are taken from (37). The solution exists and is unique because of the general fact of the existence of the Baker–Akhiezer function. It is quite difficult task in general, however, in the case $n = 1$, it is possible to write the solution explicitly in the matrix form. If

$$A = \begin{pmatrix} \dots \\ 3\bar{c}_3 \\ 2\bar{c}_2 \\ \bar{c}_1 \end{pmatrix}^T \bar{C}_{1,1}^{-1} \begin{pmatrix} \dots \\ a_3 \\ a_2 \\ a_1 \end{pmatrix}, \quad B = \begin{pmatrix} \dots \\ 3\bar{c}_3 \\ 2\bar{c}_2 \\ \bar{c}_1 \end{pmatrix}^T \bar{C}_{1,1}^{-1} \begin{pmatrix} \dots \\ a_2 \\ a_1 \\ 1 \end{pmatrix}.$$

then $\omega_1 = \frac{B}{1-A}$.

In order to apply further theory of integrable systems we need to change variables $a_n \to a_n(\mathbf{t})$, $n > 0$, $\mathbf{t} = \{t_1, t_2, \dots\}$ in the following way

$$a_n = a_n(t_1, \dots, t_n) = S_n(t_1, \dots, t_n),$$

where S_n is the n-th elementary homogeneous Schur polynomial

$$1 + \sum_{k=1}^{\infty} S_k(\mathbf{t}) z^k = \exp\left(\sum_{k=1}^{\infty} t_k z^k\right) = e^{\xi(\mathbf{t},z)}.$$

In particular,

$$S_1 = t_1, \quad S_2 = \frac{t_1^2}{2} + t_2, \quad S_3 = \frac{t_1^3}{6} + t_1 t_2 + t_3,$$

$$S_4 = \frac{t_1^4}{24} + \frac{t_2^2}{2} + \frac{t_1^2 t_2}{2} + t_1 t_3 + t_4.$$

Then the Baker–Akhiezer function corresponding to the graph W_{T_n} is written as

$$\Psi_{W_{T_n}}[g](z) = \sum_{k=-n}^{\infty} \mathscr{W}_k z^k = e^{\xi(\mathbf{t},z)} \left(1 + \sum_{k=1}^{n} \frac{\omega_k(\mathbf{t}, W_{T_n})}{z^k}\right),$$

and $\mathbf{t} = \{t_1, t_2, \dots\}$ is called the vector of generalized times. It is easy to see that

$$\partial_{t_k} a_m = 0, \quad \text{for all } m = 1, 2 \dots, k-1,$$

$\partial_{t_k} a_m = 1$ and

$$\partial_{t_k} a_m = a_{m-k}, \quad \text{for all } m > k.$$

In particular, $B = \partial_{t_1} A$. Let us denote $\partial := \partial_{t_1}$. Then in the case $n = 1$ we have

$$\omega_1 = \frac{\partial A}{1 - A}. \tag{38}$$

Now we consider the associative algebra of pseudo-differential operators $\mathscr{A} = \sum_{k=-\infty}^{n} a_k \partial^k$ over the space of smooth functions, where the derivation symbol ∂ satisfies the Leibniz rule and the integration symbol and its powers satisfy the algebraic rules $\partial^{-1}\partial = \partial\partial^{-1} = 1$ and $\partial^{-1} a$ is the operator $\partial^{-1} a = \sum_{k=0}^{\infty} (-1)^k (\partial^k a) \partial^{-k-1}$ (see, e.g., [63]). The action of the operator ∂^m, $m \in \mathbb{Z}$, is well defined over the function $e^{\xi(\mathbf{t},z)}$, where $\xi(\mathbf{t}, z) = \sum_{k=1}^{\infty} t_k z^k$, so that the function $e^{\xi(\mathbf{t},z)}$ is the eigenfunction of the operator ∂^m for any integer m, i.e., it satisfies the equation

$$\partial^m e^{\xi(\mathbf{t},z)} = z^m e^{\xi(\mathbf{t},z)}, \quad m \in \mathbb{Z}, \tag{39}$$

see, e.g., [22, 63]. As usual, we identify $\partial = \partial_{t_1}$, and $\partial^0 = 1$.

Let us introduce the dressing operator $\Lambda = \phi\partial\phi^{-1} = \partial + \sum_{k=1}^{\infty} \lambda_k \partial^{-k}$, where ϕ is a pseudo-differential operator $\phi = 1 + \sum_{k=1}^{\infty} w_k(\mathbf{t})\partial^{-k}$. The operator Λ is defined up to the multiplication on the right by a series $1 + \sum_{k=1}^{\infty} b_k \partial^{-k}$ with constant coefficients b_k. The m-th KP flow is defined by making use of the vector field

$$\partial_m \phi := -\Lambda^m_{<0}\phi, \quad \partial_m = \frac{\partial}{\partial t_m},$$

and the flows commute. In the Lax form the KP flows are written as

$$\partial_m \Lambda = [\Lambda^m_{\geq 0}, \Lambda]. \tag{40}$$

If $m = 1$, then $\partial\Lambda = [\partial, \Lambda] = \sum_{k=1}^{\infty} (\partial\lambda_k)\partial^{-k}$, which justifies the identification $\partial = \partial_{t_1}$.

Thus, the Baker–Akhiezer function $\Psi_{W_{T_n}}[g](z)$ admits the form

$$\Psi_{W_{T_n}}[g](z) = \phi \exp(\xi(\mathbf{t}, z)),$$

where ϕ is a pseudo-differential operator $\phi = 1 + \sum_{k=1}^{n} \omega_k(\mathbf{t}, W_{T_n})\partial^{-k}$. The function $\Psi_{W_{T_n}}[g](z)$ becomes the eigenfunction of the operator Λ^m, namely $\Lambda^m w = z^m w$, for $m \in \mathbb{Z}$. Besides, $\partial_m w = \Lambda^m_{>0} w$. In view of (39) we can write this function as previously,

$$\Psi_{W_{T_n}}[g](z) = \big(1 + \sum_{k=1}^{n} \omega_k(\mathbf{t}, W_{T_n}) z^{-k}\big) e^{\xi(\mathbf{t},z)}.$$

Proposition 6. *Let $n = 1$, and let the Baker–Akhiezer function be of the form*

$$\Psi_{W_{T_n}}[g](z) = e^{\xi(\mathbf{t},z)}\left(1 + \frac{\omega}{z}\right),$$

where $\omega = \omega_1$ is given by the formula (38). *Then*

$$\partial\omega = \frac{\partial^2 A}{1 - A} + \left(\frac{\partial A}{1 - A}\right)^2$$

is a solution to the KP equation with the Lax operator $L = \partial^2 - 2(\partial\omega)$.

Remark 5. Observe that the condition $\operatorname{Re} p > 0$ in $\mathbb{D}$ defines a Löwner–Kufarev cone of trajectories in the class $\mathscr{F}_0$ of univalent smooth functions which allowed us to construct Löwner–Kufarev trajectories in the neighborhood U_{H_+}, which is a cone in the Grassmannian Gr_∞. The form of solutions to the KP hierarchy is preserved along the Löwner–Kufarev trajectories. The solutions are parametrized by the initial conditions in the system (32).

Of course, one can express the Baker–Akhiezer function directly from the τ-function by the Sato formula

$$\Psi_{W_{T_n}}[g](z) = e^{\xi(\mathbf{t},z)} \frac{\tau_{W_{T_n}}(t_1 - \frac{1}{z}, t_2 - \frac{1}{2z^2}, t_3 - \frac{1}{3z^3}, \dots)}{\tau_{W_{T_n}}(t_1, t_2, t_3 \dots)},$$

or applying the *vertex operator* V acting on the Fock space $\mathbb{C}[\mathbf{t}]$ of homogeneous polynomials

$$\Psi_{W_{T_n}}[g](z) = \frac{1}{\tau_{W_{T_n}}} V \tau_{W_{T_n}},$$

where

$$V = \exp\left(\sum_{k=1}^{\infty} t_k z^k\right) \exp\left(-\sum_{k=1}^{\infty} \frac{1}{k} \frac{\partial}{\partial t_k} z^{-k}\right).$$

In the latter expression exp denotes the formal exponential series and z is another formal variable that commutes with all Heisenberg operators t_k and $\frac{\partial}{\partial t_k}$. Observe that the exponents in V do not commute and the product of exponentials is calculated by the Baker–Campbell–Hausdorff formula. The operator V is a vertex operator in which the coefficient V_k in the expansion of V is a well-defined linear operator on the space $\mathbb{C}[\mathbf{t}]$. The Lie algebra of operators spanned by $1, t_k, \frac{\partial}{\partial t_k}$, and V_k, is isomorphic to the affine Lie algebra $\hat{\mathfrak{sl}}(2)$. The vertex operator V plays a central role in the highest weight representation of affine Kac–Moody algebras [128, 176] and can be interpreted as the infinitesimal Bäcklund transformation for the Korteweg–de Vries equation [58].

The vertex operator V recovers the Virasoro algebra in the following sense. Taken in two close points $z+\lambda/2$ and $z-\lambda/2$ the operator product can be expanded in to the following formal Laurent–Fourier series

$$: V(z+\frac{\lambda}{2})V(z-\frac{\lambda}{2}) := \sum_{k\in\mathbb{Z}} W_k(z)\lambda^k,$$

where $: ab :$ stands for the bosonic normal ordering. Then $W_2(z) = T(z)$ is the stress–energy tensor which we expand again as

$$T(z) = \sum_{n\in\mathbb{Z}} L_n(\mathbf{t}) z^{n-2},$$

where the operators L_n are the Virasoro generators in the highest weight representation over $\mathbb{C}[\mathbf{t}]$. Observe that the generators L_n span the full Virasoro algebra with central extension and with the central charge 1. This can also be interpreted as a quantization of the shape evolution. We shall define a representation over the space of conformal Fock space fields based on the Gaussian free field (GFF) in the next section.

Remark 6. Let us remark that the relation of Löwner equation to the integrable hierarchies of nonlinear PDE was noticed by Gibbons and Tsarev [89]. They observed that the chordal Löwner equation plays an essential role in the classification of reductions of Benney's equations. Later Takebe, Teo, and Zabrodin [243] showed that the chordal and radial Löwner PDE serve as consistency conditions for one-variable reductions of dispersionless KP and Toda hierarchies, respectively.

Remark 7. We also mention here relations between Löwner half-plane multi-slit equations and the estimates of spectral gaps of changing length for the periodic Zakharov–Shabat operators and for Hamiltonians in KdV and nonlinear Schrödinger equations elaborated in [137–139].

13 Stochastic Löwner Evolutions, Schramm, and Connections to CFT

13.1 SLE

This section we dedicate to the stochastic counterpart of the Löwner–Kufarev theory first recalling that one of the last (but definitely not least) contributions to this growing theory was the description by Oded Schramm in 1999–2000 [230] of the stochastic Löwner evolution (SLE), also known as the Schramm–Löwner evolution.

O. Schramm

The SLE is a conformally invariant stochastic process; more precisely, it is a family of random planar curves generated by solving Löwner's differential equation with the Brownian motion as a driving term. This equation was studied and developed by Oded Schramm together with Greg Lawler and Wendelin Werner in a series of joint papers that led, among other things, to a proof of Mandelbrot's conjecture about the Hausdorff dimension of the Brownian frontier [157]. This achievement was one of the reasons Werner was awarded the Fields Medal in 2006. Sadly, Oded Schramm, born 10 December 1961 in Jerusalem, died in a tragic hiking accident on 01 September 2008 while climbing Guye Peak, north of Snoqualmie Pass in Washington.

The *(chordal) stochastic Löwner evolution* with parameter $k \geq 0$ (SLE$_k$) starting at a point $x \in \mathbb{R}$ is the random family of maps (g_t) obtained from the chordal Löwner equation (5) by letting $\xi(t) = \sqrt{k}B_t$, where B_t is a standard one-dimensional Brownian motion such that $\sqrt{k}B_0 = x$. Namely, let us consider the equation

$$\frac{dg_t(z)}{dt} = \frac{2}{g_t(z) - \xi(t)}, \qquad g_0(z) = z, \tag{41}$$

where $\xi(t) = \sqrt{k}B_t = \sqrt{k}B_t(\omega)$, $B_t(\omega)$ is the standard one-dimensional Brownian motion defined on the standard filtered probability space $(\Omega, \mathcal{G}, (\mathcal{G}_t), P)$ of Brownian motion with the sample space $\omega \in \Omega$, and $t \in [0, \infty)$, $B_0 = 0$. The solution to (41) exists as long as $g_t(z) - h(t)$ remains away from zero and we denote by T_z the first time such that $\lim_{t \to T_z - 0}(g_t(z) - h(t)) = 0$. The function g_t satisfies the hydrodynamic normalization at infinity $g_t(z) = z + \frac{2t}{z} + \dots$. Let $K_t = \{z \in \hat{\mathbb{H}} : T_z \leq t\}$ and let $\mathbb{H}_t$ be its complement $\mathbb{H} \setminus K_t = \{z \in \mathbb{H} : T_z > t\}$. The set K_t is called *SLE hull*. It is compact, $\mathbb{H}_t$ is a simply connected domain and g_t maps $\mathbb{H}_t$ onto $\mathbb{H}$. SLE hulls grow in time. The *trace* γ_t is defined as $\lim_{z \to \xi(t)} g_t^{-1}(z)$, where the limit is taken in $\mathbb{H}$. The unbounded component of $\mathbb{H} \setminus \gamma_t$ is $\mathbb{H}_t$. The Hausdorff dimension of the SLE$_k$ trace is $\min(1 + \frac{8}{k}, 2)$, see [33]. Similarly, one can define the radial stochastic Löwner evolution. The terminology comes from the fact that the Löwner trace tip tends almost surely to a boundary point in the chordal case (∞ in the half-plane version) or to the origin in the disk version of the radial case.

Chordal SLE enjoys two important properties: scaling invariance and the Markovian property. Namely,

- $g_t(z)$ and $\frac{1}{\lambda}g_{\lambda^2 t}(\lambda z)$ are identically distributed;
- $h_t(z) = g_t(z) - \xi(t)$ possesses the Markov property. Furthermore, $h_s \circ h_t^{-1}$ is distributed as h_{s-t} for $s > t$.

The SLE$_k$ depends on the choice of ω and it comes in several flavors depending on the type of Brownian motion exploited. For example, it might start at a fixed point or start at a uniformly distributed point, or might have a built in drift and so on. The parameter k controls the rate of diffusion of the Brownian motion and the behavior of the SLE$_k$ critically depends on the value of k.

The SLE_2 corresponds to the loop-erased random walk and the uniform spanning tree. The $SLE_{8/3}$ is conjectured to be the scaling limit of self-avoiding random walks. The SLE_3 is proved [52] to be the limit of interfaces for the Ising model (another Fields Medal 2010 awarded to Stanislav Smirnov), while the SLE_4 corresponds to the harmonic explorer and the GFF. For all $0 \leq k \leq 4$ SLE gives slit maps. The SLE_6 was used by Lawler, Schramm, and Werner in 2001 [157] to prove the conjecture of Mandelbrot (1982) that the boundary of planar Brownian motion has fractal dimension $4/3$. Moreover, Smirnov [240] proved the SLE_6 is the scaling limit of critical site percolation on the triangular lattice. This result follows from his celebrated proof of Cardy's formula. SLE_8 corresponds to the uniform spanning tree. For $4 < k < 8$ the curve intersects itself and every point is contained in a loop but the curve is not space-filling almost surely. For $k \geq 8$ the curve is almost sure space-filling. This phase change is due to the Bessel process interpretation of SLE, see (42).

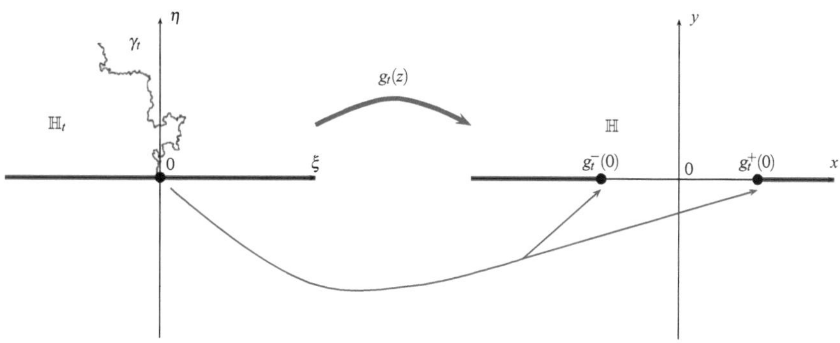

An invariant approach to SLE starts with probability measures on non-self-crossing random curves in a domain Ω connecting two given points $a, b \in \partial\Omega$ and satisfying the properties of

- *Conformal invariance*. Consider a triple (Ω, a, b) and a conformal map ϕ. If γ is a trace $SLE_k(\Omega, a, b)$, then $\phi(\gamma)$ is a trace $SLE_k(\phi(\Omega), \phi(a), \phi(b))$.;
- *Domain Markov property*. Let $\{\mathscr{F}_t\}_{t\geq 0}$ be the filtration in $\mathscr{F}$ by $\{B_t\}_{t\geq 0}$ and let g_t be a Löwner flow generated by $\xi(t) = \sqrt{k}B_t$. Then the hulls $(g_t(K_{s+t} \cap \mathbb{H}_t) - \xi_t)_{s\geq 0}$ are also generated by SLE_k and independent of the sigma-algebra $\mathscr{F}$.

The expository paper [155] is perhaps the best option to start an exploration of this fascinating branch of mathematics. Nice papers [216, 217] give up-to-date exposition of developments of SLE so we do not intend to survey SLE in detail here. Rather, we are going to show relations with CFT and other related stochastic variants of Löwner (generalized) evolution.

Here let us also mention the work by Carleson and Makarov [49] studying growth processes motivated by diffusion-limited aggregation (DLA) via Löwner's equations.

In this section we review the connections between conformal field theory (CFT) and Schramm–Löwner evolution (SLE) following, e.g., [24, 82, 130]. Indeed, SLE, being, e.g., a continuous limit of CFT's archetypical Ising model at its critical point, gives an approach to CFT which emphasizes CFT's roots in statistical physics.

Equation (41) is deterministic with a random entry and we solve it for every fixed ω. The corresponding stochastic differential equation (SDE) in the Itô form for the function $h_t(z) = g_t(z) - \xi(t)$ is

$$dh_t(z) = \frac{2}{h_t(z)}dt - \sqrt{k}dB_t, \tag{42}$$

where $-h_t/\sqrt{k}$ represents a Bessel process (of order $(4+k)/k$). For any holomorphic function $M(z)$ we have the Itô formula

$$(dM)(h_t) = -d\xi_t \mathscr{L}_{-1} M(h_t) + dt(\frac{k}{2}\mathscr{L}_{-1}^2 - 2\mathscr{L}_{-2})M(h_t), \tag{43}$$

where $\mathscr{L}_n = -z^{n+1}\partial$. From the form of (42) one can see immediately that h_t is a (time-homogeneous) diffusion, i. e., a continuous strong Markov process. The infinitesimal generator of h_t is given by $A = (\frac{k}{2}\mathscr{L}_{-1}^2 - 2\mathscr{L}_{-2})$ and this operator appears here for the first time. This differential operator makes it possible to reformulate many probabilistic questions about h_t in the language of PDE theory. If we consider $h_t(z)$ with fixed z, then (42) for h_t describes the motion of particles in the time-dependent field v with $dv = -d\xi_t\mathscr{L}_{-1} + dtA$. For instance, if we denote by $u_t(z)$ the mean function of $h_t(z)$

$$u_t(z) = \mathsf{E}h_t(z),$$

then it follows from Kolmogorov's backward equation that u_t satisfies

$$\begin{cases} \frac{\partial u_t}{\partial t} = Au_t, \\ u_0(z) = z, \end{cases} \quad z \in \mathbb{H}.$$

The kernel of the operator A describes driftless observables with time-independent expectation known as local martingales or conservation (in mean) laws of the process.

13.2 SLE and CFT

A general picture of the connections between SLE and CFT can be viewed as follows. The axiomatic approach to CFT grew up from the Hilbert sixth problem [258], and the Euclidean axioms were suggested by Osterwalder and Schrader [181].

They are centered around group symmetry, relative to unitary representations of Lie groups in Hilbert space. They define first *correlators* (complex values amplitudes) dependent on n complex variables, and a Lie group of conformal transformations of the correlators under the Möbius group. This can be extended at the infinitesimal level of the Lie algebra to invariance under infinitesimal conformal transforms, and therefore, to the so-called Ward identities. The adjoint representation of this group is given with the help of the enveloping algebra of an algebra of special operators, that act on a Fock space of *fields*. The existence of fields and relation between correlators and fields are given by a *reconstruction theorem*, see, e.g., [229]. The boundary version of this approach BCFT, i.e., CFT on domains with boundary, was developed by Cardy [48]. SLE approach starts with a family of statistical fields generated by nonrandom central charge modification of the random fields defined initially by the *GFF* and the algebra of Fock space fields, and then, defines SLE correlators, which turn to be local martingales after coupling of modified GFF on SLE random domains. These correlators satisfy the axiomatic properties of BCFT in which the infinitesimal boundary distortion leads to Ward identities involving a special boundary changing operator of conformal dimension that depends on the amplitude $\sqrt{k}$ of the Brownian motion in SLE.

N. G. Makarov

Connections between SLE and CFT were considered for the first time by Bauer and Bernard [23]. General motivation was as follows. Belavin et al. [34] defined in 1984 a class of conformal theories "minimal models," which described some discrete models (Ising, Potts, etc.) at criticality. Central theme is *universality*, i.e., the properties of a system close to the critical point are independent of its microscopic realization. Universal classes are characterized by a special parameter, central charge. Schramm's approach is based on a special evolution of conformal maps describing possible candidates for the scaling limit of interface curves. How these approaches are related? BPZ conjectured that the behavior of the system at criticality should be described by critical exponents identified as highest weights of degenerate representations of infinite dimensional Lie algebras, Virasoro in our case.

In order to make a bridge to CFT, let us address the definition of *fields* under consideration, which will be in fact *Fock space fields* and nonrandom *martingale-observables* following Kang and Makarov exposition [130], see also [221] for physical encouragement. The underlying idea is to construct the representation Fock space based on the GFF. GFF is a particular case of the Lévy Brownian motion (1947) [160] as a space extension of the classical Brownian motion. Nelson (1973) [178] considered relations of Markov properties of generalized random fields and QFT, in particular, he proved that GFF possesses the Markov property. Albeverio and Høegh-Krohn [6] and Röckner [219, 220] proved the Markov property of measure-indexed GFF and revealed relations of GFF to potential theory. Difficulty in the approach described below comes from the fact that in the definition of the Markov property, the domain of reference is chosen to be random which requires a special coupling between random fields and domains, which was realized by Schramm and Sheffield [232] and Dubédat [69].

GFF $\Phi(z)$ is defined on a simply connected domain D with the Dirichlet boundary conditions, i.e., GFF is indexed by the Hilbert space $\mathscr{E}(D)$, the completion of test functions $f \in C_0^{\infty}(D)$ (with compact support in D) equipped with the norm

$$\|f\|^2 = \int_D \int_D f(\zeta)\bar{f}(z)G(z,\zeta)d\zeta dz,$$

where $G(z,\zeta)$ is the Green function of the domain D, so $\Phi\colon \mathscr{E}(D) \to L^2(\Omega)$. For example, if $\mathbb{H} = D$, then $G(z,\zeta) = \frac{1}{2\pi}\log\left|\frac{\zeta-\bar{z}}{\zeta-z}\right|$. One can think of GFF as a generalization of the Brownian motion to complex time; however, such analogy is very rough since for a fixed $z \in D$ the expression $\Phi(z)$ is not well defined as a random variable, e.g., the correlation is $\mathsf{E}(\Phi(z)\Phi(\zeta)) = G(z,\zeta)$, but the variance does not exist in a usual sense. Instead, in terms of distributions (Φ, f) over the space $\mathscr{E}(D)$ we have covariances

$$\mathrm{Cov}((\Phi, f_1), (\Phi, f_2)) = \frac{1}{2\pi}\int_D \int_D f_1(\zeta)\bar{f_2}(z)G(z,\zeta)d\zeta dz.$$

The distributional derivatives $J = \partial\Phi$ and $\bar{J} = \bar{\partial}\Phi$ are well defined as, e.g., $J(f) = -\Phi(\partial f)$, $J\colon \mathscr{E}(D) \to L^2(\Omega)$. They are again Gaussian distributional fields.

The tensor nth symmetric product of Hilbert spaces $\mathscr{H}(D)$ we denote by $\mathscr{H}^{\odot n}$, $\mathscr{H}^{\odot 0} = \mathbb{C}$ and the Fock symmetric space is defined as the direct sum $\bigoplus_{n=0}^{\infty}\mathscr{H}^{\odot n}$. Here the sign $\odot$ of the Wick product (defined below) denotes an isomorphism to the symmetric tensor algebra multiplication. Gian Carlo Wick (1909–1992) introduced originally the product $:\cdot:\equiv \odot$ in [256] in 1950, in order to provide useful information of infinite quantities in Quantum Field Theory. In physics, Wick product is related to the normal ordering of operators over a representation space, namely, in terms of annihilation and creation operators all the creation operators appear to the left of all annihilation operators. The Wick product $\odot$ of random variables x_j is a random variable, commutative, and is defined formally as $: := 1$, $\frac{\partial}{\partial x_j}(x_1 \odot \cdots \odot x_n) = (x_1 \odot \cdots \odot x_{j-1} \odot x_{j+1} \odot \cdots \odot x_n)$, $\mathsf{E}(x_1 \odot \cdots \odot x_n) = 0$ for any $k \geq 1$. For example, $: x := \odot x = x - \mathsf{E}(x)$. However, we have to understand that $\Phi(z)$ and its derivatives are not random variables in the usual sense and must be think of as distributional random variables. Besides, $\odot\Phi = \Phi$. The mean of the field $\mathsf{E}\Phi$ is harmonic.

The *Fock space correlation functionals* are defined as span of basis correlation functionals 1 and $X(z_1) \odot \cdots \odot X(z_n)$, $z_1, \dots, z_n \in D$, $X_k = \partial^\alpha \bar{\partial}^\beta \Phi$ as well as infinite combinations (exponentials) which will play a special role in the definition of the vertex operator. We have $\partial\bar{\partial}\Phi \approx 0$, i.e., $\mathsf{E}((\partial\bar{\partial}\Phi(z)Y(\zeta)) = 0$, for any functional $Y(\zeta)$, $\zeta \neq z$. In view of this notation $\Delta\Phi \approx 0$.

The *basis Fock space fields* are formal Wick products of the derivatives of the GFF 1, Φ, $\Phi \odot \Phi$, $\partial\Phi \odot \Phi$, etc. Again, since GFF and its derivatives are understood in distributional sense, the above Wick product is formal. A *Fock space field* F_D is a linear combination of basis Fock space fields over the ring of smooth functions in D. If $F_1, \dots, F_n$ are Fock space fields and $z_1, \dots, z_n$ are distinct points of D, then $F_1(z_1) \dots F_n(z_n)$ is a correlation functional and $\mathsf{E}(F_1(z_1) \dots F_n(z_n))$ is a *correlation function* of simply correlator. The product here is thought of as a tensor product defined by the Wick formula over Feynman diagrams.

Being formally defined all these objects can be recovered through approximation by well-defined objects (scaling limit of lattice GFF) and expectations and correlators can be calculated.

We continue specifying the Markov property of domains and fields. For a shrinking deterministic subordination $D(t) \subset D(s)$ for $t < s$, the Markov property of $\{D(t)\}_{t \geq 0}$ means that domains $D(t)\big|D(s)$ and D_{t-s} coincide in law, which simply means in the Löwner case that if $\varphi_t \colon \mathbb{H}_t \to \mathbb{H}$ preserves ∞, and such that the tip $\gamma(t) \to 0$, then $\varphi_s \circ g_t$ and g_{t-s} satisfy the same equation fixing s. In the random case we have that the domains $D(t)\big|D(s)$ and D_{t-s} coincide in law in the sense of distribution, or $\mathrm{Law}(D(s)\big|D(t)) = \mathrm{Law}(D_{s-t})$. For more thorough treatment of the Markov property of domains, see [156].

In order to formulate the Markov property for fields and domains, we coming at coupling of $\{D(t)\}_{t \geq 0}$ SLE and GFF to be defined on the same probability space $(\Omega, \mathscr{G}, (\mathscr{G}_t), P)$, see [69, 231].

Let us start with a toy example of a classical real-valued field F_D defined on a simply connected domain D. Similarly to the Markov property of g_t, we say that F possesses the Markov property if for a decreasing t-dependent family of domains D_t, $F_{D_t}\big|D_s(z) = F_{D_{s-t}}$, where $0 < s < t$, $D_0 = D$. If in (41) we put $k = 0$, then the family of domains $\mathbb{H}_t$ is the family $D_t = \mathbb{H} \setminus \gamma(t)$, $\gamma(t) = (0, 2i\sqrt{t}]$. As a trivial example, construct the field $F_{D_t}(z) = \arg \varphi_t'(z)$, where $\varphi_t = z^2 + 4t^2$. The function φ_t maps D_t onto $\mathbb{C} \setminus [0, \infty)$ and F_{D_t} satisfies the Markov property.

A collection $\mathscr{F} = \{F_k\}$ of random holomorphic Fock space fields defined on D_t is said to satisfy the Markov property, if the process

$$M_t(z_1, \dots, z_n) := \mathsf{E}[F_{t1}(z_1) \dots F_{tn}(z_n)\big|D_t],$$

is a local martingale. Here $F_t(z_j)$ is push-forward of $F(z_j)$ with respect to a conformal map $D \to D_t$. That is, let $\mathrm{Aut}(D, p, q)$ is the group of conformal automorphisms of D fixing $p, q \in \partial D$ ($0, \infty$ for $\mathbb{H}$ or $\gamma(t), \infty$ for $\mathbb{H}_t$). The first requirement is that F is invariant under $\mathrm{Aut}(D, p, q)$. The second is that if $\varphi_t : D_t \to D$ is a conformal map fixing q, then primary fields F are $(\lambda, \bar{\lambda})$-forms, i.e., $F_t(z) = (\varphi_t')^{\lambda}(\overline{\varphi_t'})^{\bar{\lambda}} F(\varphi_t(z))$. A Schwarzian form F satisfies $F_t(z) = (\varphi_t')^2 F(\varphi_t(z)) + \mu S_{\varphi_t}(z)$, etc.

The space of conformal Fock space fields is a graded commutative differential algebra over the ring of nonrandom conformal fields. Conformal invariance in D is assumed for nonrandom conformal fields, and for random conformal fields this means that all correlations are conformally invariant as nonrandom conformal fields. Observe that all definitions can be given for Riemann surfaces as in [130]. The nonrandom field $M(z_1, \dots, z_n) := M_0(z_1, \dots, z_n)$ is called an *SLE martingale-observable*.

Now we want to use some Fock space fields as *states* and others as *observables* acting on states by *operator product expansion* (OPE). OPE is defined being based on the expansion for GFF

$$\Phi(\zeta)\Phi(z) = \log \frac{1}{|\zeta - z|^2} + 2 \log R_D(z) + \Phi(z) \odot \Phi(z) + o(1), \quad \text{as } \zeta \to z,$$

where $R_D(z)$ is the conformal radius of D with respect to z. The product $\Phi(\zeta)\Phi(z)$ is defined as a tensor product and is given by the Wick formula, see, e.g., [183, Sect. 4.3]. A Fock space field $F(z)$ is called holomorphic if $\mathsf{E}(F(z)Y(\zeta))$ is a holomorphic function of z for any field $Y(\zeta)$, $z \neq \zeta$. The OPE of a holomorphic field is given as a Laurent series

$$F(\zeta)Y(z) = \sum_{n \in \mathbb{Z}} C_n(z)(\zeta - z)^n, \quad \text{as } \zeta \to z.$$

Obviously, OPE is neither commutative nor associative unless one of the fields is nonrandom. The coefficients C_n are also Fock space fields. We denote $F * Y = C_0$ and $F *_{(n)} Y = C_n$ for holomorphic Fock space fields. This product satisfies the Leibniz rule.

The complex Virasoro algebra was introduced in Sect. 12.1. Let us define a special field, Virasoro field of GFF, $T(z)$ by the equality $T = -\frac{1}{2} J * J$. In particular,

$$J(\zeta)J(z) = -\frac{1}{(\zeta - z)^2} - 2T(z) + o(1), \quad \text{as } \zeta \to z \text{ in } \mathbb{H},$$

where $J(z) = \partial\Phi(z)$.

Let us define a nonrandom modification $\Phi_{(b)}$ on a simply connected domain D of the GFF $\Phi_{(0)} = \Phi$ on D parametrized by a real constant b, $\Phi_{(b)} = \Phi_{(0)} - 2b \arg \varphi'$, where $\varphi\colon D \to D$ fixing a point $\in \partial D$. Then $J_{(b)} = J_{(0)} + ib\frac{\varphi''}{\varphi'}$ is a pre-Schwarzian form and $T_{(b)} = T_{(0)} - b^2 S_{\varphi}$ is a Schwarzian form. The modified families of Fock space fields $\mathscr{F}_{(b)}$ have the central charge $c = 1-12b^2$. In order to simplify notations we omit subscript writing simply $T := T_{(b)}$, $J := J_{(b)}$, etc.

The field T becomes a stress–energy tensor in Quantum Field Theory and satisfies the equality

$$T(\zeta)T(z) = \frac{c/2}{(\zeta - z)^4} + \frac{2T(z)}{(\zeta - z)^2} + \frac{\partial T(z)}{\zeta - z} + o(1),$$

It is not a primary field and changes under conformal change of variables as $T(\zeta') = T(\zeta) - \frac{c}{12}S_{\zeta'}(\zeta)$, where $S_{\zeta'}(\zeta)$, as usual, is the Schwarzian derivative of $\zeta'(\zeta)$.

The Virasoro field has the expansion

$$T(\zeta)X(z) = \sum_{n \in \mathbb{Z}} \frac{L_n X(z)}{(\zeta - z)^{n+2}},$$

for any Fock space field $X(z)$. This way the Virasoro generators act on a field $X\colon L_n X = T *_{(-n-2)} X$ and can be viewed as operators over the space of Fock space fields. The result is again a Fock space field and one can define iteratively the field $L_{n_k} L_{n_{k-1}} \dots L_{n_1} X \equiv L_{n_k} L_{n_{k-1}} \dots L_{n_1} |X\rangle$, understanding $L_j := L_j *$ as operators acting on a "vector" X, where we adapt Dirac's notations "bra" and "ket" for vectors, operators and correlators. So we obtained a representation of the Virasoro algebra on Fock space fields

$$[L_n, L_m] = L_n * L_m - L_m * L_n = (n-m)L_{n+m} + \frac{c}{12}n(n^2-1)\delta_{n,-m}, \quad n, m \in \mathbb{Z}.$$

Let us recall that a field X is called *primary* of conformal weight $(\lambda, \bar{\lambda})$ if $X(z) = X(\phi(z))(\phi')^{\lambda}(\bar{\phi}')^{\bar{\lambda}}$. The *Virasoro primary field* $|V\rangle$ of conformal dimension λ, or $((\lambda, \bar{\lambda})$, is defined to satisfy $L_n|V\rangle = 0$, $n > 0$, $L_0|V\rangle = \lambda|V\rangle$ and $L_{-1}|V\rangle = \partial|V\rangle$. Analogously, the conjugate part of this definition is valid for the dimension $\bar{\lambda}$. Further on we omit the conjugate part because of its complete symmetry with the non-conjugate one. The *Virasoro–Verma module* $\mathscr{V}(\lambda, c)$ is constructed spanning and completing the basis vectors $L_{n_k} L_{n_{k-1}} \dots L_{n_1}|V\rangle$, where $n_k < n_{k-1} < \dots < n_1 < 0$, and $|V\rangle$ is taken to be the highest weight vector. We have a decomposition

$$\mathscr{V}(\lambda, c) = \bigoplus_{m=0}^{\infty} \mathscr{V}_m(\lambda, c),$$

where the level m space $\mathscr{V}_m(\lambda, c)$ is the eigenspace of L_0 with eigenvalue $\lambda + m$. A singular vector $|X\rangle$ by definition lies in some $\mathscr{V}_m(\lambda, c)$ and $L_n|X\rangle = 0$ for any $n > 0$. The Virasoro–Verma module is generically irreducible, having only singular-generated submodules. The Virasoro primary field becomes the highest weight vector in the representation of the Virasoro algebra $\mathfrak{vir}_{\mathbb{C}}$.

Proposition 7. *Let* $|V\rangle$ *be a Virasoro primary field in* $\mathscr{F}$ *of conformal dimension* λ *with central charge* c. *Then the field*

$$[L_{-m}+\eta_1 L_{-1}[L_{-m+1}+\eta_2 L_{-1}[L_{-m+2}+\cdots+\eta_{m-2}L_{-1}[L_{-2}+\eta_{m-1}L_{-1}^2]\ldots]|V\rangle$$

is a primary singular field of dimension $(\lambda+m)$, *if and only if* $\eta_1,\ldots,\eta_{m-1}$, c, *and* λ *satisfy a system of* m *linear equations.*

For example, if $m=2$, then the singular field is

$$[L_{-2}+\eta_1 L_{-1}^2]|V\rangle, \tag{44}$$

and

$$\begin{cases} 3+2\eta_1+4\eta_1\lambda=0, \\ c+8\lambda+12\eta_1\lambda=0, \end{cases}$$

if $m=3$, then the singular field is $[L_{-3}+\eta_2 L_{-1}L_{-2}+\eta_1\eta_2 L_{-1}^3]|V\rangle$ and

$$\begin{cases} 2+(\lambda+2)\eta_2=0, \\ 1+4(\lambda+1)\eta_1=0, \\ 5+(4\lambda+\frac{c}{2}+3)\eta_2+6(3\lambda+1)\eta_1\eta_2=0. \end{cases}$$

Now we consider the "holomorphic part" of GFF $\phi(z)=\int^z J(z)$, where $J(z)=\partial\Phi$. Of course the definition requires more work because the field constructed this way is ramified at all points. However, in correlation with a Fock space functional, the monodromy group is well-defined and finitely generated. A vertex operator is a field $\mathscr{V}_\star^\alpha(z)=e^{*i\alpha\phi(z)}$. It is a primary field of conformal dimension $\lambda=\frac{\alpha^2}{2}+\alpha\sqrt{\frac{1-c}{12}}$. Considering an infinitesimal boundary distortion $w_\varepsilon(z)=z+\frac{\varepsilon}{z-\xi}+o(\varepsilon)$ is equivalent to the insertion operator $|X(z)\rangle\to T(\xi)|X(z)\rangle$. The Ward identity implies the BCFT Cardy equation

$$(\mathscr{L}_{-1}^2-2\alpha^2\mathscr{L}_{-2})\mathsf{E}\left[\mathscr{V}_\star^\alpha(\xi)|X(z)\rangle\right]=0, \quad \text{for } \alpha\left(\alpha+\sqrt{\frac{1-c}{12}}\right)=1. \tag{45}$$

Here the representation of the Virasoro algebra is

$$\mathscr{L}_n=-(z-\xi)^{n+1}\partial-\lambda(n+1)(z-\xi)^n.$$

Merging the form of the infinitesimal generator A in (43), Virasoro primary singular field (44) and the Cardy equation (45) we arrive at the SLE numerology $\eta = -2\alpha^2$, $\alpha = \frac{\sqrt{k}}{2}$,

$$c = \frac{(6-k)(3k-8)}{2k}, \quad \lambda = \frac{6-k}{2k},$$

with the unique free parameter k.

Given a Fock space field $X(z)$, its push-forward $X_t(z)$ generically does not possess the Markov property. However, summarizing above, we can formulate the following statement.

Proposition 8. *For SLE evolution we have*

- $\mathcal{V}_{\star}^{\alpha}$ *is a Virasoro primary field;*
- $V = \mathcal{V}_{\star,t}^{\alpha}(\xi_t)X_t(z)$ *possesses the Markov property,* $\xi_t = \sqrt{k}B_t$*;*
- $M(z) = \mathsf{E}\left[\mathcal{V}_{\star}^{\alpha}(0)X(z)\right]$ *is a one-point martingale-observable;*
- *The process* $M_t(z) = \mathsf{E}\left[\mathcal{V}_{\star,t}^{\alpha}(\xi_t)X_t(z)\middle|D_t\right]$ *is a one-point martingale;*
- $(\frac{k}{2}L_{-1}^2 - 2L_{-2})|\mathcal{V}_{\star}^{\alpha}\rangle$ *is Virasoro primary singular field of level 2;*
- $(\frac{k}{2}\mathcal{L}_{-1}^2 - 2\mathcal{L}_{-2})M(z) = 0$ *is the Cardy equation for the SLE martingale observables.*

As a simple example, consider the Fock space field $V = \mathcal{V}_{\star}^{\alpha}(0)T(z)$. Then $M(z) = \mathsf{E}\left[\mathcal{V}_{\star}^{\alpha}(0)T(z)\right]$ is a one-point martingale-observable and $M(z) = 1/z^2$. If $k = 8/3$, then $M_t(z) = (h_t'(z)/h_t(z))^2$ is a local martingale.

More general construction including multi-point observables requires more technical work related to a vertex field V and boundary condition changing operators, see [130, Sect. 8.4]. Examples of martingale observables were found, e.g., by Friedrich and Werner [82] and Schramm and Sheffield [231, 232]. A radial version of SLE and relations to conformal field theory one can find in [131].

Another construction of the stress–energy tensor of CFT comes as a local observable of the conformal loop ensemble (CLE) (see, e.g., [234]) loops for any central charge, see [65, 66]. More general construction is performed on a groupoid of conformal maps of a simply connected domain, a natural generalization of the finite dimensional conformal group. The underlying manifold structure is Fréchet. Similarly to moduli (Teichmüller) spaces, the elements of the cotangent bundle are analogues of quadratic differentials, see [67]. It is shown there that the stress–energy tensor of CFT is exactly such a quadratic differential.

13.3 Generalized Löwner–Kufarev Stochastic Evolution

Another attempt to construct random interfaces different from SLE has been launched by conformal welding in [21].

We considered a setup [126] in which the sample paths are represented by the trajectories of a point (e.g., the origin) in the unit disk $\mathbb{D}$ evolving randomly under the generalized Löwner equation. The driving mechanism differs from SLE. In the SLE case the Denjoy–Wolff attracting point (∞ in the chordal case or a boundary point of the unit disk in the radial case) is fixed. In our case, the attracting point is the driving mechanism and the Denjoy–Wolff point is different from it. Relation with this model to CFT is the subject of a forthcoming paper. Let us consider the generalized Löwner evolution driven by a Brownian particle on the unit circle. In other words, we study the following initial value problem.

$$\begin{cases} \frac{d}{dt}\phi_t(z,\omega) = \frac{(\tau(t,\omega)-\phi_t(z,\omega))^2}{\tau(t,\omega)} p(\phi_t(z,\omega), t, \omega), \\ \phi_0(z,\omega) = z, \end{cases} \quad t \geq 0,\ z \in \mathbb{D},\ \omega \in \Omega. \tag{46}$$

The function $p(z,t,\omega)$ is a Herglotz function for each fixed $\omega \in \Omega$. In order for $\phi_t(z,\omega)$ to be an Itô process adapted to the Brownian filtration, we require that the function $p(z,t,\omega)$ is adapted to the Brownian filtration for each $z \in \mathbb{D}$. Even though the driving mechanism in our case differs from that of SLE, the generated families of conformal maps still possess the important time-homogeneous Markov property.

For each fixed $\omega \in \Omega$, (46) similarly to SLE, may be considered as a deterministic generalized Löwner equation with the Berkson–Porta data $(\tau(\cdot,\omega), p(\cdot,\cdot,\omega))$. In particular, the solution $\phi_t(z,\omega)$ exists, is unique for each $t > 0$ and $\omega \in \Omega$, and moreover, is a family of holomorphic self-maps of the unit disk.

The equation in (46) is an example of a so-called *random differential equation* (see, for instance, [241]). Since for each fixed $\omega \in \Omega$ it may be regarded as an ordinary differential equation, the sample paths $t \mapsto \phi_t(z,\omega)$ have continuous first derivatives for almost all ω. See an example of a sample path of $\phi_t(0,\omega)$ for $p(z,t) = \frac{\tau(t)+z}{\tau(t)-z}$, $\tau(t) = e^{ikB_t}$, $k = 5$, $t \in [0,30]$ in the figure to the left.

In order to give an explicitly solvable example let $p(z,t,\omega) = \frac{\tau(t,\omega)}{\tau(t,\omega)-z} = \frac{e^{ikB_t(\omega)}}{e^{ikB_t(\omega)}-z}$. It makes (46) linear:

$$\frac{d}{dt}\phi_t(z,\omega) = e^{ikB_t(\omega)} - \phi_t(z,\omega),$$

and a well-known formula from the theory of ordinary differential equation yields

$$\phi_t(z,\omega) = e^{-t}\left(z + \int_0^t e^s e^{ikB_s(\omega)}ds\right).$$

Taking into account the fact that $\mathsf{E}e^{ikB_t(\omega)} = e^{-\frac{1}{2}tk^2}$, we can also write the expression for the mean function $\mathsf{E}\phi_t(z,\omega)$

$$\mathsf{E}\phi_t(z,\omega) = \begin{cases} e^{-t}(z+t), & k^2 = 2, \\ e^{-t}z + \frac{e^{-tk^2/2} - e^{-t}}{1-k^2/2}, & \text{otherwise.} \end{cases} \tag{47}$$

Thus, in this example all maps ϕ_t and $\mathsf{E}\phi_t$ are affine transformations (compositions of a scaling and a translation).

In general, solving the random differential equation (46) is much more complicated than solving its deterministic counterpart, mostly because of the fact that for almost all ω the function $t \mapsto \tau(t,\omega)$ is nowhere differentiable.

If we assume that the Herglotz function has the form $p(z,t,\omega) = \tilde{p}(z/\tau(t,\omega))$, then it turns out that the process $\phi_t(z,\omega)$ has an important invariance property, that were crucial in development of SLE.

Let $s > 0$ and introduce the notation

$$\tilde{\phi}_t(z) = \frac{\phi_{s+t}(z)}{\tau(s)}.$$

Then $\tilde{\phi}_t(z)$ is the solution to the initial-value problem

$$\begin{cases} \frac{d}{dt}\tilde{\phi}_t(z,\omega) = \frac{\left(\tilde{\tau}(t,\omega) - \tilde{\phi}_t(z,\omega)\right)^2}{\tilde{\tau}(t,\omega)}\tilde{p}\left(\tilde{\phi}_t(z,\omega)/\tilde{\tau}(t)\right), \\ \tilde{\phi}_0(z,\omega) = \phi_s(z,\omega)/\tilde{\tau}(s), \end{cases}$$

where $\tilde{\tau}(t) = \tau(s+t)/\tau(s) = e^{ik(B_{s+t}-B_s)}$ is again a Brownian motion on $\mathbb{T}$ (because $\tilde{B}_t = B_{s+t} - B_s$ is a standard Brownian motion). In other words, the conditional distribution of $\tilde{\phi}_t$ given ϕ_r, $r \in [0,s]$ is the same as the distribution of ϕ_t.

By the complex Itô formula, the process $\frac{1}{\tau(t,\omega)} = e^{-ikB_t}$ satisfies the equation

$$de^{-ikB_t} = -ike^{-ikB_t}dB_t - \frac{k^2}{2}e^{-ikB_t}dt.$$

Let us denote $\frac{\phi_t(z,\omega)}{\tau(t,\omega)}$ by $\Psi_t(z,\omega)$. Applying the integration by parts formula to Ψ_t, we arrive at the following initial value problem for the Itô SDE

$$\begin{cases} d\Psi_t = -ik\Psi_t dB_t + \left(-\frac{k^2}{2}\Psi_t + (\Psi_t - 1)^2 p(\Psi_t e^{ikB_t(\omega)}, t, \omega)\right)dt, \\ \Psi_0(z) = z. \end{cases} \tag{48}$$

A numerical solution to this equation for a specific choice for $p(z,t) \equiv i$, $k = 1$, and $t \in [0,2]$, is shown in the figure to the right.

Analyzing the process $\frac{\phi_t(z,\omega)}{\tau(t,\omega)}$ instead of the original process $\phi_t(z,\omega)$ is in many ways similar to one of the approaches used in SLE theory.

The image domains $\Psi_t(\mathbb{D},\omega)$ differ from $\phi_t(\mathbb{D},\omega)$ only by rotation. Due to the fact that $|\Psi_t(z,\omega)| = |\phi_t(z,\omega)|$, if we compare the processes $\phi_t(0,\omega)$ and $\Psi_t(0,\omega)$, we note that their first hit times of the circle $\mathbb{T}_r$ with radius $r < 1$ coincide, i. e.,

$$\inf\{t \geq 0, |\phi_t(0,\omega)| = r\} = \inf\{t \geq 0, |\Psi_t(0,\omega)| = r\}.$$

In other words, the answers to probabilistic questions about the expected time of hitting the circle $\mathbb{T}_r$, the probability of exit from the disk $\mathbb{D}_r = \{z : |z| < r\}$, etc. are the same for $\phi_t(0,\omega)$ and $\Psi_t(0,\omega)$.

If the Herglotz function has the form $p(z,t,\omega) = \tilde{p}(z/\tau(t,\omega))$, then (48) becomes

$$\begin{cases} d\Psi_t = -ik\Psi_t dB_t + \left(-\frac{k^2}{2}\Psi_t + (\Psi_t - 1)^2 \tilde{p}(\Psi_t)\right) dt, \\ \Psi_0(z) = z, \end{cases} \tag{49}$$

and may be regarded as an equation of a two-dimensional time-homogeneous real diffusion written in complex form. This implies, in particular, that Ψ_t is a time-homogeneous strong Markov process. By construction, $\Psi_t(z)$ always stays in the unit disk.

Analogously to SLE we can consider random conformal Fock space fields defined on $\mathbb{D}$ changing correspondingly the definition using the Green function for $\mathbb{D}$ instead of $\mathbb{H}$. Coupling of (46) and the GFF in $\mathbb{D}$ we define the correlators $f_t(z_1,\dots,z_n) = f(z_1,\dots,z_n)\big|_{\mathbb{D}_t}$ as martingale-observables.

For a smooth function $f(z)$ defined in $\mathbb{D}$ we derive the Itô differential in the complex form

$$df(\Psi_t) = -ikdB_t(L_{-1} - \bar{L}_{-1})f(\Psi_t) + dtAf(\Psi_t),$$

where A is the infinitesimal generator of Ψ_t,

$$A = \left(-\frac{k^2}{2}z + (z-1)^2\tilde{p}(z)\right)\frac{\partial}{\partial z} - \frac{1}{2}k^2 z^2 \frac{\partial^2}{\partial z^2}$$
$$+ \left(-\frac{k^2}{2}\bar{z} + (\bar{z}-1)^2\overline{\tilde{p}(z)}\right)\frac{\partial}{\partial \bar{z}} - \frac{1}{2}k^2 \bar{z}^2 \frac{\partial^2}{\partial \bar{z}^2} + k^2|z|^2\frac{\partial^2}{\partial z \partial \bar{z}}.$$

In particular, if f is holomorphic, then

$$A = \left(-\frac{k^2}{2}z + (z-1)^2\tilde{p}(z)\right)\frac{\partial}{\partial z} - \frac{1}{2}k^2 z^2 \frac{\partial^2}{\partial z^2}. \tag{50}$$

If $f(z)$ is a martingale-observable, then $Af = 0$.

In [126] we proved the existence of a unique stationary point of Ψ_t in terms of the stochastic vector field

$$\frac{d}{dt}\Psi_t(z,\omega) = G_0(\Psi_t(z,\omega)),$$

where the Herglotz vector field $G_0(z,\omega)$ is given by

$$G_0(z,\omega) = -ikzW_t(\omega) - \frac{k^2}{2}z + (z-1)^2\tilde{p}(z).$$

Here, $W_t(\omega)$ denotes a generalized stochastic process known as *white noise*. Also nth moments were calculated and the boundary diffusion on the unit circle was considered, which corresponds, in particular, to North–South flow, see, e.g., [50].

14 Related Topics

14.1 Hele-Shaw Flows, Laplacian Growth

One of the most influential works in fluid dynamics at the end of the nineteenth century was a series of papers, see, e.g., [124] written by *Henry Selby Hele-Shaw* (1854–1941). There Hele-Shaw first described his famous cell that became a subject of deep investigation only more than 50 years later. A Hele-Shaw cell is a device for investigating two-dimensional flow (Hele-Shaw flow or Laplacian growth) of a viscous fluid in a narrow gap between two parallel plates.

This cell is the simplest system in which multi-dimensional convection is present. Probably the most important characteristic of flows in such a cell is that when the Reynolds number based on gap width is sufficiently small, the Navier–Stokes equations averaged over the gap reduce to a linear relation for the velocity similar to Darcy's law and then to a Laplace equation for the fluid pressure. Different driving mechanisms can be considered, such as surface tension or external forces

H. S. Hele-Shaw

(e.g., suction, injection). Through the similarity in the governing equations, Hele-Shaw flows are particularly useful for visualization of saturated flows in porous media, assuming they are slow enough to be governed by Darcy's law. Nowadays, the Hele-Shaw cell is used as a powerful tool in several fields of natural sciences and engineering, in particular, soft condensed matter physics, materials science, crystal growth and, of course, fluid mechanics. See more on historical and scientific account in [253].

The century-long development connecting the original Hele-Shaw experiments, the conformal mapping formulation of the Hele-Shaw flow by *Pelageya Yakovlevna Polubarinova-Kochina* (1899–1999) and *Lev Aleksandrovich Galin* (1912–1981) [83, 187, 188], and the modern treatment of the Hele-Shaw evolution based on integrable systems and on the general theory of plane contour motion, was marked by several important contributions by individuals and groups.

The main idea of Polubarinova-Kochina and Galin was to apply the Riemann mapping from an appropriate canonical domain (the unit disk $\mathbb{D}$ in most situations) onto the phase domain in order to parameterize the free boundary. The evolution equation for this map, named after its creators, allows to construct many explicit solutions and to apply methods of conformal analysis and geometric function theory to investigate Hele-Shaw flows. In particular, solutions to this equation in the case of advancing fluid give subordination chains of simply connected domains which have been studied by Löwner and Kufarev. The resulting equation for the family of functions $f(z,t) = a_1(t)z + a_2(t)z^2 + \ldots$ from $\mathbb{D}$ onto domains occupied by viscous fluid is

$$\operatorname{Re}\,[\dot{f}(\zeta,t)\overline{\zeta f'(\zeta,t)}] = \frac{Q}{2\pi}, \quad |\zeta| = 1. \tag{51}$$

P. Ya. Polubarinova-Kochina

The corresponding equation in $\mathbb{D}$ is a first-order integro-PDE

$$\dot{f}(\zeta,t) = \zeta f'(\zeta,t) \int_0^{2\pi} \frac{Q}{4\pi^2 |f'(e^{i\theta},t)|^2} \frac{e^{i\theta}+\zeta}{e^{i\theta}-\zeta} d\theta, \quad |\zeta| < 1.$$

L. A. Galin

Here Q is positive in the case of injection of negative in the case of suction. The Polubarinova–Galin and the Löwner–Kufarev equations, having some evident geometric connections, are of somewhat different nature. While the evolution of the Laplacian growth given by the Polubarinova–Galin equation is completely defined by the initial conditions, the Löwner–Kufarev evolution depends also on an arbitrary control function. The Polubarinova–Galin equation is essentially nonlinear and the corresponding subordination chains are of rather complicated nature. The treatment of classical Laplacian growth was given in the monograph [114].

The newest direction in the development of Hele-Shaw flow is related to Integrable Systems and Mathematical Physics, as a particular case of a contour dynamics. This story started in 2000 with a short note [174] by Mineev-Weinstein, Wiegmann, and Zabrodin, where the idea of the equivalence of 2D contour dynamics and the dispersionless limit of the integrable 2-D Toda hierarchy appeared for the first time. A list of complete references to corresponding works would be rather long so we only list the names of some key contributors: Wiegmann, Mineev-Weinstein, Zabrodin, Krichever, Kostov, Marshakov, Takebe, Teo et al., and some important references: [140–142, 172, 175, 243, 257]. Let us consider an "exterior" version of

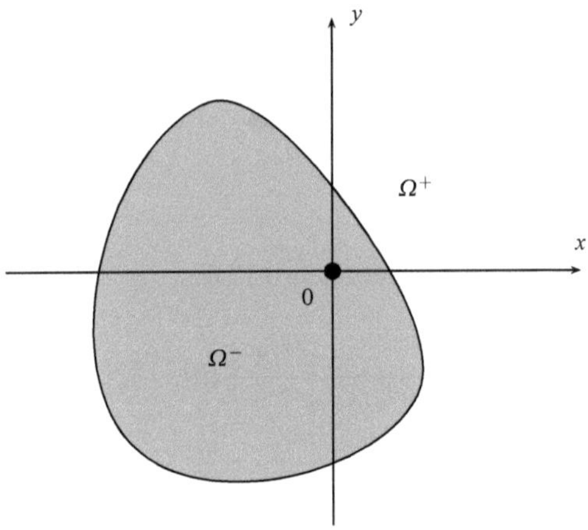

- $M_{-k} = -\int_{\Omega^+} z^{-k} d\sigma_z$;
- $M_0 = |\Omega^-|$;
- $M_k = \int_{\Omega^-} z^k d\sigma_z$;
- $k \geq 1$,
- $t = M_0/\pi$, $t_k = M_k/\pi k$ generalized times.

the process when source/sink of viscous fluid is at ∞ and the bounded domain Ω^- is occupied by the inviscid one. Then the conformal map f of the exterior of the unit disk onto Ω^+

$$f(\zeta, t) = b(t)\zeta + b_0(t) + \sum_{k=1}^{\infty} \frac{b_j(t)}{\zeta^k}, \quad b(t) > 0;$$

satisfies the analogous boundary equation

$$\operatorname{Re}\,[\dot{f}(\zeta, t)\overline{\zeta f'(\zeta, t)}] = -Q, \tag{52}$$

Following the definition of Richardson's moments let us define interior and exterior moments as in the above figure. The integrals for $k = 1, 2$ are assumed to be properly regularized. Then the properties of the moments are

- $M_0(t) = M_0(0) - Qt$ is "physical time";
- M_k are conserved for $k \geq 1$;
- M_0 and $\{M_k\}_{k \geq 1}$ determine the domain Ω^+ locally (given $\partial\Omega^+$ is smooth);

- $\{M_k\}_{k<0}$ evolve in time in a quite complicated manner;
- M_0 and $\{M_k\}_{k\geq 1}$ can be viewed as local parameters on the space of "shapes";

Suppose $\Gamma = \partial\Omega^- = \partial\Omega^+$ is analytic and $\phi(x_1, x_2, t) = 0$ is an implicit representation of the free boundary $\Gamma(t)$. Substituting $x_1 = (z+\bar{z})/2$ and $x_2 = (z-\bar{z})/2i$ into this equation and solving it for $\bar{z}$ we obtain that $\Gamma = \{z : S(z,t) = \bar{z}\}$. is given by the *Schwarz function* $S(z,t)$ which is defined and analytic in a neighborhood of Γ. The Schwarz function can be constructed by means of the Cauchy integral

$$g(z) = -\frac{1}{2\pi i}\int\limits_{\partial\Omega} \frac{\bar{\zeta}\, d\zeta}{\zeta - z}.$$

Define the analytic functions, $g_e(z)$, in Ω^+ and $g_i(z)$, in Ω^-. On $\Gamma = \partial\Omega$ the jump condition is

$$g_e(z) - g_i(z) = \bar{z}, \quad z \in \partial\Omega.$$

Then the Schwarz function $S(z)$ is defined formally by $S(z) = g_e(z) - g_i(z)$. The Cauchy integral implies the Cauchy transform of Ω^+

$$g_e(z) = -\frac{1}{2\pi i}\int\limits_{\partial\Omega} \frac{\bar{\zeta}\, d\zeta}{\zeta - z} = -\frac{1}{\pi}\int\limits_{\Omega^+} \frac{d\sigma_\zeta}{\zeta - z},$$

with the Laurent expansion

$$g_e(z) = \sum_{k=0}^{\infty} \frac{M_k}{z^{k+1}}, \quad z \sim \infty.$$

Similarly for $g_i(z)$

$$g_i(z) = -\sum_{k=1}^{\infty} M_{-k} z^{k-1}, \quad z \sim 0;$$

So formally

$$S(z) = \sum_{k=-\infty}^{\infty} \frac{M_k}{z^{k+1}} = \sum_{k=1}^{\infty} M_{-k} z^{k-1} + \frac{M_0}{z} + \sum_{k=1}^{\infty} \frac{M_k}{z^{k+1}}.$$

Let us write the logarithmic energy as

$$\mathscr{F}(\Omega^+) = -\frac{1}{\pi^2}\int_{\mathbb{C}}\int_{\mathbb{C}} \log|z-\zeta|\, d\sigma_z d\sigma_\zeta,$$

where σ_z is a measure supported in Ω^+. It is the potential for the momens

$$\frac{1}{k} M_{-k}(\Omega) = \frac{\partial \mathscr{F}(\Omega)}{\partial M_k}.$$

So (see, e.g., [174]) the moments satisfy the 2-D Toda dispersionless lattice hierarchy

$$\frac{\partial M_{-k}}{\partial t_j} = \frac{\partial M_{-j}}{\partial t_k}, \qquad \frac{\partial M_{-k}}{\partial \bar{t}_j} = \frac{\partial \bar{M}_{-j}}{\partial t_k}.$$

The function $\exp(\mathscr{F}(\Omega^+)) =: \tau(\Omega^+) = \tau(\mathbf{t})$ is the τ-function, and $t = M_0/\pi, \ldots,$ $t_k = M_k/\pi k$ are generalized times.

The real-valued τ-function becomes the solution to the Hirota equation

$$S_{f^{-1}}(z) = \frac{6}{z^2} \sum_{k,n=1}^{\infty} \frac{1}{z^{n+k}} \frac{\partial^2 \log \tau}{\partial t_k \partial t_n},$$

where $z = f(\zeta)$ is the parametric map of the unit disk onto the exterior phase domain and $S_{f^{-1}}(z)$ denotes the Schwarzian derivative of the inverse to f.

$$\frac{M_{-k}}{\pi} = \frac{\partial \log \tau}{\partial t_k}, \qquad \frac{\bar{M}_{-k}}{\pi} = \frac{\partial \log \tau}{\partial \bar{t}_k}, \qquad k \geq 1.$$

The τ-function introduced by the "Kyoto School" as a central element in the description of soliton equation hierarchies.

If $\zeta = e^{i\theta}$, $M_0(t) = M_0(0) - Qt$, then the derivatives are

$$\frac{\partial f}{\partial \theta} = i\zeta \frac{\partial f}{\partial \zeta}, \quad \frac{\partial f}{\partial t} = Q \frac{\partial f}{\partial M_0}.$$

Let $f^*(\zeta) = \overline{f(1/\bar{\zeta})}$, and let

$$\{f, g\} = \zeta \frac{\partial f}{\partial \zeta} \frac{\partial g}{\partial M_0} - \zeta \frac{\partial g}{\partial \zeta} \frac{\partial f}{\partial M_0}.$$

In view of this the Polubarinova–Galin equation (52) can be rewritten as $\{f, f^*\} = 1$. This equation is known as the *string constraint*. The equation for the τ-function with a proper initial condition provided by the string equation solves the inverse moments problem for small deformations of a simply connected domain with analytic boundary. Indeed, define the Schwarzian derivative $S_F = \frac{F'''}{F''} - \frac{3}{2}\left(\frac{F''}{F'}\right)^2$, $F = f^{-1}$. If we know the Schwarzian derivative $S(\zeta)$, we know the conformal map $w = F(\zeta) = \eta_1/\eta_2$ normalized accordingly, where η_1 and η_2 are linearly independent solutions to the Fuchs equation

$$w'' + \frac{1}{2}S(\zeta)w = 0.$$

The connection extends to the Lax–Sato approach to the dispersionless 2-D Toda hierarchy. In this setting it is shown that a Laurent series for a univalent function that provides an invertible conformal map of the exterior of the unit circle to the exterior of the domain can be identified with the Lax function. The τ-function appears to be a generating function for the inverse map. The formalism allows one to associate a notion of τ-function to the analytic curves.

The conformal map

$$f(\zeta,t) = b(t)\zeta + b_0(t) + \sum_{k=1}^{\infty} \frac{b_j(t)}{\zeta^k}, \quad b(t) > 0$$

obeys the relations

$$\frac{\partial f}{\partial t_k} = \{H_k, f\}, \quad \frac{\partial f}{\partial \bar{t}_k} = \{\bar{H}_k, f\},$$

where

- $H_k = (f^k(\zeta))_+ + \frac{1}{2}(f^k(\zeta))_0$,
- $\bar{H}_k = (\bar{f}^k(1/\zeta))_- + \frac{1}{2}(\bar{f}^k(1/\zeta))_0$.

In the paper [142], an analog of this theory for multiply-connected domains is developed. The answers are formulated in terms of the so-called Schottky double of the plane with holes. The Laurent basis used in the simply connected case is replaced by the Krichever–Novikov basis. As a corollary, analogs of the 2-D Toda hierarchy depend on standard times plus a finite set of additional variables. The solution of the Dirichlet problem is written in terms of the τ-function of this hierarchy. The relation to some matrix problems is briefly discussed.

14.2 Fractal Growth

Benoît Mandelbrot (1924, Warsaw, Poland–2010, Cambridge, Massachusetts, United States) brought to the world's attention that many natural objects simply do not have a preconceived form determined by a characteristic scale. He [168] first saw a visualization of the set named after him, at IBM's Thomas J. Watson Research Center in upstate New York.

Many of the structures in space and processes reveal new features when magnified beyond their usual scale in a wide variety of natural and industrial

B. Mandelbrot

processes, such as crystal growth, vapor deposition, chemical dissolution, corrosion, erosion, fluid flow in porous media and biological growth a surface or an interface, biological processes. A fractal, a structure coined by Mandelbrot in 1975 ("fractal" from Latin "fractus"), is a rough or fragmented geometric shape that can be subdivided into parts, each of which is (at least approximately) a reduced-size copy of the whole. Fractals are generally self-similar, independent of scale, and have (by Mandelbrot's own definition) the Hausdorff dimension strictly greater than the topological dimension. There are many mathematical structures that are fractals, e.g., the Sierpinski triangle, the Koch snowflake, the Peano curve, the Mandelbrot set, and the Lorenz attractor. One of the ways to model a fractal is the process of fractal growth that can be either stochastic or deterministic. A nice overview of fractal growth phenomena is found in [254].

Many models of fractal growth patterns combine complex geometry with randomness. A typical and important model for pattern formation is *DLA* (see a survey in [119]). Considering colloidal particles undergoing Brownian motion in some fluid and letting them adhere irreversibly on contact with another one bring us to the basics of DLA. Fix a seed particle at the origin and start another one form infinity letting it perform a random walk. Ultimately, that second particle will either escape to infinity or contact the seed, to which it will stick irreversibly. Next another particle starts at infinity to walk randomly until it either sticks to the two-particle cluster or escapes to infinity. This process is repeated to an extent limited only by modeler's patience. The clusters generated by this process are highly branched and fractal, see figure to the left.

The DLA model was introduced in 1981 by Witten and Sander [260, 261]. It was shown to have relation to dielectric breakdown [179], one-phase fluid flow in porous media [51], electro-chemical deposition [113], medical sciences [223], etc. A new conformal mapping language to study DLA was proposed by Hastings and Levitov [120, 121]. They showed that two-dimensional DLA can be grown by iterating stochastic conformal maps. Later this method was thoroughly handled in [60].

For a continuous random walk in 2-D the diffusion equation provides the law for the probability $u(z,t)$ that the walk reaches a point z at the time t,

$$\frac{\partial u}{\partial t} = \eta \Delta u,$$

where η is the diffusion coefficient. When the cluster growth rate per surface site is negligible compared to the diffusive relaxation time, the time dependence of the relaxation may be neglected (see, e.g., [261]). With a steady flux from infinity and the slow growth of the cluster the left-hand side derivative can be neglected and we have just the Laplacian equation for u. If $K(t)$ is the closed aggregate at the time t and $\Omega(t)$ is the connected part of the complement of $K(t)$ containing infinity, then the probability of the appearance of the random walker in $\mathbb{C} \setminus \Omega(t)$ is zero. Thus, the boundary condition $u(z,t)\big|_{\Gamma(t)} = 0$, $\Gamma(t) = \partial\Omega(t)$ is set. The only source of time dependence of u is the motion of $\Gamma(t)$. The problem resembles the classical Hele-Shaw problem, but the complex structure of $\Gamma(t)$ does not allow us to define the normal velocity in a good way although it is possible to do this in the discrete models.

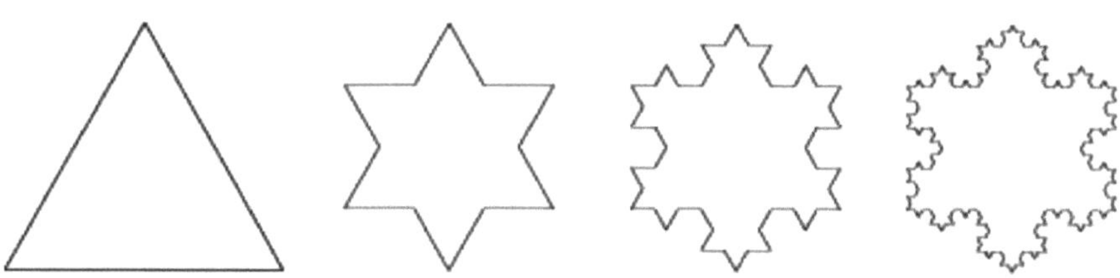

Now let us construct a Riemann conformal map $f : \mathbb{D}^- \to \hat{\mathbb{C}}$, $\mathbb{D}^- = \{z \colon |z| > 1\}$, which is meromorphic in $\mathbb{D}^-$, $f(\zeta,t) = \alpha(t)\zeta + a_0(t) + \dfrac{a_1(t)}{\zeta} + \dots$, $\alpha(t) > 0$, and maps $\mathbb{D}^-$ onto $\Omega(t)$. The boundary $\Gamma(t)$ need not even be a quasidisk, as considered earlier. While we are not able to construct a differential equation analogous to the Polubarinova–Galin one on the unit circle, the retracting Löwner subordination chain still exists, and the function $f(\zeta,t)$ satisfies the equation

$$\dot{f}(\zeta,t) = \zeta f'(\zeta,t) p_f(\zeta,t), \quad \zeta \in \mathbb{D}^-, \tag{53}$$

where $p_f(\zeta,t) = p_0(t) + p_1(t)/\zeta + \dots$ is a Carathéodory function: $\operatorname{Re} p(z,t) > 0$ for all $\zeta \in \mathbb{D}^-$ and for almost all $t \in [0,\infty)$. A difference from the Hele-Shaw problem is that the DLA problem is well posed on each level of discreteness by construction. An analogue of DLA model was treated by means of Löwner chains by Carleson and Makarov in [49]. In this section we follow their ideas as well as those from [125].

Of course, the fractal growth phenomena can be seen without randomness. A simplest example of such growth is the Koch snowflake (Helge von Koch, 1870–1924).DLA-like fractal growth without randomness can be found, e.g., in [61].

Returning to the fractal growth we want to study a rather wide class of models with complex growing structure. We note that $\alpha(t) = \operatorname{cap} K(t) = \operatorname{cap} \Gamma(t)$. Let $M(0,2\pi)$ be the class of positive measures γ on $[0,2\pi]$. The control function $p_f(\zeta,t)$ in (53) can be represented by the Riesz–Herglotz formula

$$p_f(\zeta,t) = \int_0^{2\pi} \frac{e^{i\theta}+\zeta}{e^{i\theta}-\zeta} d\gamma_t(\theta),$$

and $p_0(t) = \|\gamma_t\|$, where $\gamma_t(\theta) \in M(0,2\pi)$ for almost all $t \geq 0$ and absolutely continuous in $t \geq 0$. Consequently, $\dot{\alpha}(t) = \alpha(t)\|\gamma_t\|$. There is a one-to-one correspondence between one-parameter (t) families of measures γ_t and Löwner chains $\Omega(t)$ (in our case of growing domains $\mathbb{C} \setminus \Omega(t)$ we have only surjective correspondence).

Example 1. Suppose we have an initial domain $\Omega(0)$. If the derivative of the measure γ_t with respect to the Lebesgue measure is the Dirac measure $d\gamma_t(\theta) \equiv \delta_{\theta_0}(\theta)d\theta$, then

$$p_f(\zeta,t) \equiv \frac{e^{i\theta_0}+\zeta}{e^{i\theta_0}-\zeta},$$

and $\Omega(t)$ is obtained by cutting $\Omega(0)$ along a geodesic arc. The preimage of the endpoint of this slit is exactly $e^{i\theta_0}$. In particular, if $\Omega(0)$ is a complement of a disk, then $\Omega(t)$ is $\Omega(0)$ minus a radial slit.

Example 2. Let $\Omega(0)$ be a domain bounded by an analytic curve $\Gamma(t)$. If the derivative of the measure γ_t with respect to the Lebesgue measure is

$$\frac{d\gamma_t(\theta)}{d\theta} = \frac{1}{2\pi|f'(e^{i\theta},t)|^2},$$

then

$$p_f(\zeta,t) = \frac{1}{2\pi}\int_0^{2\pi} \frac{1}{|f'(e^{i\theta},t)|^2}\frac{e^{i\theta}+\zeta}{e^{i\theta}-\zeta}d\theta,$$

and letting ζ tend to the unit circle we obtain $\mathrm{Re}\,[\dot{f}\ \overline{\zeta f'}] = 1$, which corresponds to the classical Hele-Shaw case, for which the solution exists locally in time.

In the classical Hele-Shaw process the boundary develops by fluid particles moving in the normal direction. In the discrete DLA models either lattice or with circular patterns the attaching are developed in the normal direction too. However, in the continuous limit it is usually impossible to speak of any normal direction because of the irregularity of $\Gamma(t)$.

In [49, Sect. 2.3] this difficulty was circumvented by evaluating the derivative of f occurring in γ_t in the above Löwner model slightly outside the boundary of the unit disk.

Let $\Omega(0)$ be any simply connected domain, $\infty \in \Omega(0), 0 \notin \Omega(0)$. The derivative of the measure γ_t with respect to the Lebesgue measure is

$$\frac{d\gamma_t(\theta)}{d\theta} = \frac{1}{2\pi|f'((1+\varepsilon)e^{i\theta},t)|^2},$$

with sufficiently small positive ε. In this case the derivative is well defined.

It is worth to mention that the estimate

$$\frac{\partial \mathrm{cap}\,\Gamma(t)}{\partial t} = \dot{\alpha}(t) \lesssim \frac{1}{\varepsilon}$$

would be equivalent to the Brennan conjecture (see [191, Chap. 8]) which is still unproved. However, Theorem 2.1 [49] states that if

$$R(t) = \max_{\theta\in[0,2\pi)} |f((1+\varepsilon)e^{i\theta},t)|,$$

then

$$\limsup_{\Delta t\to 0} \frac{R(t+\Delta t) - R(t)}{\Delta t} \leq \frac{C}{\varepsilon},$$

for some absolute constant C. Carleson and Makarov [49] were, with the above model, able to establish an estimate for the growth of the cluster or aggregate given as a lower bound for the time needed to multiply the capacity of the aggregate by a suitable constant. This is an analogue of the upper bound for the size of the cluster in two-dimensional stochastic DLA given by [132].

14.3 Extension to Several Complex Variables

Pfaltzgraff in 1974 was the first one who extended the basic Löwner theory to $\mathbb{C}^n$ with the aim of giving bounds and growth estimates to some classes of univalent mappings from the unit ball of $\mathbb{C}^n$. The theory was later developed by Poreda, Graham, Kohr, Kohr, Hamada, and others (see [111] and [43]).

Since then, a lot of work was devoted to successfully extend the theory to several complex variables, and finally, it has been accomplished. The main and dramatic difference between the one-dimensional case and the higher dimensional case is essentially due to the lack of a Riemann mapping theorem or, which is the same, to the existence of the so-called Fatou–Bieberbach phenomena, that is, the existence of proper open subsets of $\mathbb{C}^n$, $n \geq 2$, which are biholomorphic to $\mathbb{C}^n$. This in turn implies that there are no satisfactory growth estimates for univalent functions on the ball (nor in any other simply connected domain) of $\mathbb{C}^n$, $n \geq 2$.

The right way to proceed is then to look at the Löwner theory in higher dimension as a discrete complex dynamical system, in the sense of random iteration, and to consider abstract basins of attraction as the analogous of the Löwner chains. In order to state the most general results, we first give some definitions and comment on that.

Most estimates in the unit disc can be rephrased in terms of the Poincaré distance, which gives a more intrinsic point of view. In higher dimension one can replace the Poincaré distance with the Kobayashi distance. First, we recall the definition of Kobayashi distance (see [134] for details and properties). Let M be a complex manifold and let $z, w \in M$. A *chain of analytic discs* between z and w is a finite family of holomorphic mappings $f_j : \mathbb{D} \to M$, $j = 1, \ldots, m$ and points $t_j \in (0, 1)$ such that

$$f_1(0) = z, f_1(t_1) = f_2(0), \ldots, f_{m-1}(t_{m-1}) = f_m(0), f_m(t_m) = w.$$

We denote by $\mathscr{C}_{z,w}$ the set of all chains of analytic discs joining z to w. Let $L \in \mathscr{C}_{z,w}$. The *length of* L, denoted by $\ell(L)$, is given by

$$\ell(L) := \sum_{j=1}^{m} \omega(0, t_j) = \sum_{j=1}^{m} \frac{1}{2} \log \frac{1+t_j}{1-t_j}.$$

We define the *Kobayashi (pseudo)distance* $k_M(z, w)$ as follows:

$$k_M(z, w) := \inf_{L \in \mathscr{C}_{z,w}} \ell(L).$$

If M is connected, then $k_M(z, w) < +\infty$ for all $z, w \in M$. Moreover, by construction, it satisfies the triangular inequality. However, it might be that $k_M(z, w) = 0$ even if $z \neq w$ (a simple example is represented by $M = \mathbb{C}$, where $k_{\mathbb{C}} \equiv 0$). In the unit disc $\mathbb{D}$, $k_{\mathbb{D}}$ coincides with the Poincaré distance.

Definition 9. A complex manifold M is said to be *(Kobayashi) hyperbolic* if $k_M(z, w) > 0$ for all $z, w \in M$ such that $z \neq w$. Moreover, M is said *complete hyperbolic* if k_M is complete.

Important examples of complete hyperbolic manifolds are given by bounded convex domains in $\mathbb{C}^n$.

The main property of the Kobayashi distance is the following: let M, N be two complex manifolds and let $f : M \to N$ be holomorphic. Then for all $z, w \in M$ it holds

$$k_N(f(z), f(w)) \leq k_M(z, w).$$

It can be proved that if M is complete hyperbolic, then k_M is Lipschitz continuous (see [18]). If M is a bounded strongly convex domain in $\mathbb{C}^n$ with smooth boundary, Lempert (see, e.g., [134]) proved that the Kobayashi distance is of class C^∞ outside the diagonal. In any case, even if k_M is not smooth, one can consider the differential dk_M as the Dini-derivative of k_M, which coincides with the usual differential at almost every point in $M \times M$.

As it is clear from the one-dimensional general theory of Löwner's equations, evolution families and Herglotz vector fields are pretty much related to semigroups and infinitesimal generators. Kobayashi distance can be used to characterize infinitesimal generators of continuous semigroups of holomorphic self-maps of complete hyperbolic manifolds. The following characterization of infinitesimal generators is proved for strongly convex domains in [41], and in general in [18]:

Theorem 16. *Let M be a complete hyperbolic complex manifold and let $H : M \to TM$ be an holomorphic vector field on M. Then the following are equivalent.*

1. *H is an infinitesimal generator,*
2. *For all $z, w \in M$ with $z \neq w$ it holds*

$$(dk_M)_{(z,w)} \cdot (H(z), H(w)) \leq 0.$$

This apparently harmless characterization contains instead all the needed information to get good growth estimates. In particular, it is equivalent to the Berkson–Porta representation formula in the unit disc.

14.4 Ld-Herglotz Vector Fields and Evolution Families on Complete Hyperbolic Manifolds

Let M be a complex manifold, and denote by $\|\cdot\|$ a Hermitian metric on TM and by d_M the corresponding integrated distance.

Definition 10. Let M be a complex manifold. A *weak holomorphic vector field of order* $d \geq 1$ on M is a mapping $G : M \times \mathbb{R}^+ \to TM$ with the following properties:

(i) The mapping $G(z, \cdot)$ is measurable on $\mathbb{R}^+$ for all $z \in M$.
(ii) The mapping $G(\cdot, t)$ is a holomorphic vector field on M for all $t \in \mathbb{R}^+$.
(iii) For any compact set $K \subset M$ and all $T > 0$, there exists a function $C_{K,T} \in L^d([0, T], \mathbb{R}^+)$ such that

$$\|G(z,t)\| \leq C_{K,T}(t), \quad z \in K, \text{ a.e. } t \in [0, T].$$

A *Herglotz vector field of order* $d \geq 1$ is a weak holomorphic vector field $G(z, t)$ of order d with the property that $M \ni z \mapsto G(z, t)$ is an infinitesimal generator for almost all $t \in [0, +\infty)$.

If M is complete hyperbolic, due to the previous characterization of infinitesimal generators, a weak holomorphic vector field $G(z, t)$ of order d if a Herglotz vector field of order d if and only if

$$(dk_M)_{(z,w)} \cdot (G(z,t), G(w,t)) \leq 0, \quad z, w \in M, z \neq w, \quad \text{a.e. } t \geq 0. \tag{54}$$

This was proved in [41] for strongly convex domains, and in [18] for the general case.

One can also generalize the concept of evolution families:

Definition 11. Let M be a complex manifold. A family $(\varphi_{s,t})_{0 \leq s \leq t}$ of holomorphic self-mappings of M is an *evolution family of order* $d \geq 1$ (or L^d-evolution family) if it satisfies the *evolution property*

$$\varphi_{s,s} = \mathsf{id}, \quad \varphi_{s,t} = \varphi_{u,t} \circ \varphi_{s,u}, \quad 0 \leq s \leq u \leq t, \tag{55}$$

and if for any $T > 0$ and for any compact set $K \subset\subset M$ there exists a function $c_{T,K} \in L^d([0, T], \mathbb{R}^+)$ such that

$$d_M(\varphi_{s,t}(z), \varphi_{s,u}(z)) \leq \int_u^t c_{T,K}(\xi) d\xi, \quad z \in K, \ 0 \leq s \leq u \leq t \leq T. \tag{56}$$

It can be proved that all elements of an evolution family are univalent (cf. [20, Proposition 2.3]).

The classical Löwner and Kufarev–Löwner equations can now be completely generalized as follows:

Theorem 17. *Let M be a complete hyperbolic complex manifold. Then for any Herglotz vector field G of order $d \in [1, +\infty]$ there exists a unique L^d-evolution family $(\varphi_{s,t})$ over M such that for all $z \in M$*

$$\frac{\partial \varphi_{s,t}}{\partial t}(z) = G(\varphi_{s,t}(z), t) \quad a.e.\ t \in [s, +\infty). \tag{57}$$

Conversely for any L^d-evolution family $(\varphi_{s,t})$ over M there exists a Herglotz vector field G of order d such that (57) *is satisfied. Moreover, if H is another weak holomorphic vector field which satisfies* (57)*, then $G(z,t) = H(z,t)$ for all $z \in M$ and almost every $t \in \mathbb{R}^+$.*

Equation (57) is the bridge between the Ld-Herglotz vector fields and L^d-evolution families. In [42] the result has been proved for any complete hyperbolic complex manifold M with Kobayashi distance of class C^1 outside the diagonal, but the construction given there only allowed to start with evolution families of order $d = +\infty$. Next, in [117] the case of L^d-evolution families has been proved for the case $M = \mathbb{B}^n$ the unit ball in $\mathbb{C}^n$. Finally, in [18], Theorem 17 was proved in full generality.

The previous equation, especially in the case of the unit ball of $\mathbb{C}^n$ and for the case $d = +\infty$, with evolution families fixing the origin and having some particular first jets at the origin has been studied by many authors, we cite here Pfaltzgraff [184, 185], Poreda [193], Graham et al. [107], Graham et al. [109] (see also [110]).

Using the so-called product formula, proved in convex domains by Reich and Shoikhet [211] (see also [212]), and later generalized on complete hyperbolic manifold in [18] we get a strong relation between the semigroups generated at a fixed time by a Herglotz vector field and the associated evolution family. Let $G(z,t)$ be a Herglotz vector field on a complete hyperbolic complex manifold M. For almost all $t \geq 0$, the holomorphic vector field $M \ni z \mapsto G(z,t)$ is an infinitesimal generator. Let (ϕ_r^t) be the associated semigroups of holomorphic self-maps of M. Let $(\varphi_{s,t})$ be the evolution family associated with $G(z,t)$. Then, uniformly on compacta of M it holds

$$\phi_t^r = \lim_{m\to\infty} \varphi_{t,t+\frac{r}{m}}^{\circ m} = \lim_{m\to\infty} \underbrace{(\varphi_{t,t+\frac{r}{m}} \circ \ldots \circ \varphi_{t,t+\frac{r}{m}})}_{m}.$$

14.5 Löwner Chains on Complete Hyperbolic Manifolds

Although one could easily guess how to extend the notion of Herglotz vector fields and evolution families to several complex variables, the concept of Löwner chains is not so easy to extend. For instance, starting from a Herglotz vector field on the unit ball of $\mathbb{C}^n$, one would be tempted to define in a natural way Löwner chains with range in $\mathbb{C}^n$. However, sticking with such a definition, it is rather hard to get a complete solution to the Löwner PDE. In fact, in case $D = \mathbb{B}^n$ the unit ball, much effort has been done to show that, given an evolution family $(\varphi_{s,t})$ on $\mathbb{B}^n$ such that $\varphi_{s,t}(0) = 0$ and $d(\varphi_{s,t})_0$ has a special form, then there exists an associated Löwner chain. We cite here the contributions of Pfaltzgraff [184, 185], Poreda [193], Graham et al. [107], Graham et al. [109], Arosio [16], Voda [255]. In the last two mentioned papers, resonances phenomena among the eigenvalues of $d(\varphi_{s,t})_0$ are taken into account.

The reason for these difficulties is due to the fact that, although apparently natural, the definition of Löwner chains as a more or less regular family of univalent mappings from the unit ball to $\mathbb{C}^n$ is not the right one. And the reason why this is meaningful in one dimension is just because of the Riemann mapping theorem, as we will explain.

Indeed, as shown before, there is essentially no difference in considering evolution families or Herglotz vector fields in the unit ball of $\mathbb{C}^n$ or on complete hyperbolic manifolds, since the right estimates to produce the Löwner equation are provided just by the completeness of the Kobayashi distance and its contractiveness properties.

The right point of view is to consider evolution families as random iteration families, and thus, the "Löwner chains" are just the charts of the abstract basins of attraction of such a dynamical system. To be more precise, let us recall the theory developed in [20]. Interesting and surprisingly enough, regularity conditions—which were basic in the classical theory for assuming the classical limiting process to converge—do not play any role.

Definition 12. Let M be a complex manifold. An *algebraic evolution family* is a family $(\varphi_{s,t})_{0\leq s\leq t}$ of univalent self-mappings of M satisfying the evolution property (55).

A L^d-evolution family is an algebraic evolution family because all elements of a L^d-evolution family are injective as we said before.

Definition 13. Let M, N be two complex manifolds of the same dimension. A family $(f_t)_{t\geq 0}$ of holomorphic mappings $f_t : M \to N$ is a *subordination chain* if for each $0 \leq s \leq t$ there exists a holomorphic mapping $v_{s,t} : M \to M$ such that $f_s = f_t \circ v_{s,t}$. A subordination chain (f_t) and an algebraic evolution family $(\varphi_{s,t})$ are *associated* if

$$f_s = f_t \circ \varphi_{s,t}, \quad 0 \leq s \leq t.$$

An *algebraic Löwner chain* is a subordination chain such that each mapping $f_t : M \to N$ is univalent. The *range* of an algebraic Löwner chain is defined as

$$\mathsf{rg}\,(f_t) := \bigcup_{t\geq 0} f_t(M).$$

Note that an algebraic Löwner chain (f_t) has the property that

$$f_s(M) \subset f_t(M), \quad 0 \leq s \leq t.$$

We have the following result which relates algebraic evolution families to algebraic Löwner chains, whose proof is essentially based on abstract categorial analysis:

Theorem 18 ([20]). *Let M be a complex manifold. Then any algebraic evolution family $(\varphi_{s,t})$ on M admits an associated algebraic Löwner chain $(f_t\colon M \to N)$.*

Moreover if $(g_t : M \to Q)$ *is a subordination chain associated with* $(\varphi_{s,t})$ *then there exist a holomorphic mapping* $\Lambda : \mathsf{rg}\,(f_t) \to Q$ *such that*

$$g_t = \Lambda \circ f_t, \quad \forall t \geq 0.$$

The mapping Λ *is univalent if and only if* (g_t) *is an algebraic Löwner chain, and in that case* $\mathsf{rg}\,(g_t) = \Lambda(\mathsf{rg}\,(f_t))$.

The previous theorem shows that the range $\mathsf{rg}\,(f_t)$ of an algebraic Löwner chain (f_t) is uniquely defined up to biholomorphisms. In particular, given an algebraic evolution family $(\varphi_{s,t})$ one can define its *Löwner range* $\mathsf{Lr}(\varphi_{s,t})$ as the biholomorphism class of the range of any associated algebraic Löwner chain.

In particular, if $M = \mathbb{D}$ the unit disc, then the Löwner range of any evolution family on $\mathbb{D}$ is a simply connected non compact Riemann surface, thus, by the uniformization theorem, the Löwner range is either the unit disc $\mathbb{D}$ or $\mathbb{C}$. Therefore, in the one-dimensional case, one can harmlessly stay with the classical definition of Löwner chains as a family of univalent mappings with image in $\mathbb{C}$.

One can also impose L^d-regularity as follows:

Definition 14. Let $d \in [1, +\infty]$. Let M, N be two complex manifolds of the same dimension. Let d_N be the distance induced by a Hermitian metric on N. An algebraic Löwner chain $(f_t : M \to N)$ is a L^d*-Löwner chain* (for $d \in [1, +\infty]$) if for any compact set $K \subset\subset M$ and any $T > 0$ there exists a $k_{K,T} \in L^d([0, T], \mathbb{R}^+)$ such that

$$d_N(f_s(z), f_t(z)) \leq \int_s^t k_{K,T}(\xi) d\xi \tag{58}$$

for all $z \in K$ and for all $0 \leq s \leq t \leq T$.

The L^d-regularity passes from evolution family to Löwner chains:

Theorem 19 ([20]). *Let* M *be a complete hyperbolic manifold with a given Hermitian metric and* $d \in [1, +\infty]$. *Let* $(\varphi_{s,t})$ *be an algebraic evolution family on* M *and let* $(f_t : M \to N)$ *be an associated algebraic Löwner chain. Then* $(\varphi_{s,t})$ *is a* L^d*-evolution family on* M *if and only if* (f_t) *is a* L^d*-Löwner chain.*

Once the general Löwner equation is established and Löwner chains have been well defined, even the Löwner–Kufarev PDE can be generalized:

Theorem 20 ([20]). *Let* M *be a complete hyperbolic complex manifold, and let* N *be a complex manifold of the same dimension. Let* $G : M \times \mathbb{R}^+ \to TM$ *be a Herglotz vector field of order* $d \in [1, +\infty]$ *associated with the* L^d*-evolution family* $(\varphi_{s,t})$. *Then a family of univalent mappings* $(f_t : M \to N)$ *is an* L^d*-Löwner chain associated with* $(\varphi_{s,t})$ *if and only if it is locally absolutely continuous on* $\mathbb{R}^+$ *locally uniformly with respect to* $z \in M$ *and solves the Löwner–Kufarev PDE*

$$\frac{\partial f_s}{\partial s}(z) = -(df_s)_z G(z, s), \quad a.e.\ s \geq 0, z \in M.$$

14.6 The Löwner Range and the General Löwner PDE in $\mathbb{C}^n$

As we saw before, given a L^d-evolution family (or just an algebraic evolution family) on a complex manifold, it is well defined the Löwner range $\mathsf{Lr}(\varphi_{s,t})$ as the class of biholomorphism of the range of any associated Löwner chain.

In practice, it is interesting to understand the Löwner range of an evolution family on a given manifold. For instance, one may ask whether, starting from an evolution family on the ball, the Löwner range is always biholomorphic to an open subset of $\mathbb{C}^n$. This problem turns out to be related to the so-called *Bedford's conjecture*. Such a conjecture states that given a complex manifold M, an automorphism $f : M \to M$ and a f-invariant compact subset $K \subset M$ on which the action of f is hyperbolic, then the stable manifold of K is biholomorphic to $\mathbb{C}^m$ for some $m \leq \dim M$. The equivalent formulation which resembles the problem of finding the Löwner range of an evolution family in the unit ball is in [78], see also [17] where such a relation is well explained.

In [16, Sect. 9.4] it is shown that there exists an algebraic evolution family $(\varphi_{s,t})$ on $\mathbb{B}^3$ which does not admit any associated algebraic Löwner chain with range in $\mathbb{C}^3$. Such an evolution family is, however, not L^d for any $d \in [1, +\infty]$.

In the recent paper [19] it has been proved the following result:

Theorem 21. *Let $D \subset \mathbb{C}^n$ be a complete hyperbolic starlike domain (for instance, the unit ball). Let $(\varphi_{s,t})$ be an L^d-evolution family, $d \in [1, +\infty]$. Then the Löwner range $\mathsf{Lr}(\varphi_{s,t})$ is biholomorphic to a Runge and Stein open domain in $\mathbb{C}^n$.*

The proof, which starts from the existence of a Löwner chain with abstract range, is based on the study of manifolds which are union of balls, using a result by Docquier and Grauert to show that the regularity hypothesis guarantees Rungeness and then one can use approximation results of Andersén and Lempert in order to construct a suitable embedding.

As a corollary of the previous consideration, we have a general solution to Löwner PDE in higher dimension, which is the full analogue of the one-dimensional situation:

Theorem 22 ([19]). *Let $D \subset \mathbb{C}^N$ be a complete hyperbolic starlike domain. Let $G : D \times \mathbb{R}^+ \to \mathbb{C}^N$ be a Herglotz vector field of order $d \in [1, +\infty]$. Then there exists a family of univalent mappings $(f_t : D \to \mathbb{C}^N)$ of order d which solves the Löwner PDE*

$$\frac{\partial f_t}{\partial t}(z) = -df_t(z)G(z,t), \quad a.a.\ t \geq 0, \forall z \in D. \tag{59}$$

Moreover, $R := \cup_{t \geq 0} f_t(D)$ is a Runge and Stein domain in $\mathbb{C}^N$ and any other solution to (59) *is of the form $(\Phi \circ f_t)$ for a suitable holomorphic map $\Phi : R \to \mathbb{C}^N$.*

In general, one can infer some property of the Löwner range from the dynamics of the evolution family. In order to state the result, let us recall what the Kobayashi pseudometric is:

Definition 15. Let M be a complex manifold. The *Kobayashi pseudometric* κ_M : $TM \to \mathbb{R}^+$ is defined by

$$\kappa_M(z; v) := \inf\{r > 0 : \exists g : \mathbb{D} \to M \text{ holomorphic } : g(0) = z, g'(0) = \frac{1}{r}v\}.$$

The Kobayashi pseudometric has the remarkable property of being contracted by holomorphic maps, and its integrated distance is exactly the Kobayashi pseudodistance. We refer the reader to [134] for details.

Definition 16. Let $(\varphi_{s,t})$ be an algebraic evolution family on a complex manifold M. For $v \in T_z M$ and $s \geq 0$ we define

$$\beta_z^s(v) := \lim_{t\to\infty} \kappa_M(\varphi_{s,t}(z); (d\varphi_{s,t})_z(v)). \tag{60}$$

Since the Kobayashi pseudometric is contracted by holomorphic mappings the limit in (60) is well defined.

The function β is the bridge between the dynamics of an algebraic evolution family $(\varphi_{s,t})$ and the geometry of its Löwner range. Indeed, in [20] it is proved that if N is a representative of the Löwner range of $(\varphi_{s,t})$ and $(f_t : M \to N)$ is an associated algebraic Löwner chain, then for all $z \in M$ and $v \in T_z M$ it follows

$$f_s^* \kappa_N(z; v) = \beta_z^s(v).$$

In the unit disc case, if $(\varphi_{s,t})$ is an algebraic evolution family, the previous formula allows to determine the Löwner range: by the Riemann mapping theorem the Löwner range is either $\mathbb{C}$ or $\mathbb{D}$. The first being non-hyperbolic, if $\beta_z^s(v) = 0$ for some $s > 0, z \in \mathbb{D}$ (v can be taken to be 1), then the Löwner range is $\mathbb{C}$, otherwise it is $\mathbb{D}$.

Such a result can be generalized to a complex manifold M. Let $\mathsf{aut}(M)$ denote the group of holomorphic automorphisms of a complex manifold M. Using a result by Fornæss and Sibony [77], in [20] it is shown that the previous formula implies

Theorem 23. *Let M be a complete hyperbolic complex manifold and assume that $M/\mathsf{aut}(M)$ is compact. Let $(\varphi_{s,t})$ be an algebraic evolution family on M. Then*

1. *If there exists $z \in M$, $s \geq 0$ such that $\beta_z^s(v) \neq 0$ for all $v \in T_z M$ with $v \neq 0$, then $\mathsf{Lr}(\varphi_{s,t})$ is biholomorphic to M.*
2. *If there exists $z \in M$, $s \geq 0$ such that $\dim_{\mathbb{C}}\{v \in T_z M : \beta_z^s(v) = 0\} = 1$, then $\mathsf{Lr}(\varphi_{s,t})$ is a fiber bundle with fiber $\mathbb{C}$ over a closed complex submanifold of M.*

15 Acknowledgements

All the authors have been supported by the ESF Networking Programme HCAA. Filippo Bracci has been supported by ERC grant "HEVO—Holomorphic Evolution Equations," no. 277691; Manuel D. Contreras and Santiago Díaz-Madrigal have been partially supported by the Ministerio de Economía y Competitividad and the European Union (FEDER), projects MTM2009-14694-C02-02 and MTM2012-37436-C02-01 and by La Consejería de Educación y Ciencia de la Junta de Andalucía; Alexander Vasil'ev has been supported by the grants of the Norwegian Research Council #204726/V30 and #213440/BG.

References

1. M. Abate, *Iteration Theory of Holomorphic Maps on Taut Manifolds* (Mediterranean Press, Rende, Cosenza, 1989)
2. M. Abate, F. Bracci, M.D. Contreras, S. Díaz-Madrigal, The evolution of Loewner's differential equations. Newslett. Eur. Math. Soc. **78**, 31–38 (2010)
3. D. Aharonov, S. Reich, D. Shoikhet, Flow invariance conditions for holomorphic mappings in Banach spaces. Math. Proc. R. Ir. Acad. **99A**(1), 93–104 (1999)
4. L.V. Ahlfors, *Conformal Invariants: Topics in Geometric Function Theory* (AMS Chelsea Publishing, Providence, RI, 2010) (Reprint of the 1973 original)
5. L.A. Aksent'ev, P.L. Shabalin, Sufficient conditions for univalence and quasiconformal extendibility of analytic functions, in *Handbook of Complex Analysis: Geometric Function Theory*, vol. 1 (North-Holland, Amsterdam, 2002), pp. 169–206
6. S. Albeverio, R. Høegh-Krohn, Uniqueness and the global markov property for euclidean fields. The case of trigonometric interactions. Commun. Math. Phys. **68**, 95–128 (1979)
7. I.A. Aleksandrov, V.V. Chernikov, Pavel Parfen'evich Kufarev. Obituary (in Russian), Uspehi Mat. Nauk **24**, 181–184 (1969)
8. I.A. Aleksandrov, S.A. Kopanev, On mutual growth of the modulus of a univalent function and of the modulus of its derivative. Sibirsk. Mat. Zh. **7**, 23–30 (1966)
9. I.A. Aleksandrov, S.A. Kopanev, The range of values of the derivative on the class of holomorphic univalent functions. Ukrain. Mat. Zh. **22**, 660–664 (1970)
10. I.A. Aleksandrov, V.I. Popov, Optimal controls and univalent functions. Ann. Univ. M. Curie-Sklodowska, Sec. A **22–24**, 13–20 (1968–1970)
11. I.A. Aleksandrov, B.Ja. Krjučkov, V.I. Popov, The first coefficients of bounded holomorphic univalent functions. Dopovīdī Akad. Nauk Ukraïn. RSR Ser. A **1**, 3–5 (1973)
12. I.A. Aleksandrov, G.A. Popova, Extremal properties of univalent holomorphic functions with real coefficients. Sibirsk. Mat. Ž. **14**, 915–926 (1973)
13. I.A. Aleksandrov, G.G. Zavozin, S.A. Kopanev, Optimal controls in coefficient problems for univalent functions. Differencial'nye Uravnenija **12**(4), 599–611 (1976)
14. I.A. Aleksandrov, *Parametric Continuations in the Theory of Univalent Functions* (Nauka, Moscow, 1976)
15. S.T. Aleksandrov, V.V. Sobolev, Extremal problems in some classes of functions, univalent in the half plane, having a finite angular residue at infinity. Siberian Math. J. **27**, 145–154 (1986). Translation from Sibirsk. Mat. Zh. **27**, 3–13 (1986)
16. L. Arosio, Resonances in Loewner equations. Adv. Math. **227**, 1413–1435 (2011)
17. L. Arosio, Basins of attraction in Loewner equations. Ann. Acad. Sci. Fenn. Math. **37**, 563–570 (2012)

18. L. Arosio, F. Bracci, Infinitesimal generators and the Loewner equation on complete hyperbolic manifolds. Anal. Math. Phys. **1**(4), 337–350 (2011)
19. L. Arosio, F. Bracci, E. Fornæss Wold, *Solving the Loewner PDE in complete hyperbolic starlike domains of* $\mathbb{C}^N$, Adv. Math. **242**, 209–216 (2013)
20. L. Arosio, F. Bracci, H. Hamada, G. Kohr, An abstract approach to Löwner chains. J. Anal. Math. **119**(1), 89–114 (2013)
21. K. Astala, P. Jones, A. Kupiainen, E. Saksman, Random curves by conformal welding. Compt. Rend. Math. **348**(5–6), 257–262 (2010)
22. O. Babelon, D. Bernard, M. Talon, *Introduction to Classical Integrable Systems* (Cambridge University Press, Cambridge, 2003)
23. M. Bauer, D. Bernard, SLE_k growth processes and conformal field theories. Phys. Lett. B **543**(1–2), 135–138 (2002)
24. M. Bauer, D. Bernard, Conformal field theories of stochastic Loewner evolutions. Comm. Math. Phys. **239**(3), 493–521 (2003)
25. R.O. Bauer, R.M. Friedrich, On radial stochastic Loewner evolution in multiply connected domains. J. Funct. Anal. **237**, 565–588 (2006)
26. R.O. Bauer, R.M. Friedrich, On chordal and bilateral SLE in multiply connected domains. Math. Z. **258**, 241–265 (2008)
27. J. Basilewitsch, Sur les théorèmes de Koebe-Bieberbach. Rec. Math. [Mat. Sbornik] N.S. **1(43)**(3), 283–291 (1936)
28. J. Basilewitsch, D'une propriété extrémale de la fonction $F(z) = \frac{z}{(1-z)^2}$. Rec. Math. [Mat. Sbornik] N.S. **12(54)**(3), 315–319 (1943)
29. I.E. Bazilevič, On a case of integrability in quadratures of the Loewner-Kufarev equation. Mat. Sb. N.S. **37**(79), 471–476 (1955)
30. I.E. Bazilevič, Generalization of an integral formula for a subclass of univalent functions. Mat. Sb. N.S. **64 (106)**, 628–630 (1964)
31. J. Becker, Löwnersche, Differentialgleichung und quasikonform fortsetzbare schlichte Funktionen. J. Reine Angew. Math. **255**, 23–43 (1972)
32. J. Becker, Ch. Pommerenke, Schlichtheitskriterien und Jordangebiete. J. Reine Angew. Math. **354**, 74–94 (1984)
33. V. Beffara, The dimension of the SLE curves. Ann. Probab. **36**(4), 1421–1452 (2008)
34. A.A. Belavin, A.M. Polyakov, A.B. Zamolodchikov, Infinite conformal symmetry in two-dimensional quantum field theory. Nuclear Phys. B **241**(2), 333–380 (1984)
35. E. Berkovich, B. Shein, *Odyssey of Fritz Noether. Epilogue*, Notes in Jewish History (2009), no. 11(114), July
36. E. Berkson, H. Porta, Semigroups of holomorphic functions and composition operators. Mich. Math. J. **25**, 101–115 (1978)
37. L. Bieberbach, Über die Koeffizienten derjenigen Potenzreihen, welche eine schlichte Abbildung des Ein- heitskreises vermitteln. S.-B. Preuss. Akad. Wiss. 940–955 (1916)
38. Bill of indictment on L.A. Vishnevskiĭ. February 15, 1938. Novosibirsk
39. K.I. Bobenko, *On the theory of extremal problems for univalent functions of class* S. Trudy Mat. Inst. Steklov. vol. 101 (Izdat. "Nauka", Moscow, 1972)
40. E. Bombieri, On the local maximum property of the Koebe function. Inv. Math. **4**, 26–67 (1967)
41. F. Bracci, M. D. Contreras, S. Díaz-Madrigal, Pluripotential theory, semigroups and boundary behaviour of infinitesimal generators in strongly convex domains. J. Eur. Math. Soc. **12**, 23–53 (2010)
42. F. Bracci, M.D. Contreras, S. Díaz-Madrigal, Evolution families and the Löwner equation I: the unit disk. J. Reine Angew. Math. **672**, 1–37 (2012)
43. F. Bracci, M.D. Contreras, and S. Díaz-Madrigal, Evolution Families and the Löwner Equation II: complex hyperbolic manifolds. Math. Ann. **344**, 947–962 (2009)
44. F. Bracci, M.D. Contreras, S. Díaz-Madrigal, Semigroups versus evolution families in the Löwner theory. J. Anal. Math. **115**(1), 273–292 (2011)

45. L. de Branges, A proof of the Bieberbach conjecture. Preprint E-5-84, Leningrad Branch of the Steklov Institute of Mathematics, LOMI, Leningrad, 1984
46. L. de Branges, A proof of the Bieberbach conjecture. Acta Math. **154**(1–2), 137–152 (1985)
47. D. Bshouty, W. Hengartner, A variation of the Koebe mapping in a dense subset of S. Can. J. Math. **39**(7), 54–73 (1987)
48. J. Cardy, Boundary conformal field theory, in *Encyclopedia of Mathematical Physics*, ed. by J.-P. Francoise, G. Naber, S. Tsun Tsou (Elsevier, Amsterdam, 2006), pp. 333–340
49. L. Carleson, N. Makarov, Aggregation in the plane and Loewner's equation. Comm. Math. Phys. **216**, 583–607 (2001)
50. A. Carverhill, Flows of stochastic dynamical systems: ergodic theory. Stochastics **14**(4), 273–317 (1985)
51. D.Y.C. Chan, B.D. Hughes, L. Paterson, Ch. Sirakoff, Simulating flow in porous media. Phys. Rev. A **38**(8), 4106–4120 (1988)
52. D. Chelkak, S. Smirnov, Universality in the 2D Ising model and conformal invariance of fermionic observables. Invent. Math. **189**(3), 515–580 (2012)
53. M.D. Contreras, S. Díaz-Madrigal, P. Gumenyuk, Löwner chains in the unit disk. Rev. Mat. Iberoam. **26**(3), 975–1012 (2010)
54. M.D. Contreras, S. Díaz-Madrigal, P. Gumenyuk, Local duality in Löwner equations. J. Nonlinear Convex Anal. (to appear). Available in http://arxiv.org/pdf/1202.2334.pdf
55. M.D. Contreras, S. Díaz-Madrigal, P. Gumenyuk, Loewner Theory in annulus I: evolution families and differential equations. Trans. Am. Math. Soc. **365**, 2505–2543 (2013)
56. M.D. Contreras, S. Díaz-Madrigal, P. Gumenyuk, Loewner Theory in annulus II: Loewner chains. Anal. Math. Phys. **1**, 351–385 (2011)
57. J. Conway, *Functions of One Complex Variable II* (Springer, New York, 1995)
58. E. Date, M. Kashiwara, T. Miwa, Vertex operators and functions: transformation groups for soliton equations II. Proc. Jpn. Acad. Ser. A Math. Sci. **57**, 387–392 (1981)
59. E. Date, M. Kashiwara, M. Jimbo, T. Miwa, Transformation groups for soliton equations. *Nonlinear Integrable Systems—Classical Theory and Quantum Theory (Kyoto, 1981)* (World Sci. Publishing, Singapore, 1983), pp. 39–119
60. B. Davidovitch, H.G.E. Hentschel, Z. Olami, I. Procaccia, L.M. Sander, E. Somfai, Diffusion limited aggregation and iterated conformal maps. Phys. Rev. E **59**(2), 1368–1378 (1999)
61. B. Davidovitch, M.J. Feigenbaum, H.G.E. Hentschel, I. Procaccia, Conformal dynamics of fractal growth patterns without randomness. Phys. Rev. E **62**(2), 1706–1715 (2000)
62. A. Dick, *Emmy Noether. 1882–1935* (Birkhäuser, Basel, 1970)
63. L. A. Dickey, *Soliton Equations and Hamiltonian Systems*, 2nd edn. Advanced Series in Mathematical Physics, vol. 26 (World Scientific, Singapore, 2003)
64. R.G. Douglas, *Banach Algebra Techniques in Operator Theory*, 2nd edn. Graduate Texts in Mathematics, vol. 179. (Springer, New York, 1998)
65. B. Doyon, Conformal loop ensembles and the stress–energy tensor: I. Fundamental notions of CLE. arXiv:0903.0372 (2009)
66. B. Doyon, Conformal loop ensembles and the stress–energy tensor: II. Construction of the stress–energy tensor. arXiv:0908.1511 (2009)
67. B. Doyon, Calculus on manifolds of conformal maps and CFT. J. Phys. A Math. Theor. **45**(31), 315202 (2012)
68. P.L. Duren, H.S. Shapiro, A.L. Shields, Singular measures and domains not of Smirnov type. Duke Math. J. **33**, 247–254 (1966)
69. J. Dubédat, SLE and the free field: partition functions and couplings. J. Am. Math. Soc. **22**(4), 995–1054 (2009)
70. D.A. Dubovikov, An analog of the Loëwner equation for mappings of strip. Izv. Vyssh. Uchebn. Zaved. Mat. **51**(8), 77–80 2007 (Russian); translation in Russian Math. (Iz. VUZ) **51**(8), 74–77 (2007)
71. P.L. Duren, M. Schiffer, The theory of the second variation in extremum problems for univalent functions. J. Anal. Math. **10**, 193–252 (1962/1963)
72. P.L. Duren, *Univalent Functions* (Springer, New York, 1983)

73. C.J. Earle, A.L. Epstein, Quasiconformal variation of slit domains. Proc. Am. Math. Soc. **129**(11), 3363–3372 (2001)
74. L.D. Faddeev, *Discretized Virasoro Algebra*. Contemporary Mathematics, vol. 391 (American Mathematical Society, Providence, RI, 2005), pp. 59–67
75. M. Fekete, G. Szegö, Eine Bemerkung über ungerade schlichte Funktionen. J. Lond. Math. Soc. **8**(2), 85–89 (1933)
76. C.H. FitzGerald, Ch. Pommerenke, The de Branges theorem on univalent functions. Trans. Am. Math. Soc. **290**, 683–690 (1985)
77. J.E. Fornæss, N. Sibony, Increasing sequences of complex manifolds. Math. Ann. 255, 351–360 (1981)
78. J.E. Fornæss, B. Stensønes, Stable manifolds of holomorphic hyperbolic maps. Int. J. Math. **15**(8), 749–758 (2004)
79. S. Friedland, M. Schiffer, Global results in control theory with applications to univalent functions. Bull. Am. Math. Soc. **82**, 913–915 (1976)
80. S. Friedland, M. Schiffer, On coefficient regions of univalent functions. J. Anal. Math. **31**, 125–168 (1977)
81. R. Friedrich, The global geometry of stochastic Loewner evolutions. *Probabilistic Approach to Geometry*. Advanced Studies in Pure Mathematics, vol. 57 (Mathematical Society of Japan, Tokyo, 2010), pp. 79–117
82. R. Friedrich, W. Werner, Conformal restriction, highest-weight representations and SLE. Comm. Math. Phys. **243**(1), 105–122 (2003)
83. L.A. Galin, Unsteady filtration with a free surface. Dokl. Akad. Nauk USSR **47**, 246–249 (1945). (in Russian)
84. R.V. Gamkrelidze, *Foundations of Optimal Control Theory* (Tbilisi University, Tbilisi, 1977)
85. P.R. Garabedian, M.A. Schiffer, A proof of the Bieberbach conjecture for the fourth coefficient. J. Rational Mech. Anal. **4**, 427–465 (1955)
86. C.S. Gardner, J.M. Greene, M.D. Kruskal, R.M. Miura, Korteweg-deVries equation and generalization. VI. Methods for exact solution. Comm. Pure Appl. Math. **27**, 97–133 (1974)
87. F.W. Gehring, E. Reich, Area distortion under quasiconformal mappings. Ann. Acad. Sci. Fenn. Ser. A I **388**, 1–14 (1966)
88. J.-L. Gervais, Infinite family of polynomial functions of the Virasoro generators with vanishing Poisson brackets. Phys. Lett. B **160**(4–5), 277–278 (1985)
89. l. Gibbons, S.P. Tsarev, Conformal maps and reductions of the Benney equations. Phys. Lett. A **258**(4–6), 263–271 (1999)
90. G.M. Goluzin, Über die Verzerrungssätze der schlichten konformen Abbildungen. Rec. Math. [Mat. Sbornik] N.S. **1(34)**(1), 127–135 (1936)
91. G.M. Goluzin, Sur les théorèmes de rotation dans la théorie des fonctions univalentes. Rec. Math. [Mat. Sbornik] N.S. **1(43)**(1), 293–296 (1936)
92. G.M. Goluzin, On the parametric representation of functions univalent in a ring. Mat. Sbornik N.S. **29(71)**, 469–476 (1951)
93. G.M. Goluzin, *Geometrical Theory of Functions of a Complex Variable*, 2nd edn. (Nauka, Moscow, 1966)
94. G. S. Goodman, Pontryagins Maximal Principle as a variational method in conformal mapping. Abstracts of Brief Scientific Communications, sec. 13 (ICM Moscow, 1966), pp. 7–8
95. G.S. Goodman, Univalent functions and optimal control, Ph.D. Thesis, Stanford University, 1968
96. V.V. Goryainov, V.Ya. Gutlyanskiĭ, *Extremal Problems in the Class* S_M, Mathematics collection (Russian) (Izdat. "Naukova Dumka", Kiev, 1976), pp. 242–246
97. V.V. Gorjainov, On a parametric method of the theory of univalent functions. Math. Notes **27**(3–4), 275–279 (1980)
98. V.V. Gorjainov, Boundary functions of a system of functionals composed of values of a univalent function and its derivative. Soviet Math. (Iz. VUZ) **26**(7), 91–94 (1982)

99. V.V. Gorjainov, *Extremals in Estimates of Functionals Depending on Values of a Univalent Function and Its Derivative*. Theory of Mappings and Approximation of Functions ("Naukova Dumka", Kiev, 1983), pp. 38–49
100. V.V. Gorjainov, The Parametric Method and Extremal Conformal Mappings. Soviet Math. Dokl. **28**(1), 205–208 (1983)
101. V.V. Goryainov, Semigroups of conformal mappings. Mat. Sb. (N.S.) **129(171)**(4), 451–472 (1986) (Russian); translation in Math. USSR Sbornik **57**, 463–483 (1987)
102. V.V. Goryajnov, Evolution families of analytic functions and time-inhomogeneous Markov branching processes. Dokl. Akad. Nauk **347**(6) 729–731 (1996); translation in Dokl. Math. **53**(2), 256–258 (1996)
103. V.V. Goryainov, *The Löwner – Kufarev equation and extremal problems for conformal mappings with fixed points on the boundary* (in Russian). Talk at the International Workshop on Complex Analysis and its Applications, Steklov Mathematical Institute, Moscow, 26 December 2011
104. V.V. Goryainov, I. Ba, Semigroups of conformal mappings of the upper half-plane into itself with hydrodynamic normalization at infinity. Ukrainian Math. J. **44**, 1209–1217 (1992)
105. V.V. Goryainov, O.S. Kudryavtseva, One-parameter semigroups of analytic functions, fixed points and the Koenigs function. Mat. Sb. **202**(7), 43–74 (2011) (Russian); translation in Sbornik: Mathematics **202**(7), 971–1000 (2011)
106. V.V. Goryainov, Semigroups of analytic functions in analysis and applications. Uspekhi Mat. Nauk **67**(6), 5–52 (2012)
107. I. Graham, H. Hamada, G. Kohr, Parametric representation of univalent mappings in several complex variables. Can. J. Math. **54**, 324–351 (2002)
108. I. Graham, G. Kohr, *Geometric Function Theory in One and Higher Dimensions* (Marcel Dekker, New York, 2003)
109. I. Graham, H. Hamada, G. Kohr, M. Kohr, Asymptotically spirallike mappings in several complex variables. J. Anal. Math. **105** 267–302 (2008)
110. I. Graham, G. Kohr, J.A. Pfaltzgraff, The general solution of the Loewner differential equation on the unit ball in $\mathbb{C}^n$. Contemp. Math. (AMS) **382**, 191–203 (2005)
111. I. Graham, G. Kohr, *Geometric Function Theory in One and Higher Dimensions* (Marcel Dekker, New York, 2003)
112. R. Greiner, O. Roth, On support points of univalent functions and a disproof of a conjecture of Bombieri. Proc. Am. Math. Soc. **129**(12), 3657–3664 (2001)
113. D. Grier, E. Ben-Jacob, R. Clarke, L.M. Sander, Morphology and Microstructure in Electrochemical Deposition of Zinc. Phys. Rev. Lett. **56**, 1264–1267 (1986)
114. B. Gustafsson, A. Vasil'ev, *Conformal and Potential Analysis in Hele-Shaw Cells* (Birkhäuser, Basel, 2006)
115. V.Ya. Gutlyanskiĭ, Parametric representation of univalent functions. Dokl. Akad. Nauk SSSR **194**, 750–753 (1970)
116. V.Ya. Gutljanskiĭ, *Parametric Representations and Extremal Problems in the Theory of Univalent Functions* (in Russian), Dissertation Dokt. Fiz.-Mat. Nauk, Mat. Inst. Akad. Nauk Ukrain. SSR, Kiev, 1972
117. H. Hamada, G. Kohr, J.R. Muir, Extension of Ld-Loewner chains to higher dimensions. J. Anal. Math. 120(1), 357–392 (2013)
118. T.E. Harris, *The Theory of Branching Processes*. Corrected reprint of the 1963 original, Dover Phoenix Editions (Dover Publications, Mineola, NY, 2002)
119. T.C. Halsey, Diffusion-Limited Aggregation: A Model for Pattern Formation. Phys. Today **53**(11), 36 (2000)
120. M.B. Hastings, Renormalization theory of stochastic growth. Phys. Rev. E **55**(1), 135–152 (1987)
121. M.B. Hastings, L.S. Levitov, Laplacian growth as one-dimensional turbulence. Phys. D **116**(1–2), 244–252 (1998)
122. W.K. Hayman, *Multivalent Functions* (Cambridge University Press, Cambridge, 1958)

123. M.H. Heins, Semigroups of holomorphic maps of a Riemann surface into itself which are homomorphs of the set of positive reals considered additively, in *E.B. Christoffel—The Influence of His Work on Mathematics and the Physical Sciences* (Bikhäuser, Basel, 1981), pp. 314–331
124. H.S. Hele-Shaw, The flow of water. Nature **58**, 33–36 (1898)
125. Yu.E. Hohlov, S.D. Howison, C. Huntingford, J.R. Ockendon, A.A. Lacey, A model for non-smooth free boundaries in Hele-Shaw flows. Q. J. Mech. Appl. Math. **47**, 107–128 (1994)
126. G. Ivanov, A. Vasil'ev, Lï£¡wner evolution driven by a stochastic boundary point. Anal. Math. Phys. **1**(4), 387–412 (2011)
127. G. Ivanov, D. Prokhorov, A. Vasil'ev, Non-slit and singular solutions to the Loewner equation. Bull. Sci. Math. **136**(3), 328–341 (2012)
128. V.G. Kac, Simple irreducible graded Lie algebras of finite growth. Math. USSR Izv. **2**, 1271–1311 (1968); Izv. Akad. Nauk USSR Ser. Mat. **32**, 1923–1967 (1968)
129. W. Kager, B. Nienhuis, L.P. Kadanoff, Exact solutions for Löwner evolutions. J. Stat. Phys. **115**(3–4), 805–822 (2004)
130. N.-G. Kang, N.G. Makarov, Gaussian free field and conformal field theory. arXiv: 1101.1024 [math.PR], 2011
131. N.-G. Kang, N.G. Makarov, Radial SLE martingale-observables. arXiv:1209.1615 [math.PR], 2012
132. H. Kesten, Hitting probabilities of random walks on $\mathbb{Z}^d$. Stoch. Proc. Appl. **25**, 165–184 (1987)
133. A.A. Kirillov, D.V. Yuriev, Representations of the Virasoro algebra by the orbit method. J. Geom. Phys. **5**(3), 351–363 (1988)
134. S. Kobayashi, *Hyperbolic Complex Spaces* (Springer, Berlin Heidelberg, 1998)
135. Y. Komatu, Über eine Verschärfung des Löwnerschen Hilfssatzes. Proc. Imp. Acad. Tokyo **18**, 354–359 (1942)
136. Y. Komatu, On conformal slit mapping of multiply-connected domains. Proc. Jpn. Acad. **26**(7), 26–31 (1950)
137. E.L. Korotyaev, P. Kargaev, Inverse problems generated by conformal mappings on complex plane with parallel slits. 2009, unpublished preprint.
138. E.L. Korotyaev, P. Kargaev, Estimates for periodic Zakharov–Shabat operators. J. Differ. Equat. **249**, 76–93 (2010)
139. E.L. Korotyaev, S. Kuksin, KdV Hamiltonian as function of actions. arXiv 1110.4475, 2011
140. I.K. Kostov, I. Krichever, M. Mineev-Weinstein, P.B. Wiegmann, A. Zabrodin, The τ-function for analytic curves. *Random Matrix Models and Their Applications*. Mathematical Sciences Research Institute Publications, vol. 40 (Cambridge University Press, Cambridge, 2001), pp. 285–299
141. I. Krichever, M. Mineev-Weinstein, P. Wiegmann, A. Zabrodin, Laplacian growth and Whitham equations of soliton theory. Phys. D **198**(1–2), 1–28 (2004)
142. I. Krichever, A. Marshakov, A. Zabrodin, Integrable structure of the Dirichlet boundary problem in multiply-connected domains. Comm. Math. Phys. **259**(1), 1–44 (2005)
143. J. Krzyż, The radius of close-to-convexity within the family of univalent functions. Bull. Acad. Polon. Sci. **10**, 201–204 (1962)
144. P.P. Kufarev, On one-parameter families of analytic functions. Rec. Math. [Mat. Sbornik] N.S. **13(55)**, 87–118 (1943)
145. P.P. Kufarev, On integrals of a simplest differential equation with movable polar singularity in the right-hand side. Tomsk. Gos. Univ. Uchen. Zap. **1**, 35–48 (1946)
146. P.P. Kufarev, Theorem on solutions to some differential equation. Tomsk. Gos. Univ. Učenye Zapiski **5**, 20–21 (1947)
147. P.P. Kufarev, A remark on integrals of Löwner's equation. Dokl. Akad. Nauk SSSR (N.S.) **57**, 655–656 (1947)
148. P.P. Kufarev, On conformal mapping of complementary regions. Dokl. Akad. Nauk SSSR (N.S.) **73**(5), 881–884 (1950)

149. P.P. Kufarev, A.E. Fales, On an extremal problem for complementary regions. Dokl. Akad. Nauk SSSR (N.S.) **81**(6), 995–998 (1950)
150. P.P. Kufarev, A.E. Fales, On an extremal problem for complementary regions. Tomsk. Gos. Univ. Ucen. Zap. **17**, 25–35 (1947)
151. P.P. Kufarev, V.V. Sobolev, L.V. Sporysheva, A certain method of investigation of extremal problems for functions that are univalent in the half-plane. Trudy Tomsk. Gos. Univ. Ser. Meh.-Mat. **200**, 142–164 (1968)
152. M.P. Kuvaev, P.P. Kufarev, An equation of Löwner's type for multiply connected regions. Tomsk. Gos. Univ. Uv. c. Zap. Mat. Meh. **25**, 19–34 (1955)
153. G.V. Kuz'mina, Moduli of families of curves and quadratic differentials. Trudy Mat. Inst. Steklov. **139**, 241 (1980)
154. M.A. Lavrent'ev, To the theory of conformal maps. Trav. Inst. Math. Stekloff **5**, 159–245 (1934)
155. G.F. Lawler, An introduction to the stochastic Loewner evolution, in *Random Walks and Geometry*, ed. by V. Kaimonovich (de Gruyter, Berlin , 2004), pp. 261–293
156. G.F. Lawler, *Conformally Invariant Processes in the Plane*, Mathematical Surveys and Monographs, vol. 114 (American Mathematical Society, Providence, RI, 2005)
157. G. Lawler, O. Schramm, W. Werner, Values of Brownian Intersection Exponents. I. Half-plane Exponents. Acta Math. **187**(2), 237–273 (2001); Values of Brownian intersection exponents. II. Plane exponents. Acta Math. **187**(2), 275–308 (2001); Values of Brownian intersection exponents. III. Two-sided exponents. Ann. Inst. H. Poincaré. Probab. Statist. **38**(1), 109–123 (2002)
158. N.A. Lebedev, On parametric representation of functions regular and univalent in a ring (in Russian). Dokl. Akad. Nauk SSSR (N.S.) **103**, 767–768 (1955)
159. N.A. Lebedev, *The Area Principle in the Theory of Univalent Functions* (Izdat. "Nauka", Moscow, 1975)
160. P. Lévy, *Processus Stochastiques et Mouvement Brownian* (Gauthier-Villars, Paris, 1947)
161. E.P. Li, On the theory of univalent functions on a circular ring. Dokl. Akad. Nauk SSSR (N.S.) **92**, 475–477 (1953)
162. J. Lind, A sharp condition for the Löwner equation to generate slits. Ann. Acad. Sci. Fenn. Math. **30**(1), 143–158 (2005)
163. J. Lind, D. Marshall, S. Rohde, Collisions and spirals of Loewner traces. Duke Math. J. **154**(3), 527–573 (2010)
164. J.E. Littlewood, R.E.A.C. Paley, A proof that an odd schlicht function has bounded coefficients. J. Lond. Math. Soc. **7**, 167–169 (1932)
165. K. Löwner, Untersuchungen über schlichte konforme Abbildungen des Einheitskreises. Math. Ann. **89**, 103–121 (1923)
166. K. Löwner, Über monotone Matrixfunktionen. Math. Z. **38**, 177–216 (1934)
167. Ch. Loewner, Conservation laws in compressible fluid flow and associated mapping. J. Ration. Mech. Anal. **2**, 537–561 (1953)
168. B. Mandelbrot, Fractal aspects of the iteration of $z \to \lambda z(1-z)$ for complex λ and z, in *Non-Linear Dynamics*, ed. by Robert H.G. Helleman. Annals of the New York Academy of Sciences, vol. 357 (New York Academy of Sciences, New York, 1979), pp. 249–259
169. I. Markina, D. Prokhorov, A. Vasil'ev, Sub-Riemannian geometry of the coefficients of univalent functions. J. Funct. Anal. **245**(2), 475–492 (2007)
170. I. Markina, A. Vasil'ev, Virasoro algebra and dynamics in the space of univalent functions. Contemp. Math. **525**, 85–116 (2010)
171. I. Markina, A. Vasil'ev, Evolution of smooth shapes and integrable systems. ArXiv: 1108.1007 [math-ph], 2011, pp. 25
172. A. Marshakov, P. Wiegmann, A. Zabrodin, Integrable structure of the Dirichlet boundary problem in two dimensions. Comm. Math. Phys. **227**(1), 131–153 (2002)
173. D.E. Marshall, S. Rohde, The Löwner differential equation and slit mappings. J. Am. Math. Soc. **18**(4), 763–778 (2005)

174. M. Mineev-Weinstein, P.B. Wiegmann, A. Zabrodin, Integrable structure of interface dynamics. Phys. Rev. Lett. **84**(22), 5106–5109 (2000)
175. M. Mineev-Weinstein, A. Zabrodin, Whitham-Toda hierarchy in the Laplacian growth problem. J. Nonlinear Math. Phys. **8**, 212–218 (2001)
176. R.V. Moody, A new class of Lie algebras. J. Algebra **10**, 211–230 (1968)
177. J. Muñoz Porras, F. Pablos Romo, Generalized reciprocity laws. Trans. Am. Math. Soc. **360**(7), 3473–3492 (2008)
178. E. Nelson, The free Markov field. J. Funct. Anal. **12**, 221–227 (1973)
179. L. Niemeyer, L. Pietronero, H.J. Wiessmann, Fractal dimension of dielectric breakdown. Phys. Rev. Lett. **52**(12), 1033–1036 (1984)
180. J. O'Connor, E.F. Robertson, *Charles Loewner*, MacTutor History of Mathematics archive, University of St Andrews
181. K. Osterwalder, R. Schrader, Axioms for Euclidean Green's functions. Comm. Math. Phys. **31**, 83–112 (1973); **42**, 281–305 (1975)
182. E. Peschl, Zur Theorie der schlichten Funktionen. J. Reine Angew. Math. **176**, 61–94 (1936)
183. M.E. Peskin, D.V. Schroeder, *An Introduction to Quantum Field Theory* (Perseus Books, Cambridge, 1995)
184. J.A. Pfaltzgraff, Subordination chains and univalence of holomorphic mappings in $\mathbb{C}^n$. Math. Ann. **210**, 55–68 (1974)
185. J.A. Pfaltzgraff, Subordination chains and quasiconformal extension of holomorphic maps in $\mathbb{C}^n$. Ann. Acad. Sci. Fenn. Ser. A I Math. **1**, 13–25 (1975)
186. G. Pick, Über die konforme Abbildung eines Kreises auf ein schlichtes und zugleich beschränktes Gebiet. S.-B. Kaiserl. Akad. Wiss. Wien, Math.-Natur. Kl. Abt. II a **126**, 247–263 (1917)
187. P.Ya. Polubarinova-Kochina, On a problem of the motion of the contour of a petroleum shell. Dokl. Akad. Nauk USSR **47**(4), 254–257 (1945)
188. P.Ya. Polubarinova-Kochina, Concerning unsteady motions in the theory of filtration. Prikl. Matem. Mech. **9**(1), 79–90 (1945)
189. Ch. Pommerenke, Über die Subordination analytischer Funktionen. J. Reine Angew. Math. **218**, 159–173 (1965)
190. Ch. Pommerenke, *Univalent Functions, with a Chapter on Quadratic Differentials by G. Jensen* (Vandenhoeck & Ruprecht, Göttingen, 1975)
191. Ch. Pommerenke, *Boundary Behaviour of Conformal Maps* (Springer, Berlin, 1992)
192. L.S. Pontryagin, V.G. Boltyanskii, R.V. Gamkrelidze, E.F. Mishchenko, *The Mathematical Theory of Optimal Processes* (A Pergamon Press Book. The Macmillan Co., New York, 1964)
193. T. Poreda, On generalized differential equations in Banach spaces. Diss. Math. **310**, 1–50 (1991)
194. V.I. Popov, The range of a system of functionals on the class S. Trudy Tomsk. Gos. Univ. Ser. Meh.-Mat. **182**(3), 106–132 (1965)
195. V.I. Popov, L.S. Pontrjagin's maximum principle in the theory of univalent functions. Dokl. Akad. Nauk SSSR **188**, 532–534 (1969)
196. N.V. Popova, Connection between the Löwner equation and the equation $\frac{dw}{dt} = \frac{1}{w-\lambda(t)}$. Izv. Acad. Sci. Belorussian SSR **6**, 97–98 (1954)
197. A. Pressley, G. Segal, *Loop Groups*. Oxford Mathematical Monographs (Oxford Science Publications, The Clarendon Press, Oxford University Press, New York, 1986)
198. D.V. Prokhorov, The geometric characterization of functions from Bazilevič subclasses. Sov. Math. (Iz. VUZ) **19**(2), 106–107 (1975)
199. D.V. Prokhorov, Optimal control method in an extremal problem on the class of univalent functions. Dokl. Akad. Nauk SSSR **275**(4), 798–801 (1984)
200. D.V. Prokhorov, Value sets for systems of functionals in the classes of univalent functions. Mat. Sb. **181**(12), 1659–1677 (1990)
201. D.V. Prokhorov, The value set of initial coefficients for bounded univalent typically real functions. Sib. Mat. Zh. **32**(5), 132–141 (1990)

202. D.V. Prokhorov, *Reachable Set Methods in Extremal Problems for Univalent Functions* (Saratov University, Saratov, 1993)
203. D.V. Prokhorov, Even coefficient estimates for bounded univalent functions. Ann. Polon. Math. **58**(3), 267–273 (1993)
204. D.V. Prokhorov, Coefficients of functions close to the identity function. Complex Variables Theor. Appl. **33**(1–4), 255–263 (1997)
205. D.V. Prokhorov, Coefficients of holomorphic functions. Complex analysis and representation theory, 2. J. Math. Sci. (New York) **106**(6), 3518–3544 (2001)
206. D.V. Prokhorov, A.A. Nikulin, Asymptotic estimates for the coefficients of bounded univalent functions. Sibirsk. Mat. Zh. **47**(5), 1128–1138 (2006)
207. D. Prokhorov, A. Vasil'ev, Univalent functions and integrable systems. Comm. Math. Phys. **262**(2), 393–410 (2006)
208. D. Prokhorov, A. Vasil'ev, Optimal control in Bombieri's and Tammi's conjectures. Georgian Math. J. **12**(4), 743–761 (2005)
209. D. Prokhorov, A. Vasil'ev, Singular and tangent slit solutions to the Löwner equation, in *Analysis and Mathematical Physics*, Trends in Mathematics (Birkhäuser, Basel, 2009), pp. 455–463
210. M. Reed, B. Simon, *Methods of Modern Mathematical Physics. I. Functional Analysis*, 2nd edn. (Academic Press, New York, 1980), p. 400
211. S. Reich, D Shoikhet, Metric domains, holomorphic mappings and nonlinear semigroups. Abstr. Appl. Anal. **3**(1–2), 203–228 (1998)
212. S. Reich, D. Shoikhet, *Nonlinear Semigroups, Fixed Points, and Geometry of Domains in Banach Spaces* (Imperial College Press, London, 2005)
213. M.S. Robertson, A remark in the odd schlicht functions. Bull. Am. Math. Soc. **42**(6), 366–370 (1936)
214. M.S. Robertson, Applications of the subordination principle to univalent functions. Pacific J. Math. **11**, 315–324 (1961)
215. R.M. Robinson, Bounded univalent functions. Trans. Am. Math. Soc. **52**, 426–449 (1942)
216. S. Rohde, O. Schramm, Basic properties of SLE. Ann. Math. **161**(2), 883–924 (2005)
217. S. Rohde, Oded Schramm: from circle packing to SLE. Ann. Probab. **39**(5), 1621–1667 (2011)
218. M. Rosenblum, J. Rovnyak, *Topics in Hardy Classes and Univalent Functions* (Birkhäuser Verlag, Basel, 1994)
219. M. Röckner, Markov property of generalized fields and axiomatic potential theory. Math. Ann. **264**, 153–177 (1983)
220. M. Röckner, Generalized Markov fields and Dirichlet forms. Acta Applicandae Math. **3**, 285–311 (1985)
221. I. Rushkin, E. Bettelheim, I.A. Gruzberg, P. Wiegmann, Critical curves in conformally invariant statistical systems. J. Phys. A Math. Theor. **40**, 2165–2195 (2007)
222. M. Sato, Y. Sato, Soliton equations as dynamical systems on infinite-dimensional Grassmann manifold. *Nonlinear Partial Differential Equations in Applied Science Tokyo, 1982*. North-Holland Mathematics Studies, vol. 81 (North-Holland, Amsterdam, 1983), pp. 259–271
223. A. Savakis, S. Maggelakis, Models of shrinking clusters with applications to epidermal wound healing. Math. Comput. Model. **25**(6), 1–6 (1997)
224. G. Segal, G. Wilson, Loop groups and equations of KdV type. Publ. Math. IHES **61**, 5–65 (1985)
225. A.C. Schaeffer, D.C. Spencer, The coefficients of schlicht functions. II **12**, 107–125 (1945)
226. A.C. Schaeffer, D.C. Spencer, *Coefficient Regions for Schlicht Functions*. American Mathematical Society Colloquium Publications, vol. 35 (American Mathematical Society, New York, NY, 1950)
227. M. Schiffer, D.C. Spencer, *Functionals of Finite Riemann Surfaces* (Princeton University Press, Princeton, NJ, 1954)
228. M. Schiffer, O. Tammi, On the fourth coefficient of bounded univalent functions. Trans. Am. Math. Soc. **119**(1965), 67–78

229. M. Schottenloher, *A Mathematical Introduction to Conformal Field Theory*, 2nd edn. Lecture Notes in Physics, vol. 759 (Springer, Berlin, 2008)
230. O. Schramm, Scaling limits of loop-erased random walks and uniform spanning trees. Isr. J. Math. **118**, 221–288 (2000)
231. O. Schramm, S. Sheffield, Contour lines of the two-dimensional discrete Gaussian free field. Acta Math. **202**(1), 21–137 (2009)
232. O. Schramm, S. Sheffield, A contour line of the continuum Gaussian free field. arXiv: 1008.2447, 2010
233. D.-S. Shah, Le-le Fan, On the parametric representation of quasi-conformal mappings. Sci. Sin. **11**, 149–162 (1962)
234. S. Sheffield, Exploration trees and conformal loop ensembles. Duke Math. J. **147**(1), 79–129 (2009)
235. T. Sheil-Small, Some remarks on Bazilevič functions. J. Anal. Math. **43**, 1–11 (1983/1984)
236. B.M. Shein, *Private communication*, 2012
237. L. Siewierski, Sharp estimation of the coefficients of bounded univalent functions close to identity. Diss. Math. **86**, 1–153 (1971)
238. A.G. Siskakis, Semigroups of composition operators and the Cesàro operator on $H^p(D)$, PhD Thesis, University of Illinois, 1985
239. A.G. Siskakis, Semigroups of composition operators on spaces of analytic functions, a review. Contemp. Math. **213**, 229–252 (1998)
240. S. Smirnov, Critical percolation in the plane: conformal invariance, Cardy's formula, scaling limits. C. R. Acad. Sci. Paris Sér. I Math. **333**, 239–244 (2001)
241. T.T. Soong, *Random Differential Equations in Science and Engineering*. Mathematics in Science and Engineering, vol. 103 (Academic Press [Harcourt Brace Jovanovich Publishers], New York, 1973)
242. Lad. Špaček, Contribution à la théorie des fonctions univalentes. Čas. Mat. Fys. **62**(2), 12–19 (1932)
243. T. Takebe, L.-P. Teo, A. Zabrodin, Löwner equations and dispersionless hierarchies. J. Phys. A **39**(37), 11479–11501 (2006)
244. O. Tammi, On the maximalization of the coefficient a_3 of bounded schlicht functions. Ann. Acad. Sci. Fennicae. Ser. A. I. Math. Phys. **149**, 1–14 (1953)
245. O. Tammi, On optimizing parameters of the power inequality for a_4 in the class of bounded univalent functions. Ann. Acad. Sci. Fennicae. Ser. A. I. Math. Phys. **560**, 1–24 (1973)
246. O. Tammi, *Extremal Problems for Bounded Univalent Functions*, Lecture Notes in Mathematics, vol. 646 (Springer, New York, 1978)
247. O. Tammi, *Extremal Problems for Bounded Univalent Functions, II*, Lecture Notes in Mathematics, vol. 913 (Springer, New York, 1982)
248. F. Tricomi, Sull'iterazione delle funzioni di linee. Giorn. Math. Battaglini **55**, 1–8 (1917)
249. J. Tsai, The Loewner driving function of trajectory arcs of quadratic differentials. J. Math. Anal. Appl. **360**, 561–576 (2009)
250. A. Vasil'ev, Mutual change of initial coefficients of univalent functions. Matemat. Zametki **38**(1), 56–65 (1985)
251. A. Vasil'ev, Value set $\{f(r_1), f(r_2)\}$ for the class of univalent functions with real coefficients. Sib. Mat. Zh. **27**(6), 28–32 (1986)
252. A. Vasil'ev, Optimal control methods in an extremal problem on a class of solutions of the Loewner-Kufarev equation. Diff. Uravneniya **26**(3), 386–392 (1990)
253. A. Vasil'ev, From the Hele-Shaw experiment to integrable systems: a historical overview. Complex Anal. Oper. Theor. **3**(2), 551–585 (2009)
254. T. Vicsek, *Fractal Growth Phenomena* (World Scientific Publishing, Singapore, 1989)
255. M. Voda, Solution of a Loewner chain equation in several variables. J. Math. Anal. Appl. **375**(1), 58–74 (2011)
256. G.C. Wick, The evaluation of the collision matrix. Phys. Rev. **80**(2), 268–272 (1950)
257. P.B. Wiegmann, A. Zabrodin, Conformal maps and integrable hierarchies. Comm. Math. Phys. **213**(3), 523–538 (2000)

258. A.S. Wightman, *Hilberts Sixth Problem: Mathematical Treatment of the Axioms of Physics*, ed. by F.E. Browder. Proceedings of Symposia in Pure Mathematics (part 1), vol. 28 (American Mathematical Society, Providence, RI, 1976), pp. 241–268
259. E. Witten, Quantum field theory, Grassmannians, and algebraic curves. Comm. Math. Phys. **113**(4), 529–600 (1988)
260. T.A. Witten, Jr., L.M. Sander, Diffusion-limited aggregation, and kinetic critical phenomenon. Phys. Rev. Lett. **47**(2), 1400–1403 (1981)
261. T.A. Witten, L.M. Sander, Diffusion-limited aggregation. Phys. Rev. B **27**(9), 5686–5697 (1983)
262. N.J. Zabusky, M.D. Krushkal, Interaction of ï£¡solitonsï£¡ in a collisionless plasma and initial states. Phys. Rev. Lett. **15**, 240–243 (1965)
263. V.E. Zaharov, L.D. Faddeev, The Korteweg-de Vries equation is a fully integrable Hamiltonian system. Funkcional. Anal. i Prilozh. **5**(5), 18–27 (1971)

The Schwarz Lemma: Rigidity and Dynamics

Mark Elin, Fiana Jacobzon, Marina Levenshtein, and David Shoikhet

Abstract The Schwarz Lemma has given impetus to developments in several areas of complex analysis and mathematics in general. We survey some investigations related to its three parts (invariance, rigidity, and distortion) that began early in the twentieth century and are still being carried out. We consider only functions analytic in the unit disk. Special attention is devoted to the Boundary Schwarz Lemma and to applications of the Schwarz–Pick Lemma and the Boundary Schwarz Lemma to modern rigidity theory and complex dynamics.

1 Introduction

The Schwarz Lemma is one of the most quoted and central results in all of complex function theory, and there is hardly a result that has been as influential. It is difficult to overestimate the significance of this lemma which gave a great push to the development of geometric function theory, fixed point theory of holomorphic mappings, hyperbolic geometry, and many other fields of analysis.

The Schwarz Lemma concerns holomorphic self-mappings of the unit disk in the complex plane that have a fixed point. It consists of three conclusions. The first one gives general sharp estimates of the values of bounded analytic functions on the open unit disk of the complex plane. Geometrically, the estimates of the values of an analytic self-mapping of the disk given by the Schwarz and Schwarz–Pick Lemmas provide the invariance of the hyperbolic disks around the interior fixed point under this mapping.

M. Elin (✉) · F. Jacobzon · M. Levenshtein · D. Shoikhet
Department of Mathematics, ORT Braude College, P.O. Box 78, 21982 Karmiel, Israel
e-mail: mark_elin@braude.ac.il; fiana@braude.ac.il; marlev@braude.ac.il; davs@braude.ac.il

A. Vasil'ev (ed.), *Harmonic and Complex Analysis and its Applications*, Trends in Mathematics, DOI 10.1007/978-3-319-01806-5_3,

The second conclusion (the estimate of the derivative) served as the origin for various distortion theorems, while the third one establishes a rigidity property of a holomorphic self-mapping to be a rotation if it coincides with the rotation up to the first order at the origin. This uniqueness principle and its various modifications lead to solutions of many extremal problems for different classes of holomorphic functions.

Whereas the Schwarz and Schwarz–Pick Lemmas establish general properties of holomorphic self-mappings of the open unit disk with an interior fixed point, a variety of boundary versions of these lemmas in the spirit of Julia, Wolff and Carathéodory as well as later rigidity results provide analogous properties in the case of boundary fixed points.

In the 150 year period since their discovery, numerous extensions and generalizations of the Schwarz and Schwarz–Pick Lemmas have appeared. Ideas generated by these now classical results continue to attract mathematicians to this day. Dozens of books and papers have been devoted to these results (see, for example, [1, 8, 10, 33, 93, 112] and references therein). Various applications of developments of these lemmas can be found in several areas of classical analysis as well as in new areas such as composition operators (see, for example, [37, 57, 129]), $\overline{\partial}$-problems (see, for example, [86]), semigroup theory [66, 131], multi-valued functions [61, 79, 111], the theory of regular functions over quaternions and octonions (see [73]).

In this survey we do not examine all the known extensions of the Schwarz and Schwarz–Pick Lemmas. Many important aspects such as the famous Schwarz–Ahlfors Lemma [5, 6, 112, 113], its generalizations for Riemann surfaces and Kähler manifolds (see, for example, [93, 143]), the Schwarz–Pick inequalities for multiply connected domains [15] as well as finite-dimensional and infinite-dimensional Schwarz and Schwarz–Pick Lemmas (see, for example, [1, 60, 127] and [57]) are not considered here. Instead, we deal mainly with extensions and generalizations of the Schwarz and Schwarz–Pick Lemmas and Julia–Wolff–Carathéodory Theorem for functions of one complex variable holomorphic in the open unit disk. In particular, we devote attention to the development of the invariance and rigidity aspects of the Schwarz Lemma as well as estimates for derivatives in both the interior and boundary cases, emphasizing a rigidity nature of the existence of extremal functions in the sharp estimates for the boundary derivatives at contact points. We also present some applications to the theory of complex dynamical systems.

We start with a brief historical sketch of the development of the Schwarz Lemma in these directions that took place in the beginning of the twentieth century in Sect. 2. It also includes an explanation of the Schwarz–Pick Lemma and Julia–Wolff–Carathéodory Theorem in terms of the hyperbolic metric, the geometric Landau–Toeplitz Theorem, and contributions of famous mathematicians such as Littlewood, Rogosinski, Dieudonné, Beckenbach, and Löwner.

Section 3 is devoted to generalizations of the first and the second conclusions of the Schwarz Lemma. In particular, we consider geometric versions of the Schwarz Lemma originated by the Landau–Toeplitz Theorem, Mercer's and Beardon's approaches, the hyperbolic derivative, and multi-point Schwarz–Pick lemmas.

In Sect. 4, we retrace the progress in the study of inequalities including angular derivatives, from the early results by Unkelbach and Herzig and remarkable Cowen–Pommerenke inequalities to advanced modifications of the Julia Lemma and the Julia–Wolff–Carathéodory Theorem.

The rigidity theory, i.e., the development of the uniqueness part of the Schwarz Lemma is discussed in Sect. 5. These investigations were initiated by the breakthrough by Burns and Krantz which established the first boundary rigidity principle. Other boundary versions have since been found give conditions on the boundary behavior of a holomorphic function under which it coincides with a given mapping.

In the last section, we consider infinitesimal versions of the Schwarz Lemma and the Julia–Wolff–Carathéodory Theorem for one-parameter continuous semigroups on the open unit disk and their generators.

We believe that the Schwarz Lemma and its boundary versions will serve as a basis for many further investigations and that its modern enhancements such as the rigidity theory as well as its applications to complex dynamics will open new areas of research.

2 Short Historical Overview

2.1 Schwarz Lemma

We start with the classical Schwarz Lemma. Henceforth denote by $\mathrm{Hol}(D, E)$ the set of all functions holomorphic on a set D which take values in E and let $\mathrm{Hol}(D) := \mathrm{Hol}(D, D)$. In particular, $\mathrm{Hol}(\Delta)$ is the set of all holomorphic self-mappings of the open unit disk $\Delta := \{z : |z| < 1\}$ in the complex plane $\mathbb{C}$.

Theorem 2.1 (The Schwarz Lemma). *Suppose* $F \in \mathrm{Hol}(\Delta)$ *and* $F(0) = 0$. *Then either*

$$|F'(0)| < 1 \tag{1}$$

and

$$|F(z)| < |z| \tag{2}$$

for all $z \in \Delta \setminus \{0\}$, *or* $|F'(0)| = 1$ *and* F *is the rotation* $F(z) = F'(0)z$.

The original version of the Schwarz Lemma applies to holomorphic functions from the unit disk into itself that fix the origin. If, however, F fixes some other point $\zeta \in \Delta$, the Schwarz Lemma applies to the function $G := m_\zeta \circ F \circ m_\zeta$, where $m_\zeta(z) := \dfrac{\zeta - z}{1 - \bar{\zeta} z}$ is the involutive automorphism of the unit disk. Differentiating G at 0 yields $|F'(\zeta)| \le 1$, and equality holds only in the case F is either the identity or an elliptic automorphism of Δ.

The Schwarz Lemma, although unpretentious, has turned out to be fruitful for deeper understanding of self-mappings as well as iterates of holomorphic mappings and their asymptotic behavior. For example, an immediate consequence is the following rigidity property: *If* $F(\zeta) = \zeta$ *and* $F'(\zeta) = 1$ *for an arbitrary point* $\zeta \in \Delta$ *(i.e.,* F *coincides with the identity mapping up to first order at* ζ*), then* F *is the identity*. Moreover, if a holomorphic self-mapping F of Δ fixes two points $\tau, \zeta \in \Delta$, the Schwarz Lemma, applied to the composite mapping $m_\zeta \circ F \circ m_\zeta$ (which fixes 0 and $m_\zeta(\tau)$), we have again $F(z) = z$ for all $z \in \Delta$.

The Schwarz Lemma was thus named by Constantin Carathéodory [44] in honor of Hermann Amandus Schwarz, who is the eponym of various other analytical objects, including the Cauchy–Schwarz inequality, the Schwarz–Christoffel formula, the Schwarzian derivative, and the Schwarz reflection principle.

H.A. Schwarz (1843–1921) was born in Hermsdorf, Silesia (now Jerzmanowa, Poland) and died in Berlin. He was married to Marie Kummer, a daughter of the mathematician Ernst Eduard Kummer; they had six children. Schwarz originally studied chemistry at Gewerbeinstitut (later called the Technical University of Berlin), but Kummer and Weierstraß persuaded him to switch to Mathematics. He continued to study in Berlin, where he was supervised by Weierstraß, until 1864 when he was awarded his doctorate. His doctoral thesis was examined by Kummer. Between 1867 and 1869 he worked in Halle, then in Zürich. From 1875 he worked at Göttingen University, studying function theory, differential geometry, and the calculus of variations. In 1892, he became a member of the Berlin Academy of Science and a professor at the University of Berlin, where his students included Lipót Fejér, Paul Koebe, and Ernst Zermelo.

It should be mentioned that Schwarz himself arrived at the Schwarz Lemma in his study of schlicht functions and originally proved it for these functions only. The now canonical proof of the Schwartz Lemma appeared in Carathéodory's 1907 paper [43], where he attributed this proof to the German mathematician Erhard Schmidt.

For the sake of completeness, we formulate here a consequence of Theorem 2.1 due to Ernst Lindelöf (see [102, p. 11]).

Corollary 2.1 (Lindelöf's Inequality). *If* $F \in \mathrm{Hol}(\Delta)$, *then it satisfies the estimate*

$$|F(z)| \le \frac{|z| + |F(0)|}{1 + |z||F(0)|}, \quad z \in \Delta.$$

We note in passing that an application of Lindelöf's inequality to the function $F \circ m_\zeta$ shows that every function $F \in \mathrm{Hol}(\Delta)$ satisfies

$$|F(z)| \le \frac{|F(\zeta)| + |m_\zeta(z)|}{1 + |F(\zeta)||m_\zeta(z)|} \tag{3}$$

for all $z, \zeta \in \Delta$.

Actually, the Schwarz Lemma contains three assertions. The first one is not only an estimate of the value of a self-mapping but also an invariance result. The second one is an estimate of the value of the derivative at the unique interior fixed point of F. The last assertion is, essentially, a rigidity result; namely, if $F(0) = 0$ and $F'(0) = 1$, then $F(z) = z$ for all $z \in \Delta$. Each of these assertions has opened new areas in geometric function theory and has been developed and improved in different ways.

2.2 Schwarz–Pick Lemma

Suppose that ζ is an arbitrary point of the open unit disk (not necessary fixed by F). The Schwarz Lemma applied to the function $m_{F(\zeta)} \circ F \circ m_\zeta$ gives $|m_{F(\zeta)} \circ F \circ m_\zeta| \le |z|$ for all $z \in \Delta$, or, since m_ζ is an involution, $|m_{F(\zeta)} \circ F(z)| \le |m_\zeta(z)|$. This generalized version of the Schwarz inequality is invariant under conformal automorphisms of the open unit disk and is known as the Schwarz–Pick Lemma.

Theorem 2.2 (The Schwarz–Pick Lemma). *Suppose* $F \in \mathrm{Hol}(\Delta)$ *and* $\zeta \in \Delta$. *Then*

$$\left| \frac{F(\zeta) - F(z)}{1 - \overline{F(\zeta)}F(z)} \right| \le \left| \frac{\zeta - z}{1 - \overline{\zeta} z} \right| \quad \textit{for all} \quad z \in \Delta, \tag{4}$$

and

$$|F'(\zeta)| \le \frac{1 - |F(\zeta)|^2}{1 - |\zeta|^2}. \tag{5}$$

Moreover, equality either in (4) at some $z \in \Delta$ *or in (5) holds if and only if* F *is the identity or an automorphism which fixes* ζ.

Fig. 1 The invariance of non-Euclidean disks

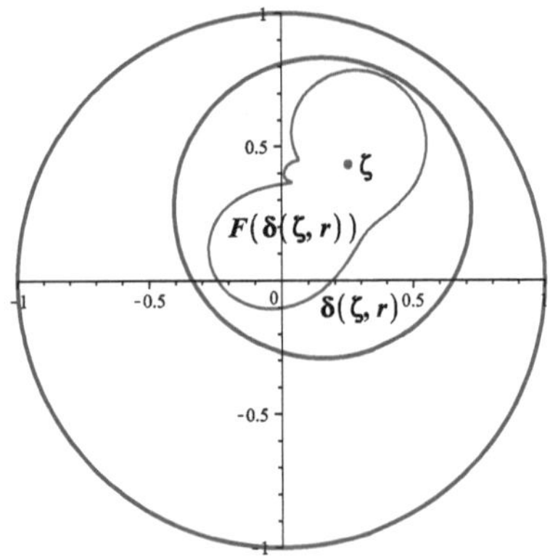

Constantin Carathéodory mentioned the idea of composing self-mappings with automorphisms of the disk in [44]. Nevertheless, inequality (4) first appears in the work [88] by Gaston Julia as part of a preliminary section in which he collects auxiliary results.

Inequality (4) is commonly called the "Schwarz–Pick Lemma" in honor of Georg Alexander Pick who wrote an influential paper [115] on the subject in 1916, two years before Julia's work. In the case $\zeta \in \Delta$ is a fixed point of F, Pick's observation provides a simple geometric interpretation of (4), namely, that a non-Euclidean disk $\delta(\zeta, r) := \{z : |m_\zeta(z)| < r\}$ with non-Euclidean center ζ is invariant under F for each $0 < r < 1$, i.e., is mapped by F into itself (see Fig. 1).

G.A. Pick (1859–1942) was born in Vienna, Austria into a Jewish family. In 1877, at the age of seventeen and one year before entering the University of Vienna, he published a mathematics paper. Leo Königsberger, a chairman at the University of Vienna, became Pick's supervisor. Three years later, Pick wrote

his dissertation "Über eine Klasse abelscher Integrale." He was appointed as ordinary professor (full professor) in 1892 at the German University of Prague. His mathematical work was extremely broad and his 67 papers range across many topics such as linear algebra, invariant theory, integral calculus, potential theory, functional analysis, and geometry. Terms such as "Pick matrices," "Pick–Nevanlinna interpolation," and the "Schwarz–Pick lemma" are used widely to the present day. Pick supervised about 20 doctoral students, the most famous being Karl Löwner. Pick was the driving force behind the appointment of Einstein to a chair in the faculty of mathematical physics at the German University of Prague in 1911. Pick was elected a member of the Czech Academy of Sciences and Arts, but was expelled when Nazis took over Prague. He was sent to Theresienstadt concentration camp on July 13, 1942, where he died two weeks later.

Inequality (4) implies that in the case $F \in \mathrm{Hol}(\Delta)$ fixes a point $\zeta \in \Delta$, all the disks $\delta(\zeta, r)$, $r \in (0,1)$, are F-invariant. Inequality (5), in turn, is an estimate of the first derivative at each point $\zeta \in \Delta$. If ζ is the fixed point of F, this inequality coincides with the conclusion of the Schwarz Lemma $|F'(\zeta)| \le 1$. Once again, the last assertion of Theorem 2.2 has a rigidity character.

Pick's innovation is not so much inequality (4) itself, but rather his observation that holomorphic self-mappings of the unit disk decrease the hyperbolic (non-Euclidean) distance, the so-called Poincaré metric.

2.3 *Poincaré Metric: Hyperbolic Geometry in the Unit Disk*

In 1882 Poincaré discovered that the open unit disk Δ equipped with the metric

$$\lambda(z)|dz| = \frac{|dz|}{1-|z|^2}$$

can be regarded as the hyperbolic plane. This metric is conformally invariant, i.e., $\lambda(z)|dz| = \lambda(\zeta)|d\zeta|$ for any automorphism $\zeta = M(z)$ of Δ. Now, for two different points $z, w \in \Delta$ and each curve $\gamma : [0,1] \mapsto \Delta$ joining z and w having a piecewise continuous derivative γ', we define the length L_γ of γ by

$$L_\gamma := \int_0^1 \lambda(\gamma)|\gamma'(t)|dt.$$

The length of the shortest such γ is called *the Poincaré hyperbolic distance between z and w*, i.e.,

$$\rho(z,w) := \inf_\gamma L_\gamma.$$

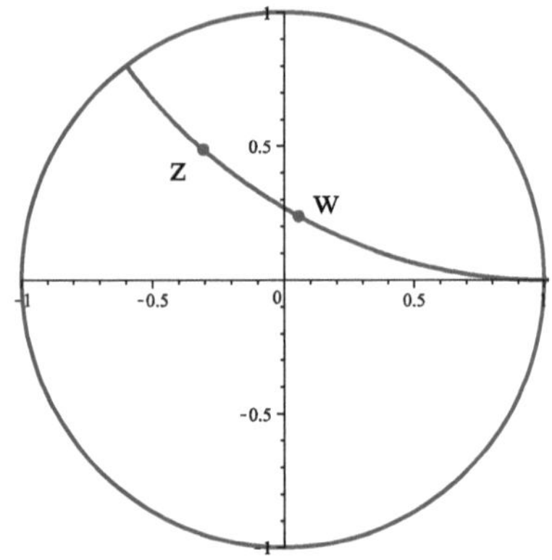

Fig. 2 The geodesic through the points z and w

For each pair of points z and w in Δ, there exists a unique geodesic segment joining z and w whose length is $\rho(z, w)$). If z, 0, and w are colinear, this geodesic is a line segment. If z, 0, and w are not colinear, this geodesic is the arc of the circle in Δ which passes through z and w and is orthogonal to $\partial\Delta$, the boundary of Δ (see Fig. 2). A simple calculation shows that

$$\rho(z, w) := \frac{1}{2} \log \frac{1 + |m_w(z)|}{1 - |m_w(z)|}, \quad z, w \in \Delta. \tag{6}$$

The function ρ defines a metric on Δ which agrees with the standard topology on Δ. The pair (Δ, ρ) is a complete metric space.

Jules Henri Poincaré (1854–1912) was born in Nancy, France, where his father was Professor of Medicine at the University. Henri graduated the École Polytechnique in 1875 and continued his studies at the École des Mines. He received

his doctorate in mathematics from the University of Paris in 1879 under the supervision of Charles Hermite. Poincaré was appointed to a chair in the Faculty of Science at the Sorbonne in 1881. In parallel to his academic carrier, he worked at the Ministry of Public Services and eventually, in 1893, became chief engineer of the Corps de Mines and inspector general in 1910. In 1886 he was nominated as a chairman of the department of mathematical physics and probability, and as a chairman at the École Polytechnique. He held these chairs in Paris until his premature death at the age of 58.

Poincaré was preoccupied by many aspects of mathematics, physics, and philosophy and is often described as the Last Universalist in mathematics. He made contributions to numerous branches of mathematics, celestial mechanics, fluid mechanics, the special theory of relativity, and the philosophy of science.

Despite its name, the Poincaré hyperbolic geometry on the disk was introduced as a model of hyperbolic space by Eugenio Beltrami [25].

E. Beltrami (1835–1900) was born in Cremona in Lombardy, which was then a part of the Austrian Empire and now part of Italy. He studied mathematics at University of Pavia. In 1862, he was appointed professor at the University of Bologna.

Throughout his life, Beltrami had various professorial jobs at universities in Pisa, Rome and Pavia. From 1891 until the end of his life, he lived in Rome. He became the president of the Accademia dei Lincei in 1898 and a senator of the Kingdom of Italy in 1899. In 1868, he published two memoirs dealing with consistency and interpretations of non-Euclidean geometry of Bolyai and Lobachevsky.

The Schwarz–Pick inequality (4) implies that each holomorphic self-mapping F of Δ is nonexpansive with respect to the Poincaré hyperbolic distance (ρ-nonexpansive), i.e.,

$$\rho(F(z), F(w)) \leq \rho(z, w), \quad z, w \in \Delta. \tag{7}$$

For more detailed representation of the hyperbolic geometry on the open unit disk Δ, see, for example, [75, 78, 131] and [92].

2.4 Landau–Toeplitz Theorem

One of the earliest generalizations of the Schwarz Lemma is due to Edmund Landau and Otto Toeplitz, who in [97] obtained a result similar to the Schwarz Lemma. In their work, the diameter of the image set assumes the role of maximum modulus of the function. To state their result precisely, for $F \in \mathrm{Hol}(\Omega, \mathbb{C})$ we let

$$\mathrm{Diam}\, F(\Omega) := \sup_{z,w\in\Omega} |F(z) - F(w)|\,.$$

Theorem 2.3 (Landau–Toeplitz Theorem, 1907). *Let* $F \in \mathrm{Hol}(\Delta, \mathbb{C})$ *and* $\mathrm{Diam}\, F(\Delta) = 2$. *Then*

$$\mathrm{Diam}\, F(r\Delta) \leq 2r \tag{8}$$

for every $0 < r < 1$, *and*

$$|F'(0)| \leq 1. \tag{9}$$

Moreover, equality in (8) for some $r \in (0, 1)$, *or in (9) holds if and only if* F *is an Euclidean isometry* $a + cz$ *for some constants* $a, c \in \mathbb{C}$ *and* $|c| = 1$.

Remark 2.1. Quoting [40], we note that the growth estimate on the diameter (8) should be viewed in analogy with the classical growth bound in the Schwarz Lemma. Notice, however, that Theorem 2.3 covers the case $F(\Delta)$ is an equilateral triangle of side-length 2 which is, of course, not contained in a disk of radius 1 and the case $F(\Delta)$ is contained in the so-called Reuleaux triangle that is obtained from the equilateral triangle by joining adjacent vertices by a circular arc having center at the third vertex (see Fig. 3).

E. Landau (1877–1938) was born in Berlin to a wealthy Jewish family. He studied mathematics at the University of Berlin and received his doctorate in 1899. He taught at the University of Berlin from 1899 until 1909 and held a chairmanship at the University of Göttingen from 1909. Beginning from the 1920s, Landau played an active role in establishing of the Mathematics Institute at the Hebrew University of Jerusalem.

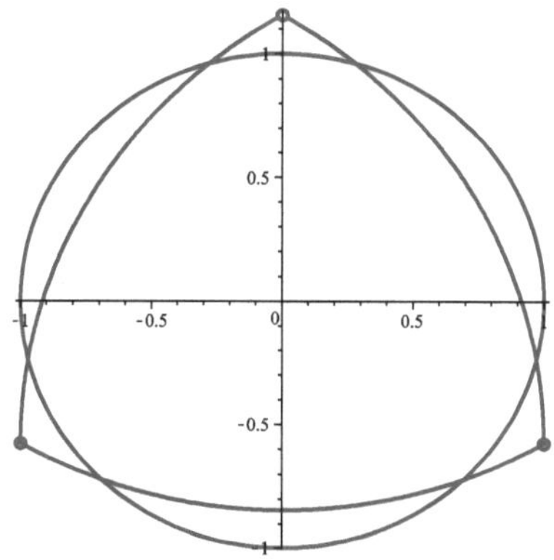

Fig. 3 The unit circle and the Reuleaux triangle

In 1927 Landau and his family emigrated to Palestine, and he began teaching at the Hebrew University. In 1933 Landau returned to Göttingen. He remained there until he was forced out by the Nazi regime in 1933. Thereafter, he lectured only outside of Germany. In 1934, he moved to Berlin, where he died in early 1938 of natural causes. G.H. Hardy wrote that none was ever more passionately devoted to mathematics than Landau.

O. Toeplitz (1881–1940) was born in Breslau, studied mathematics at the University of Breslau, and was awarded a doctorate in algebraic geometry in 1905. In 1906, he arrived at Göttingen University and remained there for seven years. Member of the mathematics faculty included David Hilbert, Felix Klein, and Hermann Minkowski. From 1913 Toeplitz worked at the University of Kiel. In 1933, the Civil Service Law came into effect and professors of Jewish origin were removed from teaching. Toeplitz was dismissed in 1935. In 1939, he emigrated to Palestine, where he was scientific advisor to the rector of the Hebrew University of Jerusalem. He died in Jerusalem from tuberculosis one year later.

To emphasize the deep connection between the Schwarz Lemma and Landau–Toeplitz Theorem, we consider for a given mapping $F \in \mathrm{Hol}(\Delta, \mathbb{C})$, the function

$$\mathrm{Rad}\, F(r\Delta) := \sup_{|z|<r} |F(z) - F(0)|$$

(see [41]). A geometrical interpretation of $\mathrm{Rad}\, F(r\Delta)$ is as the radius of the smallest disk centered at $F(0)$ which contains $F(r\Delta)$. Using this notion, the Schwarz Lemma can be reformulated as follows.

Suppose $F \in \mathrm{Hol}(\Delta)$. *The function* $\phi(r) := \frac{1}{r}\,\mathrm{Rad}\, F(r\Delta)$ *is strictly increasing for* $0 < r < 1$, *except when* F *is linear, in which case* ϕ *is constant.*

There are many versions of the Schwarz Lemma dealing with other geometric quantities related to images. Some versions are presented in Sect. 3.2. Here we introduce only a result by George Pólya and Gábor Szegö (see, for example, [116]):

For $F \in \mathrm{Hol}(\Delta, \mathbb{C})$, *define the width of* $F(r\Delta)$, $r \in (0, 1)$, *by*

$$W(F(r\Delta)) = \sup_{z,w \in r\Delta} |\mathrm{Re}\, F(z) - \mathrm{Re}\, F(w)|, \quad 0 < r \le 1.$$

If $W(F(\Delta)) = \frac{\pi}{2}$, *then* $|F'(0)| \le 1$ *with equality when* $F(z) = \arctan z$.

2.5 Boundary Versions of the Schwarz Lemma

The Schwarz–Pick Lemma reveals properties of a holomorphic self-mapping F of Δ with an interior fixed point $\zeta \in \Delta$. In particular, an immediate consequence of the Schwarz–Pick Lemma is that such a mapping (if it is not the identity) has at most one fixed point in Δ, and the hyperbolic disks centered at this point are F-invariant. The natural question arises: what can be said about a fixed point free mapping F? More precisely,

(i) If F has no interior fixed point, does there exist a boundary fixed point τ?
(ii) If so, do there exist disks attached τ that are invariant under F?

These questions are answered by the Julia–Wolff–Carathéodory Theorem. It combines conclusions of the celebrated Julia theorem [89] proved in 1920, Wolff's boundary version of the Schwarz Lemma [140] proved in 1926, and Carathéodory's contribution [45] proved in 1929 (see also [46]).

Julia's result answers question (ii). He carried out most of his investigations in the upper half-plane $\Pi = \{w : \operatorname{Im} w > 0\}$ and proved that if a function $f \in \operatorname{Hol}(\overline{\Pi})$ satisfies $f(\tau) = \tau_1$ for two boundary points $\tau, \tau_1 \in \partial\Pi$, then

$$\operatorname{Im} \frac{1}{f(w) - \tau_1} < \operatorname{Im} \frac{1}{(w - \tau) f'(\tau)}, \quad w \in \Pi, \tag{10}$$

or, in the case of the boundary fixed point, when $\tau = \tau_1 = 0$,

$$\operatorname{Im} \frac{1}{f(w)} < \operatorname{Im} \frac{1}{w f'(0)}, \quad w \in \Pi. \tag{11}$$

Note that Julia's inequality does not actually require the function f to be defined on the boundary. Translating this result to the unit disk $\Delta = \{z : |z| < 1\}$ and passing to sequences, Julia's result can be reformulated as follows.

Theorem 2.4 (Julia's Lemma). *Let $F \in \operatorname{Hol}(\Delta)$, and suppose that there exists a sequence $\{z_n\}_{n=1}^{\infty}$ converging to a boundary point $\zeta \in \partial\Delta$ such that $\lim_{n\to\infty} F(z_n) = \eta \in \partial\Delta$ and $\lim_{n\to\infty} \frac{1-|F(z_n)|}{1-|z_n|} = \alpha < \infty$. Then for each $z \in \Delta$,*

$$\frac{|1 - F(z)\overline{\eta}|^2}{1 - |F(z)|^2} \leq \alpha \frac{|1 - z\overline{\zeta}|^2}{1 - |z|^2}. \tag{12}$$

Equality in (12) holds for some $z \in \Delta$ if and only if F is an automorphism of Δ.

Moreover, the radial limit $F(\zeta) := \lim_{r\to 1^-} F(r\zeta)$ exists and equals η.

Note that the existence of the radial limit of $F \in \operatorname{Hol}(\Delta)$ at a boundary point $\zeta \in \partial\Delta$ is equivalent to the existence of the angular (non-tangential) limit $\angle \lim_{z\to\zeta} F(z)$ taken as $z \to \zeta$ in each angular region in Δ with vertex at ζ (see, for example, [117]). If this limit has modulus 1, the point $\zeta \in \partial\Delta$ is called a *contact point* of F.

Given $\zeta \in \partial\Delta$ and $k > 0$, consider the disk

$$D(\zeta, k) := \left\{ z \in \Delta : \frac{|1 - z\overline{\zeta}|^2}{1 - |z|^2} < k \right\} \tag{13}$$

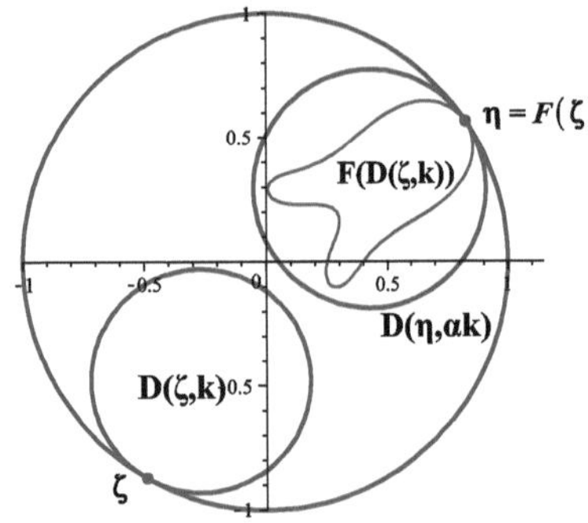

Fig. 4 The image of a horodisk in another one

tangent to the boundary at the point ζ (usually called the horodisk). A geometrical meaning of inequality (12) is that $F(D(\zeta,k)) \subseteq D(\eta,\alpha k)$, see Fig. 4.

Gaston Julia (1893–1978) was born in Sidi bel Abbès, Algeria and he was raised in Algeria. He went to Paris on a scholarship, but World War I interrupted his education. While serving on the front lines in 1915, he suffered a severe injury during an attack, losing his nose. He finished his dissertation during the long recuperation from his injury. Subsequent operations failed to repair the damage to his nose and for the rest of his life he wore a leather strap across his face to hide the missing nose. In 1918, he married Marianne Chausson and they had six children. Julia was one of the forefathers of the modern dynamical system theory and is best remembered for what is now called the Julia set.

In particular, Julia's Lemma implies that if there exists a boundary fixed point ζ of F with $\alpha \leq 1$, then all the disks $D(\zeta,k)$ are F-invariant (see Fig. 5).

Julius Wolff proved that in the case F has no interior fixed point, such a boundary point does indeed exist; thus, he answered question (i) affirmatively.

Fig. 5 Invariant horodisk

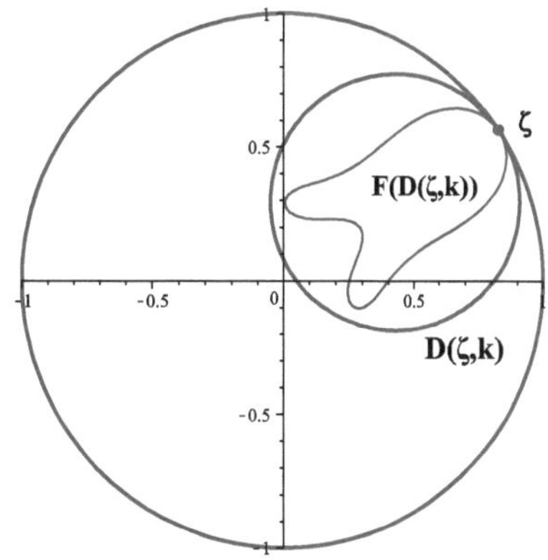

Theorem 2.5 (Wolff Theorem). *Suppose $F \in \mathrm{Hol}(\Delta)$ has no fixed point in Δ. Then there is a unique boundary fixed point $\zeta \in \partial\Delta$ such that for all $z \in \Delta$,*

$$\frac{|1 - F(z)\overline{\zeta}|^2}{1 - |F(z)|^2} \leq \frac{|1 - z\overline{\zeta}|^2}{1 - |z|^2}. \tag{14}$$

Note that the original version of this theorem also asserts that for each $z \in \Delta$, the sequence of iterates $F_1(z) := F(z)$, $F_n(z) := F(F_{n-1}(z))$, $n = 1, 2, \ldots$, converges to ζ. This result was proved independently by Arnaud Denjoy and is thus usually named the Denjoy–Wolff Theorem. The attractive fixed point of F is called the Denjoy–Wolff point.

J. Wolff (1882–1945) was born in Bergen-Belsen. He studied mathematics and physics in Amsterdam, where he obtained his doctorate supervised by J. Korteweg. In 1917, he was appointed lecturer at the University of Groningen, and in 1922 at the University at Utrecht. In addition, he was a math advisor to the life insurance company "Eigen Hulp" in The Hague. Julius' family was featured on the Barneveld list as well as on the Gerzon list. The lists were of those working for the clothing

firm Gerzon who were deemed useful for the war industry by the occupier. Those on the lists were transported with the Gerzon group from Westerbork to Bergen-Belsen, where they were all murdered.

From the geometrical point of view, inequality (14) means that for all $k > 0$, the disks $D(\zeta, k)$ are invariant under F.

The Wolff Theorem can be interpreted as a direct analogue of the Schwarz–Pick Lemma, where the role of the fixed point is taken over by a point on the unit circle. The result is the key to all deep results related to sequences of iterates of a holomorphic self-mapping F of Δ.

Although both Julia's Lemma and Wolff's Theorem imply the existence of invariant horodisks sharing a common boundary point, it is not clear how their hypotheses are related. Julia's Lemma states the existence of the angular limit of F at ζ. In 1929, C. Carathéodory [45] proved that under Julia's hypotheses, the angular derivative $F'(\zeta)$ exists (see also [83]).

Theorem 2.6 (Julia–Carathéodory Theorem). *For $F \in \mathrm{Hol}(\Delta)$ and $\zeta \in \partial\Delta$, the following statements are equivalent.*

(i) $\displaystyle \liminf_{z \to \zeta, z \in \Delta} \frac{1 - |F(z)|}{1 - |z|} = \alpha < \infty;$

(ii) $\displaystyle \angle \lim_{z \to \zeta} \frac{F(z) - \eta}{z - \zeta} =: F'(\zeta)$ *exists for some* $\eta \in \partial\Delta;$

(iii) $\displaystyle \angle \lim_{z \to \zeta} F'(z) = F'(\zeta)$ *exists and* $\displaystyle \angle \lim_{z \to \zeta} F(z) = \eta \in \partial\Delta.$

Moreover, $\alpha > 0$, the boundary points η in (ii) and (iii) are the same, and $F'(\zeta) = \alpha\zeta\overline{\eta}$.

Simultaneously, Edmund Landau and Georges Valiron in [98] established an essentially equivalent result for self-mappings of the right half-plane.

A point $\zeta \in \partial\Delta$ at which any of the equivalent statements of this theorem holds is said to be a *regular contact point* of F. It is also known that $\angle \lim_{z \to \zeta} (\zeta - z)^{n-1} F^{(n)}(z) = 0$ for all $n = 2, 3, \ldots$.

The Julia–Carathéodory Theorem implies that in the settings of the Wolff Theorem, i.e., when F has no interior fixed point, there exists a boundary point ζ such that the angular derivative $F'(\zeta)$ exists and is a positive real number not exceeding 1. In this case, the mapping F is said to be *of parabolic type* if $F'(\zeta) = 1$, and *of hyperbolic type* if $0 < F'(\zeta) < 1$.

Constantin Carathéodory (1873–1950) was born in Berlin. After graduating from the high school Athénée Royal d'Ixelles in 1891, he began training as a military engineer. In 1900, he entered the University of Berlin where Frobenius and Schwarz were professors. Carathéodory frequently attended Frobenius's lectures and Schwarz' colloquium, worked on the calculus of variations, and was much influenced by both Hilbert and Klein. Carathéodory received his doctorate in 1904 from Göttingen University under supervision of Hermann Minkowski. In 1910, Carathéodory was appointed to the Chair of Higher Mathematics at the Technical

University of Breslau, in 1913 to professor at Göttingen, and in 1918 at the University of Berlin. He established a second university in Smyrna at the request of Greek government. In 1928, he became the first visiting lecturer of the American Mathematical Society. After time spent as a visiting professor at Harvard, he held the chair in Munich until he retired in August 1938. Carathéodory made significant contributions to the calculus of variations, the theory of point set measure, and the theory of functions of a real variable. He contributed to partial differential equations and calculus of variations, to the theory of functions of several variables, to conformal representations of simply connected regions, and developed a theory of boundary correspondence. He also made contributions to thermodynamics, the special theory of relativity, mechanics, and geometrical optics.

Julia's Lemma and the Julia–Carathéodory Theorem do not require F to be fixed point free. At the same time, if F has a boundary fixed point ζ (in the sense that $\angle \lim\limits_{z \to \zeta} F(z) = \zeta$) such that $F'(\zeta) =: \alpha \leq 1$, then

$$\frac{|1 - F(z)\overline{\zeta}|^2}{1 - |F(z)|^2} \leq \alpha \frac{|1 - z\overline{\zeta}|^2}{1 - |z|^2}, \tag{15}$$

i.e., for all $k > 0$, $D(\zeta, k) \subseteq D(\zeta, \alpha k)$, and F necessarily has no interior fixed point.

Theorem 2.6 asserts that in the case F (neither the identity nor an elliptic automorphism) has an interior fixed point, then the angular derivative at each boundary contact point ζ (if it exists) satisfies the inequality $|F'(\zeta)| > 1$. In particular, if $\zeta \in \partial\Delta$ is a fixed point of F, then $F'(\zeta) > 1$.

It is worth mentioning that there is an essential difference between the Schwarz Lemma and its boundary versions. Suppose, for example, that F fixes zero and $|F'(0)| \leq q < 1$. The last inequality does not ensure that the image $F(r\Delta)$, $r \in (0, 1]$, lies in the disk of radius rq or of any other radius smaller than r. At the same time, the inclusion $F(r\Delta) \subset rq\Delta$ for at least one $r \in (0,1]$ implies that $|F'(0)| \leq q$. On the contrary, if $F \in \mathrm{Hol}(\Delta)$ has a boundary fixed point ζ, then the inequality $F'(\zeta) = q < 1$ implies that the image of each horodisk $F(D(\zeta,k))$, $k > 0$, is contained in $D(\zeta,\alpha k)$. However, the inverse implication fails: the inclusion $F(D(\zeta,k)) \subseteq D(\zeta,\alpha k)$ does not imply that $F'(\zeta) \leq \alpha$.

Note that early boundary generalizations of the Schwarz Lemma do not extend its rigidity part. This aspect was investigated much later and is discussed in Sect. 5 below.

2.6 First Refinements

This subsection is devoted to the earliest improvements and generalizations of the Schwarz and Schwarz–Pick Lemmas which are now considered as classical by themselves.

We start with an assertion which, similarly to Schmidt's proof of the Schwarz Lemma, can be proved using the maximum modulus principle.

Theorem 2.7. *Let $F \in \mathrm{Hol}(\Delta)$ with $F(z_1) = \ldots = F(z_n) = 0$. Then*

$$|F(z)| \leq \prod_{k=1}^{n} |m_{z_k}(z)|, \quad z \in \Delta,$$

where each point in the product is taken according to its multiplicity.

The case of equality in Theorem 2.7 was studied by Lehto. He proved that equality holds for some z with $F(z) \neq 0$ if and only if F is an inner function. In this case, equality holds for all z outside of a subset of Δ having logarithmic capacity zero.

Setting one of the points $z_1, \ldots, z_n$

Theorem 2.7 improves inequality (2) of the Schwarz Lemma since it includes the case in which one of the given points equals zero. Moreover, Theorem 2.7 in the case in which several points coincide with zero implies the following assertion (see, for example, [78]).

Theorem 2.8 (Generalized Schwarz Lemma). *Let k be a positive integer and $F \in \mathrm{Hol}(\Delta)$. If $F(0) = F'(0) = \cdots = F^{(k-1)}(0) = 0$, then*

(i) $\dfrac{1}{k!}|F^{(k)}(0)| \leq 1$*;*

(ii) $|F(z)| \leq |z|^k$ *for all $z \in \Delta$.*

Equality in (i) or in (ii) at a point $0 \neq z \in \Delta$ holds if and only if $F(z) = cz^k$ in Δ with $|c| = 1$.

Another classical generalization of the Schwarz Lemma is due to John Edensor Littlewood (see, for example, [103, Theorem 214]). It concerns multivalent self-mappings of Δ fixing the origin.

Suppose that $w \in F(\Delta)$ *and* $\{z_1, \dots z_n\} \subset F^{-1}(w)$ *(points of the preimage of* F *taken according to their multiplicity). Then*

$$|F(z_1)| \leq \prod_{k=1}^{n} |z_k|. \tag{16}$$

J.E. Littlewood (1885–1977) was a British mathematician, best known for the results achieved in collaboration with G. H. Hardy. He was born in Rochester in Kent, England. He lived in Wynberg in Cape Town from 1892 to 1900, where his father was a headmaster. The younger Littlewood then attended St Paul's School in London for three years, where he was taught by F.S. Macaulay who is now best known for his contributions to ideal theory. Littlewood studied at Trinity College, Cambridge, and was the Senior Wrangler in the Mathematical Tripos of 1905. He was elected a Fellow of Trinity College in 1908 and, apart from three years as Richardson Lecturer in the University of Manchester, spent his entire career at the University of Cambridge. He was appointed Rouse Ball Professor of Mathematics in 1928, and retired in 1950. He was elected a Fellow of the Royal Society in 1916, awarded the Royal Medal in 1929, the Sylvester Medal in 1943 and the Copley Medal in 1958. He was president of the London Mathematical Society from 1941 to 1943 and was awarded the De Morgan Medal in 1938 and the Senior Berwick Prize in 1960.

The next assertion may also be viewed as a sharpened form of the Schwarz Lemma. It turns out that if, in addition to the hypotheses of the lemma, the value $F'(0)$ is known, we can find, for a given z, a disk smaller than the disk of radius $|z|$ which contains the value $F(z)$. A result which refines inequality (2) was established by Werner Wolfgang Rogosinski in 1934 [126]. In what follows, let

$$B(c,r) := \{z \in \mathbb{C} : |z - c| < r\}. \tag{17}$$

Theorem 2.9 (Rogosinski's Lemma). *Let $|\lambda| < 1$. For the set of all functions $F \in \mathrm{Hol}(\Delta)$ with $F(0) = 0$ and $F'(0) = \lambda$, the range of values of $F(z)$ is the closed disk $\overline{B(c,r)}$, where*

$$c = \frac{z\lambda(1-|z|^2)}{1-|z\lambda|^2} \qquad \text{and} \qquad r = \frac{|z|^2(1-|\lambda|^2)}{1-|z\lambda|^2}.$$

W.W. Rogosinski (1894–1964) was born in Breslau. In 1913, he attended the University of Breslau, but his studies were interrupted by World War I, in which Rogosinski served as a medic. Later, he continued his studies at the University of Freiburg and then at the University of Göttingen under Edmund Landau's supervision. His dissertation, "New Application of Pfeiffer's method for Dirichlet's divisor problem," written in 1922, caused a stir. In 1923, he moved to Koenigsberg, first as a lecturer before becoming an associate professor in 1928. He worked for five years with Richard Brauer, Gábor Szegö, and Kurt Reidemeister. In 1936, after the nazi takeover, he was dismissed from his professorship. Hardy and Littlewood invited him to come to Cambridge, where he moved with his wife and child. In 1947, he was appointed professor at Newcastle University and in 1948 to Head of Department. From 1959, he worked at the Mathematical Institute at Aarhus. He died in Aarhus after a long illness.

Rogosinski's Lemma has an interesting geometric consequence. Consider the set of all self-mappings of the open unit disk such that $F(0) = 0$, $\arg F'(0) = \phi$ and $|F'(0)| \le r < 1$. Then the range of values of $F(z)$ (for fixed $z \in \Delta$) is the closed domain containing the disk $|\zeta| \le r|z|^2$ and bounded by the half-circle $|\zeta| = r|z|^2$ and the two circular arcs connecting the point $re^{i\phi}z$ to $re^{i(\phi+\pi/2)}z|z|$ and $re^{i(\phi-\pi/2)}z|z|$, respectively, and tangent to the circle at these points (see Fig. 6, cf., [64]). In particular, if $\mathrm{Re}\left(\frac{z_0F'(z_0)}{F(z_0)}\right) < 0$ at some point $z_0 \in \Delta$, then $|F(z_0)| \le |z_0|^2$.

The above results improve the "invariance" part of the Schwarz Lemma.

Fig. 6 The circle of radius $|z_0|$ and the range of $F(z_0)$ with vertex at the point $z_0 e^{i \arg F'(0)}$

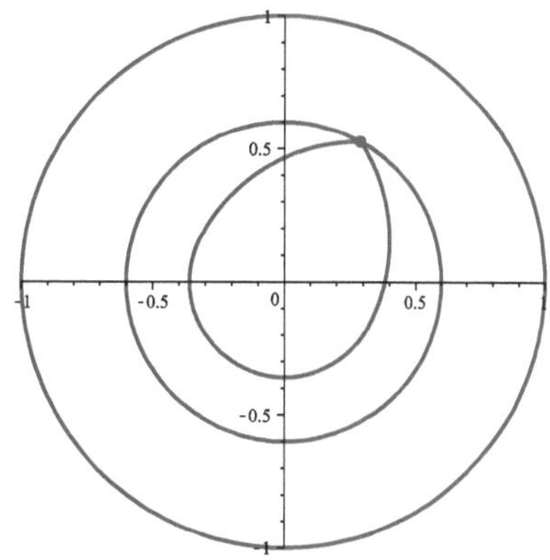

We now turn to estimates of the derivative $F'(z)$. The range of values of the derivative given by the Schwarz–Pick Lemma, namely,

$$|F'(z)| \leq \frac{1 - |F(z)|^2}{1 - |z|^2} \tag{18}$$

can also be specified when $F(0) = 0$ and the value $F(z)$ is given for some $z \neq 0$. The first such improvement of the Schwarz inequality (1) was established by Jean Alexandre Dieudonné in 1931 [59].

J.A. Dieudonné (1906–1992) was born and brought up in Lille. In 1924, he was accepted by the École Normale Supérieure, where he was inspired by Émile Picard, Jacques Hadamard, Élie Cartan, Paul Montel, Arnaud Denjoy, and Gaston Julia. His doctoral studies were supervised by Montel. In 1934, Dieudonné was one of the group of normaliens convened by Weil, which became known as "Bourbaki." He served in the French Army during World War II and then taught in Clermont-Ferrand until the liberation of France. In 1953, after holding professorships at the

University of São Paulo (1946–47), the University of Nancy (1948–1952), and the University of Michigan (1952–53), he joined the Department of Mathematics at Northwestern University before returning to France as a founding member of the Institut des Hautes Études Scientifiques. In 1964, he moved to the University of Nice to found the Department of Mathematics. He retired in 1970. He was elected as a member of the Académie des Sciences in 1968.

Theorem 2.10 (Dieudonné's Lemma). *Let $z, w \in \Delta$. For the set of all functions $F \in \mathrm{Hol}(\Delta)$ with $F(0) = 0$ and $F(z) = w$, the range of values of $F'(z)$ is the closed disk $\overline{B(c,r)}$, where*

$$c = \frac{w}{z} \qquad \textit{and} \qquad r = \frac{|z|^2 - |w|^2}{|z|(1-|z|^2)}.$$

As a consequence, Dieudonné obtained the following assertion.

Corollary 2.2. *If $F \in \mathrm{Hol}(\Delta, \Delta)$ and $F(0) = 0$, then*

$$|F'(z)| \le \begin{cases} 1, & |z| \le \sqrt{2} - 1, \\ \dfrac{(1+|z|^2)^2}{4|z|(1-|z|^2)}, & \sqrt{2} - 1 < |z| < 1. \end{cases}$$

The bound is sharp.

Another refinement of the distortion part of the Schwarz Lemma is due to Edvin F. Beckenbach [22]. We formulate it in an equivalent form.

Theorem 2.11. *If f is holomorphic in Δ and*

$$\int_0^1 |f'(tz)|dt \le \frac{1}{|z|} \qquad \textit{for all} \quad z \in \Delta \setminus \{0\},$$

then

$$\int_0^1 |f'(tz)|dt \le 1 \qquad \textit{for all} \qquad z \in \Delta.$$

In particular, $|f'(0)| \le 1$. Moreover, equality in either of these inequalities holds if and only if $|f'(z)| \equiv 1$.

In this context, the next result by Julia (see [90]) can also be considered an analog of the Schwarz Lemma.

If $f \in \mathrm{Hol}(\Delta, \mathbb{C})$ then for $0 < p < \infty$, the function

$$\Phi_p(r) = \frac{\int_0^{2\pi} |f(re^{it})|^p dt}{r^p}, \qquad 0 < r < 1,$$

is increasing.

This result is in the spirit of the Schwarz Lemma interpreted in terms of the increasing function $\phi(r) = \frac{1}{r} \operatorname{Rad} F(r\Delta)$ (cf., Sect. 2.4). Such an interpretation leads to many recent generalizations of the Schwarz Lemma, some of which are presented in Sect. 3.2 below.

In the case that a self-mapping F is univalent, a simple manipulation of the Schwarz Lemma immediately implies that the derivative $F'(0)$ has a lower bound.

Theorem 2.12. *Let F be a univalent analytic function mapping Δ onto an open set $\Omega \subset \mathbb{C}$, with $F(0) = 0$. If Ω contains the disk $\{w : |w| < r\}$, then $|F'(0)| \geq r$, with equality if and only if $F(z) = re^{i\theta}z$.*

It has already been mentioned that if F has an interior fixed point and a boundary regular contact point ζ, then $F'(\zeta) > 1$. Regarding boundary versions of the Schwarz Lemma, in 1923, Karl Löwner [104] proved deformation theorems that can be considered early continuous versions of the last fact. His results can be formulated as follows.

Theorem 2.13 (Löwner Theorem). *Let $F \in \operatorname{Hol}(\Delta)$ satisfy $F(0) = 0$. Assume that F maps an arc $A \subset \partial\Delta$ of length s onto an arc $F(A) \subset \partial\Delta$ of length σ. Then $\sigma \geq s$ with equality if and only if either $s = \sigma = 0$ or F is just a rotation.*

Further discussion of early and contemporary results connected to the boundary behavior of holomorphic self-mappings, in particular involving angular derivatives, can be found in Sect. 4.1 and subsequent sections.

3 Generalizations of the Schwarz and Schwarz–Pick Lemmas

In this section, we present results which generalize both the invariance part of the Schwarz and Schwarz–Pick Lemmas as well as estimates on derivatives of a holomorphic self-mapping of the unit disk. Although these generalizations give more precise bounds, they usually require additional assumptions. We also consider analogous estimates for hyperbolic derivatives and hyperbolic divided differences (including multi-point Schwarz–Pick Lemma).

Regarding the rigidity principle, notice that in most of the estimates in this section, equality holds if and only if F has a specific form. In many different situations, extremal functions for the Schwarz–Pick type inequalities are either affine functions, automorphisms, or more generally, inner functions.

Section 3.2 presents geometric versions of the Schwarz Lemma which give different ways of measuring the image $F(r\Delta)$, $0 < r < 1$, using various geometric characteristics, including diameter, capacity, width, and boundary length. These versions originate from the Landau–Toeplitz Theorem 2.3. Some of these results also contain extremal functions as a rigidity part.

3.1 Strengthened Forms of the Schwarz Lemma

We begin with a consequence of the Schwarz Lemma obtained by Lawrence A. Harris in [81] for a holomorphic self-mapping of Δ not necessarily fixing the origin.

Corollary 3.1. *Every $F \in \mathrm{Hol}(\Delta)$ satisfies*

$$|F(z) - F(0)| \le |z| \frac{1 - |F(0)|^2}{1 - |F(0)||z|}.$$

Obviously, in the case $F(0) = 0$, Corollary 3.1 reduces to the original Schwarz inequality.

Sharpened estimates of $F(z)$ and $F'(z)$ were established by Peter R. Mercer in 1997 [106] under the assumption that the images of two points are known. We formulate his results as follows.

Theorem 3.1. *Let $F \in \mathrm{Hol}(\Delta)$, $F(0) = 0$ and $F(z_0) = w_0$ $(z_0 \neq 0)$. Denote $\sigma = \dfrac{w_0}{z_0}$. Then for all $z \in \Delta$,*

(i) $F(z) \in \overline{B(c_1, r_1)}$, where

$$c_1 = \frac{z\sigma(1 - |m_{z_0}(z)|^2)}{1 - |\sigma|^2 |m_{z_0}(z)|^2} \qquad \text{and} \qquad r_1 = |z m_{z_0}(z)| \cdot \frac{1 - |\sigma|^2}{1 - |\sigma|^2 |m_{z_0}(z)|^2}.$$

In particular,

$$|F(z)| \le |z| \cdot \frac{|\sigma| + |m_{z_0}(z)|}{1 + |\sigma||m_{z_0}(z)|}$$

and this inequality is sharp.

(ii) $F'(0) \in \overline{B(c_2, r_2)}$, where

$$c_2 = \sigma \cdot \frac{1 - |z_0|^2}{1 - |w_0|^2} \qquad \text{and} \qquad r_2 = |z_0| \cdot \frac{1 - |\sigma|^2}{1 - |w_0|^2}.$$

One can see that the Rogosinski Lemma 2.9 and the Dieudonné Lemma 2.10 are the limiting cases of Theorem 3.1 (i) as $z_0 \to 0$ and $z \to z_0$, respectively. Later, a particular case of Theorem 3.1, which follows also by the Rogosinski Lemma 2.9, was reproved by Robert Osserman [114, Lemma 2].

In the same paper [106], Mercer obtained an estimate of $|F(z)|$ for a holomorphic self-mapping of Δ that fixes the origin, where, instead of the value at an additional point z_0, he used the values of the first and second derivatives of F at zero.

Theorem 3.2. *Let $F \in \mathrm{Hol}(\Delta)$ and $F(0) = 0$. For a point $z \in \Delta$, set $a := |z| \dfrac{b + |z|}{1 + b|z|}$ and $b := \dfrac{|F''(0)|}{2(1 - |F'(0)|^2)}$. Then*

$$|F(z)| \le |z|\frac{a + |F'(0)|}{1 + a|F'(0)|}.$$

This inequality is sharp. Moreover, $b \le 1$, i.e., the sharpened Cauchy estimate $\frac{|F''(0)|}{2} \le (1 - |F'(0)|^2)$ holds.

An improvement of the Schwarz inequality (2) for univalent self-mappings of the disk was proved by Shinji Yamashita in 1997 [142] under the additional assumption that the value of the derivative at zero is given: *Let $F \in \mathrm{Hol}(\Delta)$ be univalent, $F(0) = 0$ and $F'(0) = \alpha$. Then for all $z \in \Delta$,*

$$|F(z)| \le |z|\frac{4|\alpha|}{\left(1 - |z| + \sqrt{(1 - |z|)^2 + 4|\alpha z|}\right)^2}. \tag{19}$$

Yamashita also described the class of univalent functions for which equality holds in (19).

Recently, Xiaojun Huang and Ling Chen studied the following natural question (see [84]). Is there an analog of the Schwarz Lemma for meromorphic functions? Their answer is affirmative and in their proof they use winding numbers and homotopy.

Let $n(r, F)$ be the number of poles and $n(r, 1/F)$ the number of zeros of $F : \Delta \mapsto \Delta$ in $r\Delta$, each counted according to multiplicity.

Theorem 3.3. *Let k be a positive integer. Suppose that F is a meromorphic function on Δ such that*

(a) $F(0) = F'(0) = \ldots = F^{(k-1)}(0) = 0$ and
(b) there exists a real number $0 < r_0 < 1$ such that $|F(z)| \le 1$ for all $z \in \{z : r_0 < |z| < 1\}$.

The following assertions hold.

(i) If $w \in \Delta$ is neither pole nor zero of F and the function $G_w(z) = z^k - \frac{w^k F(z)}{F(w)}$ satisfies $n(1, 1/G_w) - n(1, G_w) \ne k$, then $|F(z)| \le |z|^k$. Moreover, the equality $|F(w)| = |w|^k$ holds at a point $w \ne 0$ if and only if $F(z) = cz^k$ with $|c| = 1$.
(ii) If $F^{(k)}(0) \ne 0$ and the function $G(z) = z^k - \frac{k!F(z)}{F^{(k)}(0)}$ satisfies $n(1, 1/G) - n(1, G) \ne k$, then $|F^{(k)}(0)| \le k!$. Moreover, $|F^{(k)}(0)| = k!$ if and only if $F(z) = cz^k$ with $|c| = 1$.

In [63], Vladimir Dubinin obtained refinements and supplements to the Schwarz inequalities for functions with free domain of analyticity (i.e., not necessarily analytic in the whole unit disk). Namely, he obtained conditions that imply $|F(z)| \le c|z|^n$, $|F'(0)| \le 1$, as well as $F'(\zeta) \ge 0$ at some boundary point ζ.

3.2 Geometric Versions of the Schwarz Lemma

Recently, geometric versions of the Schwarz–Pick Lemma and its various modifications have generated considerable interest. We now describe some of these developments. Following Robert B. Burckel, Donald E. Marshall and Pietro Poggi-Corradini [40], we start with a consequence of Landau–Toeplitz Theorem 2.3: *Suppose F is analytic in Δ. If the function $\frac{\operatorname{Diam} F(r\Delta)}{r}$ is constant, then F is linear.*

In the same paper [40], a related modulus growth estimate was given as follows.

Theorem 3.4. *Suppose F is analytic in Δ and* Diam $F(\Delta) \le 2$. *Then for all $z \in \Delta$,*

$$|F(z) - F(0)| \le \frac{2|z|}{1 + \sqrt{1 - |z|^2}}. \tag{20}$$

Moreover, equality in (20) holds at some point in $\Delta \setminus \{0\}$ if and only if F is a linear fractional transformation of the form

$$F(z) = c\frac{z - b}{1 - \bar{b}z} + a \tag{21}$$

for some constants $a \in \mathbb{C}$, $b \in \Delta \setminus \{0\}$ and $c \in \partial\Delta$.

Note that in the particular case $F \in \operatorname{Hol}(\Delta)$, the inequality of Harris (see Corollary 3.1) is stronger than (20), but the latter inequality is applicable to a wider class of functions. In addition, it should be mentioned that in contrast to the Schwarz Lemma, if F is a linear fractional transformation of the form (21), then equality in (20) occurs at the point $z = 2b/(1 + |b|^2)$ only.

Since the origin does not play a special role in (20), the inequality can be rewritten in the following symmetric form.

$$|F(z) - F(w)| \le \operatorname{Diam} F(\Delta) \tanh\left(\frac{\rho(z, w)}{2}\right) \qquad \text{for all } z, w \in \Delta.$$

In [134], Alexander Solynin established the following stronger form of the Landau–Toeplitz theorem for meromorphic functions.

Theorem 3.5. *Let $F(z) = cz^p + c_1 z^{p+1} + \ldots$ be meromorphic in Δ. Suppose that the length of every straight line segment embedded into the Riemann surface of F^{-1}, that passes through P_0 (where $F^{-1}(P_0) = 0$), does not exceed* 2. *Then $|c| \le 1$ with equality if and only if $F(z) = cz^p$ for some c, $|c| = 1$.*

In [30], Dimitrios Betsakos discovered a multi-point extension of Theorem 3.4 for the case of multivalent functions (cf., the Littlewood inequality (16)). As above, $z_1(w), \ldots, z_n(w)$ denote preimages of a point $w \in F(\Delta)$ under F.

Theorem 3.6. *Let* $F \in \mathrm{Hol}(\Delta, \mathbb{C})$ *satisfy* $\mathrm{Diam}\, F(\Delta) = 2$ *and* $w \in F(\Delta)$. *Then*

$$\frac{4|w - F(0)|}{4 + |w - F(0)|^2} \le \prod_{j=1}^{n} |z_j(w)|. \quad w \in F(\Delta) \setminus \{F(0)\}.$$

Equality holds for some $w_0 \in F(\Delta) \setminus \{F(0)\}$ *if and only if there exist an* $a \in \Delta \setminus \{0\}$ *and an inner function* h *with* $h(0) = 0$ *such that*

$$F(z) = m_a(h(z)) + a + w_0, \quad z \in \Delta.$$

In [41], Robert B. Burckel, Donald E. Marshall, David Minda, Pietro Poggi-Corradini, and Thomas J. Ransford explored other geometric quantities that can be used to measure the size of the image of an analytic function, namely, n-diameter, capacity, area, and perimeter. They established corresponding analogs of the Schwarz Lemma.

For $E \subset \mathbb{C}$ and $n \ge 2$, let $\mathrm{Diam}_n(E) := \sup \left(\prod_{1 \le j < k \le n} |\zeta_j - \zeta_k| \right)^{\frac{2}{n(n-1)}}$, where the supremum is taken over all n-tuples of points in E. We call $\mathrm{Diam}_n(E)$ the *n-diameter* of E. The limit $\mathrm{Diam}_\infty(E) := \lim_{n \to \infty} \mathrm{Diam}_n(E)$ is well defined and is called the *transfinite diameter* of E. It coincides with the logarithmic capacity $\mathrm{Cap}(E)$.

Theorem 3.7 (Diameter Schwarz' Lemma). *Suppose* $F \in \mathrm{Hol}(\Delta, \mathbb{C})$ *and either* n *is a natural number,or* $n = \infty$. *The functions* ϕ_n *defined by*

$$\phi_n(r) := \frac{\mathrm{Diam}_n(F(r\Delta))}{r},$$

are increasing and log-convex. *Moreover, they are strictly increasing for* $0 < r < 1$ *except in the special case that* F *is linear. Consequently, if* $\mathrm{Diam}_n(F(\Delta)) = \mathrm{Diam}_n(\Delta)$, *then*

$$\mathrm{Diam}_n(F(r\Delta)) \le \mathrm{Diam}_n(\Delta) r \tag{22}$$

for all $r \in (0, 1)$, *and*

$$|F'(0)| \le 1. \tag{23}$$

Moreover, equality either in (22) for some $0 < r < 1$, *or in (23) holds if and only if* $F(z) = a + cz$ *for some constants* $a \in \mathbb{C}$, $c \in \partial\Delta$.

It is easy to see that the condition Area $F(\Delta) = \pi$ is weaker than the condition Diam $F(\Delta) = 2$. Thus, the following analog of Schwarz's Lemma is stronger than the Landau–Toeplitz Theorem 2.3.

Theorem 3.8 (Area Schwarz' Lemma). *Suppose F is analytic in Δ. Then the function $\phi_{\text{Area}}(r) := (\pi r^2)^{-1} \operatorname{Area} F(r\Delta)$ is strictly increasing for $0 < r < 1$, except when F is linear, in which case ϕ_{Area} is constant. Consequently, if $\operatorname{Area} F(\Delta) = \pi$, then*

$$\operatorname{Area} F(r\Delta) \leq \pi r^2 \tag{24}$$

for all $r \in (0, 1)$, and

$$|F'(0)| \leq 1. \tag{25}$$

Moreover, equality either in (24) for some $0 < r < 1$, or in (25) holds if and only if $F(z) = a + cz$ for some constants $a \in \mathbb{C}$, $c \in \partial\Delta$.

As the authors of [41] noted, the Perimeter Schwarz' Lemma is a weaker result because log-convexity is not concluded and extra conditions on the image must be imposed. In the next theorem we combine [116, Problem 309] with a result in [41].

Theorem 3.9. *Let $F \in \operatorname{Hol}(\Delta, \mathbb{C})$. The function $r \mapsto r^{-1} \operatorname{Length} \partial(F(r\Delta))$ is strictly increasing, unless F is linear. If, in addition, f is univalent and $\partial F(\Delta)$ is a Jordan curve of Euclidean length at most 2π, then*

$$\operatorname{Length} \partial(F(r\Delta)) \leq 2\pi r \tag{26}$$

for all $0 < r < 1$, and

$$|F'(0)| \leq 1. \tag{27}$$

Moreover, equality either in (26) for some $0 < r < 1$, or in (27) holds if and only if $F(z) = a + cz$ for some $a \in \mathbb{C}$, $c \in \partial\Delta$.

Corollary 3.2. *Suppose Ω is a simply connected region in $\mathbb{C}$ and $\partial\Omega$ is a Jordan curve of Euclidean length at most 2π. If $f \in \operatorname{Hol}(\Delta, \Omega)$, then $|f'(0)| \leq 1$ and equality holds if and only if $f(z) = a + cz$ for some $a \in \mathbb{C}$, $c \in \partial\Delta$.*

Note that the last Corollary also follows from Beckenbach's Theorem 2.11.

In [30], Betsakos obtained the following assertion which deals with the width of the image and strengthens the result of Pólya and Szegö quoted in Sect. 2.4.

Theorem 3.10. *Let $F \in \operatorname{Hol}(\Delta, \mathbb{C})$ satisfy $W(F(\Delta)) = \frac{\pi}{2}$. The following assertions hold.*

(a) $W(F(r\Delta)) \leq 2 \arctan r$ *for all $r \in (0, 1)$. Equality holds for some $r \in (0, 1)$ if and only if there exist $b \in \mathbb{C}$ and $\theta \in [0, 2\pi)$ such that $F(z) = \arctan(e^{i\theta} z) + b$.*

(b) $\tanh |w - F(0)| \leq \prod_k |z_k(w)|$ *for all $w \in F(\Delta) \setminus \{F(0)\}$, where $z_k(w)$ are preimages of $w \in F(\Delta)$. Equality holds for some $w_0 \in F(\Delta) \setminus \{F(0)\}$ if and only if there exists an inner function h with $h(0) = 0$ such that*

$$F(z) = \arctan(m_{iu}(h(z))) + t$$

with $u = \tanh(\operatorname{Im} w_0) \in (-1, 1)$ *and* $t = \operatorname{Re} w_0 \in \mathbb{R}$. *In this case, the image* $F(\Delta)$ *is a vertical strip and equality holds for all* $w \neq F(0)$ *on the mid-line of this strip.*

(c) $|F(z) - F(0)| \leq \operatorname{arctanh}|z|$ *for all* $z \in \Delta$. *Equality holds at a point* $z \in \Delta \setminus \{0\}$ *with* $\arg z = \theta$ *if and only if there exist* $t \in \mathbb{R}$ *and* $u \in (-1, 1)$ *such that*

$$F(z) = \pm \operatorname{arctanh}\left(m_{iu}(-ie^{-i\theta}z)\right) + t.$$

If F *has this form, then equality holds for all* $z \in \{re^{i\theta} : -1 < r < 1\}$.

Tom Carroll and Jesse Ratzkin continued this investigation. In particular, in [47] they answered affirmatively the question posed in [41] concerning the validity of inequalities similar to those above for the first eigenvalue λ of the Laplacian with Dirichlet boundary data.

Theorem 3.11. *Let* f *be a conformal mapping of* Δ. *The function*

$$\Phi_\lambda(r) = \frac{\lambda(f(r\Delta))}{\lambda(r\Delta)}, \quad 0 < r < 1,$$

is strictly decreasing, unless f *is linear (in which case* Φ_λ *is constant).*

Some analogs of the monotonicity property for the capacity of a condenser were established by Betsakos and Pouliasis in [31]. A *condenser* in the complex plane $\mathbb{C}$ is a pair (D, K), where D is a Greenian open subset of $\mathbb{C}$ and K is a compact subset of D. Let h be the solution of the generalized Dirichlet problem on $D \setminus K$ with boundary values 0 on ∂D and 1 on ∂K. The function h is called the *equilibrium potential* of the condenser (D, K). The *capacity* of (D, K) is defined by

$$\operatorname{Cap}(D, K) = \int_{D \setminus K} |\nabla h|^2.$$

The *equilibrium energy* of (D, K) is defined by the extended real number

$$I(D, K) = \frac{2\pi}{\operatorname{Cap}(D, K)}.$$

We set

$$C_2(D, K) = e^{I(D,K)}.$$

Theorem 3.12. *Let* $f \in \operatorname{Hol}(\Delta, \mathbb{C})$ *be such that the image* $f(\Delta)$ *is a Greenian domain. Then the function*

$$\Phi_C(r) := \frac{C_2\left(f(\Delta), f(\overline{r\Delta})\right)}{r}, \quad r \in (0, 1),$$

is increasing. If Φ_C *is not strictly increasing, there exists* $d \in (0,1]$ *such that* Φ_C *is constant on* $(0,d)$, Φ_C *is strictly increasing on* $(d,1)$, *and* f *is univalent on* $d\Delta$.

Let $D \subset \mathbb{C}$ be a Greenian domain with the Green function $G_D(x,y)$ and $z_0 \in D$. The inner radius $R(D,z_0)$ of D at z_0 is

$$R(D,z_0) = \exp\left[\lim_{z\to z_0}\left(G_D(z,z_0) - \log\frac{1}{z-z_0}\right)\right].$$

A simple property of the inner radius is that if D is simply connected and f maps Δ conformally onto D with $f(0) = z_0$, then $R(D,z_0) = |f'(0)|$. The following monotonicity property for the inner radius $R(f(r\Delta), f(0))$ was proved in [31].

Theorem 3.13. *Let* $f \in \mathrm{Hol}(\Delta,\mathbb{C})$. *Then the function*

$$\Phi_R(r) := \frac{R(f(r\Delta), f(0))}{r}, \quad 0 < r < 1,$$

is increasing. Moreover, if Φ_R *is not strictly increasing, there exists an* $s \in (0,1]$ *such that* Φ_R *is constant on* $(0,s)$ *and strictly increasing on* $(s,1)$, *and* f *is univalent on* $s\Delta$.

In [29], Betsakos proposed another approach to proving general volume, capacity, and diameter versions of the Schwarz Lemma for quasiregular mappings in higher dimensions. His approach is based on conformal invariants (the conformal capacity of condensers, the Green function, the modulus metric), and their behavior under suitable types of symmetrization. There are also different variations of the Schwarz Lemma for the sharp constant in the Sobolev embedding, for principal frequency and other characteristics (see, for example, [48] and references therein).

3.3 Sharpened Estimates of the First Derivative

Now we turn to the question as to how the Schwarz–Pick inequality (5)

$$\frac{|F'(z)|}{1-|F(z)|^2} \le \frac{1}{1-|z|^2}$$

can be strengthened, given additional information on a self-mapping F.

In [77], Gennadii Mikhailovich Goluzin refined the Schwarz–Pick inequality, using first Taylor coefficients. Namely, he considered holomorphic self-mappings F of Δ that have a Taylor expansion of the form

$$F(z) = c_0 + \sum_{k=n}^{\infty} c_k z^k, \quad n \ge 1. \tag{28}$$

We formulate his results as follows.

Theorem 3.14. *Let* $F \in \mathrm{Hol}(\Delta)$ *admit representation (28). Then*

$$\frac{|F'(z)|}{1-|F(z)|^2} \le \frac{n|z|^{n-1}}{1-|z|^{2n}}, \quad z \in \Delta.$$

Moreover, if equality holds at a point $z_0 \in \Delta$, *then* $F(z) = \gamma m_{z_0^n}(z^n)$ *with* $|\gamma| = 1$.

Theorem 3.15. *Let* $F \in \mathrm{Hol}(\Delta)$ *admit representation (28). Then*

$$\frac{|F'(z)|}{1-|F(z)|^2} \le \frac{\phi'(r)}{1-\phi(r)^{2n}} \quad z \in \Delta,$$

where $r = |z|$, $\phi(r) = r^n \frac{r+|\gamma_n|}{1+r|\gamma_n|}$ *with* $\gamma_n = \frac{c_n}{1-|c_0|^2}$. *Moreover, equality holds only for the function*

$$F(z) = \frac{c_0 + \Phi_n(z)}{1+\overline{c_0}\Phi_n(z)}, \qquad \text{where} \quad \Phi_n(z) = z^n \frac{z+\gamma_n}{1+z\overline{\gamma_n}}.$$

In particular, for $n = 1$, we get

$$\frac{|F'(z)|}{1-|F(z)|^2} \le \frac{1}{1-r^2} \cdot \frac{2r + |\gamma_1| + r^2|\gamma_1|}{1+r^2+2r|\gamma_1|}.$$

In 1994, Yamashita [141] using a different approach obtained a strengthened version of inequality (5) in the form

$$|F'(z)| \le \frac{1-|F(z)|^2}{1-|z|^2} \cdot L, \quad z \in \Delta,$$

where

$$L := \frac{\gamma(1+|z|^2)+2|z|}{1+|z|^2+2\gamma|z|} \; (\le 1) \quad \text{with} \quad \gamma := \frac{|F'(0)|}{1-|F(0)|^2}.$$

The last inequality coincides with Goluzin's in Theorem 3.15 for $n = 1$. Yamashita proved that equality holds at a point $z_0 \neq 0$ if and only if F is either an automorphism of Δ or the product of two automorphisms.

In his next paper in 1997 [142], Yamashita extended these results as follows.

Theorem 3.16. *Let* $F \in \mathrm{Hol}(\Delta)$ *admit representation (28). Then*

$$|F'(z)| \le \frac{1-|F(z)|^2}{1-|z|^2} \cdot L_1$$

for all $z \in \Delta$, *where* $L_1 := \dfrac{(1-r^2)|\Phi_n'(r)|}{1-|\Phi_n(r)|^2} - R_n(r)$ *with* $r = |z|$ *and*

$$\delta_n = \begin{cases} 1, & |\gamma_n| = 1 \\ \dfrac{|c_{n+1}|}{(1-|c_0|^2)(1-|\gamma_n|^2)}, & |\gamma_n| < 1 \end{cases},$$

$$\Phi_n(z) = \frac{z(z+\delta_n)}{1+\delta_n z},$$

$$R_n(r) = \frac{r^n(1-\Phi_0(r))(1-\Gamma(r,\Psi))}{1-\Phi_n(r)}.$$

Moreover, equality holds at some point if and only if one of the following is valid.

(i) $F(z) = T(z^n)$, *where* T *is an automorphism of* Δ. *In this case equality holds everywhere in* Δ.
(ii) $F(z) = T(z^n S(z))$, *where* T *and* S *are automorphisms of* Δ. *In this case, equality holds at each point* z *such that* $\arg z = \arg a + \pi$, *where* $a = S^{-1}(0)$.

Returning to the settings of the Schwarz Lemma, i.e., $F(0) = 0$, we mention that in [21], Alan F. Beardon and David Minda completed the Schwarz Lemma as well as Dieudonné's lemma with the following assertion: *Suppose that* $F \in \mathrm{Hol}(\Delta)$ *with* $F(0) = 0$. *If at some point* $z \in \Delta$ *the function* F *satisfies* $|F(z)| \le 1 - |z| - |z|^2$, *then* $|F'$ï£¡$(z)| \le 1$.

3.4 Upper Bounds for Higher Order Derivatives

The Schwarz–Pick inequality (5) for the first derivative has been generalized to the derivatives of arbitrary order. In 1920, O. Szaśz obtained sharp estimates of $|F^{(n)}(z)|$ independent of $F(z)$ for odd n; their nature is different from (5). The following result was proved in 1985 by Stephan Ruscheweyh [128].

Theorem 3.17. *Let* $F \in \mathrm{Hol}(\Delta)$ *and* $z \in \Delta \setminus \{0\}$. *Then*

$$\frac{|F^{(n)}(z)|}{1-|F(z)|^2} \le \frac{n!(1+|z|)^{n-1}}{(1-|z|^2)^n}. \tag{29}$$

Later on, estimates for higher derivatives were studied by many authors (see, for example, [26,58,74,145]). In particular, James Milne Anderson and James Rovnyak in [12] used a Hilbert space method to derive Ruscheweyh's Theorem 3.17 and specified it as follows.

Theorem 3.18. *Let* $F \in \mathrm{Hol}(\Delta, \overline{\Delta})$.

(i) For $n = 1$, *equality in (29) holds for some* $z \in \Delta$ *if and only if it holds for all* $z \in \Delta$, *and this occurs if and only if*

$$F(z) = \gamma m_c(z)$$

for some $\gamma \in \partial\Delta$ *and* $c \in \Delta$.

(ii) *If* $z = 0$, *then for any* $n \geq 1$, *equality in (29) holds if and only if* $F(z) = \gamma m_c(z^n)$.

(iii) *If* $z \in \Delta \setminus \{0\}$ *and* $n \geq 2$, *equality in (29) holds only for a constant of absolute value one.*

A similar problem was studied by Barbara D. MacCluer, Karel Stroethoff, and Ruhan H. Zhao in [105] using characterizations of boundedness and compactness of weighted composition operators acting on Bloch-type spaces. They stated more generalized estimates.

Theorem 3.19. *Let* $F \in \mathrm{Hol}(\Delta)$ *and* $\alpha, \beta > 0$. *For each natural* n,

$$(a) \qquad \sup_{z\in\Delta} \frac{|F'(z)|\left(1-|z|^2\right)^{\beta}}{\left(1-|F(z)|^2\right)^{\alpha}} < \infty \Longrightarrow \sup_{z\in\Delta} \frac{\left|F^{(n)}(z)\right|\left(1-|z|^2\right)^{\beta+n-1}}{\left(1-|F(z)|^2\right)^{\alpha}} < \infty$$

and

$$(b) \qquad \lim_{|z|\to 1^-} \frac{|F'(z)|\left(1-|z|^2\right)^{\beta}}{(1-|F(z)|^2)^{\alpha}} = 0 \Longrightarrow \lim_{|z|\to 1^-} \frac{\left|F^{(n)}(z)\right|\left(1-|z|^2\right)^{\beta+n-1}}{\left(1-|F(z)|^2\right)^{\alpha}} = 0.$$

If $\beta > \alpha > 0$, *then the implications converse to (a) and (b) also hold.*

To describe further developments, we rewrite inequality (29) in the form

$$\tau_{F,n}(z) \leq n!(1+|z|)^{n-1},$$

where $\tau_{F,n}(z) := \dfrac{(1-|z|^2)^n |F^{(n)}(z)|}{1-|F(z)|^2}$.

In 2004 [74], Pratibha Ghatage and Dechao Zheng proved the following result.

Theorem 3.20. *Let* $F \in \mathrm{Hol}(\Delta)$. *For each* $n \in \mathbb{N}$, *the function* $\tau_{F,n}(z)$ *is Lipschitz continuous with respect to the pseudo-hyperbolic metric* $|m_z(w)|$, *i.e.,*

$$|\tau_{F,n}(z) - \tau_{F,n}(w)| \leq C_n |m_z(w)|$$

for all $z, w \in \Delta$. *Here* C_n *is a positive constant depending only on* n.

In 2008 [58], Shaoyu Dai and Yifei Pan proved that $C_n \leq (n+1)! \cdot 2^{n+2}$ and, in doing so, completed Theorem 3.20.

3.5 Estimates in the Hyperbolic Metric

We now discuss some refined versions of the Schwarz–Pick inequality

$$\left|m_{F(w)}(F(z))\right| \le |m_w(z)|, \quad z, w \in \Delta. \tag{30}$$

As already mentioned, each holomorphic self-mapping of the open unit disk is nonexpansive relative to the Poincaré hyperbolic metric, and (30) can be rewritten in the form.

$$\rho(F(z), F(w)) \le \rho(z, w), \quad z, w \in \Delta. \tag{31}$$

Moreover, if F is not an automorphism, the inequality in (31) is strict. We will now see now that this contractive property can be strengthened.

We begin with the definition of the hyperbolic derivative $F^*(z_0)$ of a holomorphic self-mapping F of Δ at a point $z_0 \in \Delta$. Following [20] and [17], we define the hyperbolic difference quotient (hyperbolic divided difference) $\Delta_{z_0} F(z)$ of F by

$$\Delta_{z_0} F(z) = \frac{m_{F(z_0)}(F(z))}{m_{z_0}(z)}, \quad z \in \Delta \setminus \{z_0\}. \tag{32}$$

The operator Δ_{z_0} is invertible in the following sense: *If* $G \in \mathrm{Hol}(\Delta, \overline{\Delta})$ *and* $z_1, w_1 \in \Delta$, *then there exists a unique function* $F \in \mathrm{Hol}(\Delta)$ *such that* $\Delta_{z_1} F(z) = G(z)$ *for all* $z \in \Delta$, *and* $F(z_1) = w_1$. *This function is given by* $F(z) = m_{w_1}(m_{z_1}(z)G(z))$ (see [19] and [17]).

Inequality (30) implies that $|\Delta_{z_0} F(z)| \le 1$ for all $z \in \Delta \setminus \{z_0\}$; hence, the function $\Delta_{z_0} F$ has a removable singularity at the point z_0. Its limit at z_0 is called the hyperbolic derivative of F at z_0 and it is denoted by $F^*(z_0)$. An explicit computation shows that

$$F^*(z_0) := \lim_{z \to z_0} \Delta_{z_0} F(z) = \frac{F'(z_0)\left(1 - |z_0|^2\right)}{1 - |F(z_0)|^2}. \tag{33}$$

Note that from this point of view, the results presented in Sect. 3.3 are just estimates of the hyperbolic derivative.

In 1992 [19], Alan F. Beardon and Thomas K. Carne established the following quantitative characteristic of the strong contractibility.

Theorem 3.21. *Let* $F \in \mathrm{Hol}(\Delta)$. *Then for all* $z, w \in \Delta$,

$$\rho(F(z), F(w)) \le \frac{1}{2} \log\big(\cosh(2\rho(z, w)) + |F^*(w)| \sinh(2\rho(z, w))\big).$$

Since $|F^*(w)| \le 1$, the right-hand side of the above inequality is at most $\rho(z, w)$ and we get an improvement of inequality (31). Moreover, if F is not an isometry, then $|F^*(w)| < 1$, and Theorem 3.21 shows the influence of values of F^* on the contracting effect of F in the hyperbolic distance ρ.

We note in passing that Theorem 3.21 is well suited for proving that if F (not an automorphism of Δ) has a fixed point $\tau \in \Delta$, then the iterates of F converge to τ uniformly on each closed hyperbolic disk centered at τ and, hence, locally uniformly on Δ.

Recall that the Schwarz–Pick Lemma implies that a self-mapping F having an interior fixed point which coincides with an automorphism up to the first order at this point should coincide with this automorphism identically. Thus, it is natural to ask how close is a self-mapping to the automorphism if we weaken the above requirement preserving arguments of the derivatives only. The next result of Beardon and Minda [21] answers this question.

Theorem 3.22. *Let* $F \in \mathrm{Hol}(\Delta)$ *satisfy* $F(0) = 0$ *and* $F'(0) = \alpha \in [0, 1]$. *Then for all* $z \in \Delta$,

$$\begin{aligned} &\rho(0, z) - \frac{1}{2}\log\big(\cosh(2\rho(0, z)) + \alpha \sinh(2\rho(0, z))\big) \\ &\leq \rho(F(z), z) \\ &\leq \rho(0, z) + \frac{1}{2}\log\big(\cosh(2\rho(0, z)) - \alpha \sinh(2\rho(0, z))\big). \end{aligned}$$

Both inequalities are sharp.

In the same paper, the authors also gave the sharp Euclidean bounds

$$\frac{(1-\alpha)|z|(1-|z|)}{1+\alpha|z|} \leq |F(z) - z| \leq \frac{(1-\alpha)|z|(1+|z|)}{1-\alpha|z|}.$$

The following strengthened version of the Schwarz–Pick inequality is due to Mercer. In his 1999 paper [107], he estimated the hyperbolic distance $\rho(F(z), F(w))$ for a holomorphic self-mapping F of Δ under the assumption that the value of the hyperbolic derivative F^* at a point $\tau \in \Delta$ is given.

Theorem 3.23. *Let* $F \in \mathrm{Hol}(\Delta)$ *and* $\tau \in \Delta$. *For all* $z, w \in \Delta$,

$$\rho\left(F(z), F(w)\right) \leq \rho(z, w) + \frac{1}{2}\log\left[1 - (1 - A)\frac{2|m_z(w)|}{1 + |m_z(w)|^2}\right], \tag{34}$$

where

$$A = \begin{cases} \dfrac{|F^*(\tau)| + |m_\tau(w)|}{1 + |F^*(\tau)||m_\tau(w)|}, & \textit{if} \quad \rho(z, \tau) \leq \rho(z, w), \\ \dfrac{|F^*(\tau)| + |m_\tau(z)|}{1 - |F^*(\tau)||m_\tau(z)|}, & \textit{if} \quad \rho(\tau, w) \leq \rho(z, w), \\ \mu(w, z, \tau), & \textit{otherwise}. \end{cases}$$

with

$$\mu(w,z,\tau)=\frac{(|m_w(z)|-|F^*(\tau)|)(u^2+1)+2u(|F^*(\tau)||\phi_w(z)|-1)}{2u(|m_w(z)|-|F^*(\tau)|)+(|F^*(\tau)||m_w(z)|-1)(u^2+1)}$$

and

$$u:=\max\{|m_\tau(z)|,|m_\tau(w)|\}.$$

Since $0<A<1$, Theorem 3.23 strengthens the Schwarz–Pick inequality (31).

Remark 3.1. Setting $\tau=w$ in the first case, or $\tau=z$ in the second, one can obtain by some manipulations in the right-hand side of (34) the conclusion of Theorem 3.21 as a corollary (see also [19]). This result is also implied by a (sharper) Rogosinski's Lemma 2.9.

3.6 Schwarz–Pick Inequalities for the Hyperbolic Derivative

As mentioned above, the hyperbolic derivative F^* of a holomorphic self-mapping F of Δ satisfies $|F^*(z)|<1$ for all $z\in\Delta$ (unless F is an automorphism of Δ). Thus the function F^* itself is a self-mapping of the disk. However, it is not analytic and, consequently, does not necessarily satisfy the Schwarz–Pick inequality. For example, for the function $F(z)=z^2$, $F^*(z)=\dfrac{2z}{1+|z|^2}$ and $F^*(0)=0$. A simple calculation shows that for all $z\in\Delta,\ z\neq 0$,

$$\rho(F^*(0),F^*(z))=2\rho(0,z)>\rho(0,z).$$

However, since $F^*(\Delta)\subset\Delta$, we can measure the hyperbolic distance between two values of the hyperbolic derivatives. It turns out that analogs of the Schwarz–Pick inequalities for the hyperbolic derivatives do exist. The first such result was established by Beardon in 1997 [18].

Theorem 3.24. *If $F\in\mathrm{Hol}(\Delta)$ is not an automorphism and $F(0)=0$, then*

$$\rho(F^*(0),F^*(z))\le 2\rho(0,z). \tag{35}$$

Equality holds for each z for the function $F(z)=z^2$.

In 2002, Luis Bernal-González and María C. Calderón-Moreno (see [28]) proved that inequality (35) holds for hyperbolic derivatives of higher order. In addition, they established that

(a) *equality in (35) holds for some $z\in\Delta, z\neq 0$, if and only if it holds for all points on a diameter of Δ, and if and only if $\frac{F(z)}{z}$ is an automorphism of the disk;*

(b) *equality in (35) holds for all* $z \in \Delta$ *if and only if it holds for two nonzero points lying in two distinct diameters of* Δ*, and if and only if* $F(z) = e^{i\theta} z^2$.

It seems natural to relax the condition $F(0) = 0$ in order to furnish a more complete analog of the Schwarz–Pick Lemma for F^*. Unfortunately, in general, the inequality $\rho(F^*(z), F^*(w)) \leq 2\rho(z, w)$ does not hold. Hakki Turgay Kaptanoğlu [91] obtained the next result as a particular case of his multidimensional extension of Theorem 3.24.

Theorem 3.25. *If* $F \in \mathrm{Hol}(\Delta)$ *with* $F(z_i) = w_i$, $i = 1, 2$, *then*

$$\rho\left(\frac{m_{z_1}(z_2)}{m_{z_2}(z_1)} F^*(z_1), \frac{m_{w_1}(w_2)}{m_{w_2}(w_1)} F^*(z_2)\right) \leq 2\rho(z_1, z_2).$$

Another approach to removing the normalization $F(0) = 0$ was presented by Mercer in 2006.

Theorem 3.26 ([109]). *If* $F \in \mathrm{Hol}(\Delta)$ *(not an automorphism of* Δ*), then for each pair of distinct points* $z_1, z_2 \in \Delta$,

$$|\rho(F^*(z_1), F^*(z_2)) - \rho(\sigma_1, \sigma_2)| \leq 2\rho(z_1, z_2),$$

where $\sigma_1 = m_{w_2}(w_1) m_{z_2}^{-1}(z_1)$ *and* $\sigma_2 = m_{w_1}(w_2) m_{z_1}^{-1}(z_2)$ *with* $w_1 = F(z_1)$ *and* $w_2 = F(z_2)$.

Note, that if $F(0) = 0$, then for $z_2 = 0$ we have $\sigma_1 = \sigma_2$, and so inequality (35) is included in this result.

A related fact was obtained by Yamashita [141] as a consequence of his results mentioned in Sect. 3.3. Namely, he proved that *if* F *is not an automorphism, then for all* $z_1, z_2 \in \Delta$,

$$\rho\left(F^*(z_1), 0\right) - \rho\left(0, F^*(z_2)\right) \leq 2\rho(z_1, z_2). \tag{36}$$

If equality in (36) holds for a pair $z_1 \neq z_2$*, then* F *is the product of two automorphisms. In this case, for a fixed* $a \in \Delta$*, the equality*

$$\rho\left(F^*(z), 0\right) - \rho\left(0, F^*(a)\right) = 2\rho(z, a)$$

holds alternatively

(A) *at each* $z \in \Delta$ *if* $F(z) = T(X(z)^2)$*, where* T *and* X *are automorphisms with* $X(a) = 0$*, or*
(B) *at each* z *of the part of the geodesic* $X^{-1}\{-rb/|b| : 0 \leq r < 1\}$ *and at no other point if* $F(z) = T(X(z)S(X(z)))$*, where* S, T, X *are automorphisms with* $S(b) = 0$, $b \neq 0$*, and* $X(a) = 0$.

Later, Mercer [109] proved the following assertion which implies inequality (36) as a consequence.

Theorem 3.27. *If $F \in \mathrm{Hol}(\Delta)$, then for all $z_1, z_2 \in \Delta$,*

$$m_{-a}(b) \leq |F^*(z_2)| \leq -m_{-a}(c), \tag{37}$$

where $a = |m_{z_1}(z_2)|$, $b = m_a(|F^*(z_1)|)$ and $c = -m_{-a}(|F^*(z_1)|)$.

If F is not an automorphism of Δ, then evaluating the increasing function $t \mapsto \frac{1}{2}\log\frac{1+t}{1-t}$ at either side of inequality (37), yields (see [109])

$$\rho(F^*(z_1), 0) - 2\rho(z_1, z_2) \leq \rho(F^*(z_2), 0) \leq \rho(F^*(z_1), 0) + 2\rho(z_1, z_2),$$

which coincides with (36).

As suggested in [144], Beardon's Theorem 3.24 can be strengthened using Mercer's Theorem 3.23.

Theorem 3.28. *Suppose $F \in \mathrm{Hol}(\Delta)$ is not an automorphism and $F(0) = 0$. Then*

$$\rho(F^*(0), F^*(z)) \leq 2\rho(0, z) - 2B_0,$$

where

$$B_0 = -\frac{1}{2}\log\left[1 - (1 - A_0)\frac{2|z|}{(1+|z|)^2}\right],$$

with

$$A_0 = \begin{cases} \dfrac{|F^*(\tau)| + |\tau|}{1 + |F^*(\tau)||\tau|}, & \text{if} \quad \rho(z,\tau) \leq \rho(z,0), \\ \dfrac{|F^*(\tau)| + |m_\tau(z)|}{1 + |F^*(\tau)||m_\tau(z)|}, & \text{if} \quad \rho(\tau,0) \leq \rho(z,0), \\ \mu(z,\tau), & \text{otherwise}, \end{cases}$$

where

$$\mu(z,\tau) = \frac{(|z| - |F^*(\tau)|)(u^2+1) + 2u(|F^*(\tau)||z| - 1)}{2u(|z| - |F^*(\tau)|) + (|F^*(\tau)||z| - 1)(u^2+1)}$$

and

$$u := \max\{|m_\tau(z)|, |\tau|\}.$$

We end this section with an estimate of the hyperbolic derivative of univalent self-mappings of the disk:

$$|F^*(z)| \leq \frac{|F'(0)|(1+|z|)^2}{(1-|z|)^2 + 4|F'(0)z|}.$$

This inequality was obtained in [118].

3.7 Multi-point Schwarz–Pick Lemma

The original Schwarz–Pick Lemma states that a holomorphic self-mapping of the open unit disk is nonexpansive with respect to the hyperbolic metric in Δ, i.e., it gives a relationship between two points $z, w \in \Delta$ and their images $F(z)$ and $F(w)$ under a holomorphic self-mapping F of Δ. This section is devoted to analogous relationships for an arbitrary number of points in terms of the hyperbolic divided differences. What is crucial in these investigations is that, by the Schwarz–Pick Lemma, for each $F \in \mathrm{Hol}(\Delta)$ the function $\Delta_{z_1} F$ either belongs to $\mathrm{Hol}(\Delta)$ for all $z_1 \in \Delta$, or is a unimodular constant. Thus Δ_{z_1} can be interpreted as an operator acting from $\mathrm{Hol}(\Delta)$ to $\mathrm{Hol}(\Delta, \overline{\Delta})$. As a consequence, multi-point versions enable us to improve the estimate of the hyperbolic distance $\rho(F(z), F(w))$ given by the Schwarz–Pick Lemma in the case images of F at some additional points are known.

In [20], Beardon and Minda established the following three-point Schwarz–Pick Lemma.

Theorem 3.29. *Suppose that $F \in \mathrm{Hol}(\Delta)$ is not an automorphism of Δ. For all $z, u, z_1 \in \Delta$,*

$$\rho(\Delta_{z_1} F(z), \Delta_{z_1} F(u)) \le \rho(z, u).$$

Equality holds if and only if F is a Blaschke product of degree two.

As a consequence, they proved that if $F \in \mathrm{Hol}(\Delta),\ F(0) = 0,\ F'(0) \ge 0$, then $\mathrm{Re}\, F'(z) > 0$ for $|z| \le \dfrac{F'(0)}{1 + \sqrt{1 - F'(0)^2}}$, so that F is univalent in this disk.

Other consequences are a generalization of Theorem 3.24.

$$\rho\left(|F^*(z_1)|, |F^*(z_2)|\right) \le 2\rho(z_1, z_2) \tag{38}$$

and the following refinement of Theorem 3.21:

Corollary 3.3. *Suppose that $F \in \mathrm{Hol}(\Delta)$. For all $z, u \in \Delta$ and v on the closed geodesic arc joining z and u,*

$$\rho\,(F(z), F(u)) \le \frac{1}{2} \log\bigl(\cosh\,(2\rho(z, u)) + |F^*(v)| \sinh\,(2\rho(z, u))\bigr).$$

Equality holds if and only if F is a Blaschke product of degree two and the points z and u lie on a hyperbolic geodesic ray emanating from the critical point c of F, and either $v = z$ is between c and u, or $v = u$ is between c and z.

In addition, Beardon and Minda gave "hyperbolic" explanations of Dieudonné's Lemma and Rogosinski's Lemma. It is worth mentioning that the inequality in Theorem 3.29 is equivalent to an inequality obtained earlier by Mercer in terms of the Euclidean distance (see [108, Lemma 3.2]).

For further considerations, we define the hyperbolic difference quotient of higher order following [17] and [50]. For a sequence (finite or infinite) of pairwise distinct points $\{z_j\}_{j=0,1,\dots} \subset \Delta$, we define the hyperbolic divided difference $\Delta^k_{z_1,\dots,z_k}$ by

$$\Delta^k_{z_1,\dots,z_k} F(z) = \left(\Delta_{z_k} \circ \Delta_{z_{k-1}} \circ \dots \Delta_{z_1}\right) F(z).$$

For brevity, we also write $\Delta^k F(z) = \Delta^k_{z_1,\dots,z_k} F(z)$ and $\gamma_k = \Delta^k F(z_k)$.

Line Baribeau, Patrice Rivard, and Elias Wegert [17] generalized Theorem 3.29 as follows (see also [50]).

Theorem 3.30. *Let $F \in \mathrm{Hol}(\Delta)$ and $z_1, \dots, z_k, z, u$ be points in Δ. Then*

$$|\Delta^k_{z_1,\dots,z_k} F(z)| \le 1, \quad \text{for all } z \in \Delta. \tag{39}$$

Equality holds for a point $z \in \Delta$ if and only if F is a Blaschke product of degree k. Moreover, if F is not a Blaschke product of degree $\le k-1$, then

$$\rho\left(\Delta^k_{z_1,\dots,z_k} F(z), \Delta^k_{z_1,\dots,z_k} F(u)\right) \le \rho(z,u), \quad z \in \Delta. \tag{40}$$

Equality holds for a point $z \ne u$ precisely when F is a Blaschke product of degree $(k+1)$.

It is shown in [17] that the Schur algorithm for finding a solution of the classical Nevanlinna–Pick interpolation problem can be reformulated in terms of hyperbolic divided differences. Following the analogy that hyperbolic divided differences operate on Blaschke products in the same way that ordinary divided differences act on polynomials, the authors of that work showed that this reformulation of the Schur algorithm is analogous to the Newton algorithm for polynomial interpolation.

As another application of Theorem 3.30, we present a result from [50] by Kyung Hyun Cho, Seong-A Kim, and Toshiyuki Sugawa, which generalizes many known estimates including Theorem 3.1.

Theorem 3.31. *Let $z, z_1, \dots, z_k$ be given points in Δ. Denote $\tau_j := m_{z_j}(z)$ for $j = 1, \dots, k$.*

(i) Suppose that $F \in \mathrm{Hol}(\Delta)$ is not a Blaschke product of degree at most $k-1$. For $j = 1, \dots k$, let $F_j(z) = \Delta^j_{z_1,\dots,z_j} F(z)$, $\gamma_j = F_j(z_j)$ and $A_j(z) = m_{\gamma_j}(z\tau_j)$. Then $F(z) \in (A_1 \circ \dots \circ A_k)(\overline{\Delta})$. If, in addition, F is not a Blaschke product of degree k, then $F(z) \in (A_1 \circ \dots \circ A_k)(\Delta)$.

(ii) Conversely, suppose that points $\gamma_1, \dots, \gamma_k \in \Delta$ are given. Let F_j and A_j be as above. Then for each $w \in (A_1 \circ \dots \circ A_k)(\overline{\Delta})$ there exists a function $F \in \mathrm{Hol}(\Delta)$ with $F(z) = w$ such that $\gamma_j = F_j(z_j)$ for $j = 1, \dots, k$.

This theorem implies the following simple estimate which generalizes Lindelöf's inequality: *Let $F \in \mathrm{Hol}(\Delta)$ and $z_0 \in \Delta$. Then*

$$\max\left\{\frac{|F(z_0)| - |m_{z_0}(z)|}{1 - |m_{z_0}(z) F(z_0)|}, 0\right\} \le |F(z)| \le \frac{|F(z_0)| + |m_{z_0}(z)|}{1 + |m_{z_0}(z) F(z_0)|}$$

for all $z \in \Delta$. If $z \neq z_0$, equality holds on the right-hand side only if F is an automorphism.

With the aid of Theorem 3.31, the Schwarz–Pick inequality (31) can be refined as follows. We retain the notations of Theorem 3.31 and set

$$T_j(x) = (|\tau_j|x + |\gamma_j|)/(1 + |\gamma_j \tau_j|x), \; R(x) = (1 + |\tau_1|x)/(1 - |\tau_1|x)$$

and $R_j = R \circ T_1 \circ \ldots \circ T_j$ for $j \geq 1$.

Theorem 3.32 (see [50]). *Let $z_1, \ldots, z_k, z, u \in \Delta$ and $F \in \mathrm{Hol}(\Delta)$. Suppose that F is not a Blaschke product of degree at most k. Then*

$$\rho(F(z), F(u)) \leq \frac{1}{2} \log R_n(1) \leq \frac{1}{2} \log R_{n-1}(1) \leq \ldots$$
$$\cdots \leq \frac{1}{2} \log R_1(1) \leq \frac{1}{2} \log R(1) = \rho(z, u).$$

In fact, Theorem 3.30 contains generalizations of some earlier results. Following [50], we consider two particular cases: (1) $|\Delta^3_{z,0,0} F(z)| \leq 1$ under the condition $F(0) = 0$ and (2) $|\Delta^2_{z,z_0} F(z)| \leq 1$. Both cases can be described explicitly as the following refinements of Dieudonné's Lemma.

Theorem 3.33 (Generalized Dieudonné's Lemma). *Let $F \in \mathrm{Hol}(\Delta)$ and $z_0 \in \Delta$. Set $c = F'(0)$ and $w_0 = F(z_0)$.*

(i) If $F(0) = 0$, then

$$\left| F'(z)(1 - |c|^2) - \frac{2F(z)}{z} + \overline{c} \cdot \left(\frac{F(z)}{z} \right)^2 + c \right|$$
$$\leq \frac{1}{1 - |z|^2} \left(|z - \overline{c} F(z)|^2 - \left| \frac{F(z)}{z} - c \right|^2 \right)$$

for all $z \in \Delta$. In particular, if $c (= F'(0)) = 0$, then

$$\left| F'(z) - \frac{2F(z)}{z} \right| \leq \frac{|z|^4 - |F(z)|^2}{|z|^2 (1 - |z|^2)}.$$

(ii) For any point $z \in \Delta$,

$$\left| F'(z) - \frac{F(z) - w_0}{z - z_0} \cdot \frac{1 - \overline{w_0} F(z)}{1 - |w_0|^2} \cdot \frac{1 - |z_0|^2}{1 - \overline{z_0} z} \right| \tag{41}$$
$$\leq \frac{1}{1 - |z|^2} \left(\frac{|1 - \overline{w_0} F(z)|^2}{1 - |w_0|^2} \cdot \left| \frac{z - z_0}{1 - \overline{z_0} z} \right| - \frac{|F(z) - w_0|^2}{1 - |w_0|^2} \cdot \left| \frac{1 - \overline{z_0} z}{z - z_0} \right| \right),$$

where equality in (41) *holds if and only if* F *is a Blaschke product of degree at most* 2.

Inequality (41) was obtained earlier by Mercer [108, Lemma 3.3]. It reduces to the original Dieudonné's Lemma when $z_0 = F(z_0) = 0$.

To describe further developments, we follow Patrice Rivard [125] and define the higher order hyperbolic derivatives by using the higher order hyperbolic divided differences. Namely, *for* $F \in \mathrm{Hol}(\Delta, \overline{\Delta})$ *and* $n \geq 1$, *the hyperbolic derivative of order* n *of* F *at a point* $z \in \Delta$ *is defined by*

$$H^n F(z) := \Delta^n_{z,\ldots,z} F(z).$$

The next result is the Schwarz–Pick Lemma for higher order hyperbolic derivatives. It generalizes Beardon's Theorem 3.24 as well as inequality (38).

Theorem 3.34. *Let* $F \in \mathrm{Hol}(\Delta, \overline{\Delta})$, $u, v \in \Delta$ *and* $n \geq 1$. *Then*

$$\rho(H^n F(u), H^n F(v)) \leq 2\rho(u, v) + \rho(\Delta^n_{u,\ldots,u} F(v), \Delta^n_{v,\ldots,v} F(u)) \tag{42}$$

and

$$\rho(|H^n F(u)|, |H^n F(v)|) \leq 2\rho(u, v) + \rho(|\Delta^n_{u,\ldots,u} F(v)|, |\Delta^n_{v,\ldots,v} F(u)|). \tag{43}$$

Moreover,

(i) equality in (42) *holds for a pair of distinct points* u *and* v *if and only if* F *is a Blaschke product of degree* $(n + 1)$ *and also if* $u, v, (\Delta^n_{u,\ldots,u})^{-1} \Delta^n_{v,\ldots,v} F(u)$ *and* $(\Delta^n_{u,\ldots,u})^{-1} H^n F(v)$ *lie in this order on the same hyperbolic geodesic, and*
(ii) equality in (43) *holds for a pair of distinct points* u *and* v *if and only if* F *is a Blaschke product of degree* $(n + 1)$ *and if* $H^n F(u)$, $\Delta^n_{u,\ldots,u} F(v)$, $\Delta^n_{v,\ldots,v} F(u)$ *and* $H^n F(v)$ *lie in this order on a hyperbolic geodesic ray emanating zero.*

4 Inequalities Involving Angular Derivatives

4.1 Early Results of Unkelbach and Herzig

It was already mentioned that by the Julia–Wolff–Carathéodory Theorem, if $F \in \mathrm{Hol}(\Delta)$ is not an automorphism and has an interior fixed point and a boundary regular fixed point $\zeta \in \partial\Delta$, then $F'(\zeta) > 1$. But no quantitative bound on how much each hyperbolic disk around fixed point is shrunk follows from the standard Schwarz–Pick inequality.

Perhaps, the first estimates quantifying this shrinkage were obtained in 1938 by Helmut Unkelbach [137] and in 1940 by Alfred Herzig [83]. We formulate their results in a form compatible with the context (see also [33, 95, 114]).

Theorem 4.1. *Let $F \in \mathrm{Hol}(\Delta)$ be not the identity mapping which satisfies $F(0) = F'(0) = \ldots = F^{(n-1)}(0) = 0$ with $\frac{F^{(n)}(0)}{n!} = A$, $0 \le |A| < 1$. Let $z = 1$ be a boundary regular fixed point of F. Then*

(i) the following inequality holds:

$$F'(1) \ge n + \frac{1 + |A|^2 - 2\,\mathrm{Re}\,A}{1 - |A|^2} \ge n + \frac{1 - |A|}{1 + |A|}. \tag{44}$$

Equality in (44) *holds only if*

$$F(z) = z^n \cdot \frac{(1 - A)z + A - |A|^2}{\left(\overline{A} - |A|^2\right) z + 1 - \overline{A}}.$$

(ii) If $\mathrm{Re}\,A \ge 0$*, then*

$$F'(1) \ge n + |\sin \phi| \tag{45}$$

with $\phi = \arg A$*. Equality in* (45) *holds for*

$$F(z) = \begin{cases} e^{i\phi} z^n \cdot \dfrac{z + ia}{za + i}, & \mathrm{Im}\,A \ge 0, \\ e^{i\phi} z^n \cdot \dfrac{za - i}{z - ia}, & \mathrm{Im}\,A \le 0, \end{cases}$$

where $a = \tan\left(\frac{\pi}{4} - \frac{\phi}{2}\right)$.

(iii) If $\mathrm{Re}\,A \le 0$*, then* $F'(1) \ge n + 1$ *with equality for* $F(z) = z^{n+1}$.

Moreover, Unkelbash who investigated the case $n = 1$, also established that the functions given in assertions (ii) and (iii) are the only extremal functions for which equality holds.

More than 60 years later, the inequality $F'(1) \ge \dfrac{2}{1 + |F'(0)|}$ was rediscovered by Robert Osserman in [114] and is now sometimes called Osserman's inequality.

As an application of Theorem 4.1, consider a holomorphic self-mapping of the unit disk extended continuously to an arc $A \subset \partial\Delta$ and that maps A into $\partial\Delta$. This puts us in the settings of the Löwner Theorem 2.13. It was shown by Herzig [83] (see also Robert Osserman [114] and Steven G. Krantz [95]), that inequality (44) allows a quantitative strengthening of Löwner Theorem 2.13. We formulate it as follows.

Corollary 4.1. *Let $F \in \mathrm{Hol}(\Delta)$ satisfy hypotheses of Theorem 4.1. Assume that F extends continuously to an arc $A \subset \partial\Delta$ of length s and maps it onto an arc $F(A) \subset \partial\Delta$ of length σ. Then*

$$\sigma \ge s\left(n + \frac{1 + |A|^2 - 2\,\mathrm{Re}\,A}{1 - |A|^2}\right)$$

with equality only for

$$F(z) = \lambda z^n \cdot \frac{(1-A)z + A - |A|^2}{\left(\overline{A} - |A|^2\right)z + 1 - \overline{A}}, \quad |\lambda| = 1.$$

For the sake of completeness, we present one more result that can be considered a boundary analog of Theorem 4.1. More precisely, whereas Theorem 4.1 deals with the case that F has two fixed points—interior and boundary, the next assertion gives an analogous estimate in the case F has two boundary contact points.

Theorem 4.2. *Let* $F \in \mathrm{Hol}(\Delta)$. *Assume that* $\zeta_1, \zeta_2 \in \partial\Delta$ *with* $\zeta_1 \neq \zeta_2$ *and* $|F(\zeta_1)| = |F(\zeta_2)| = 1$. *If the angular derivatives* $F'(\zeta_1)$ *and* $F'(\zeta_2)$ *exist finitely, then*

$$|F'(\zeta_1)F'(\zeta_2)| \geq \left|\frac{F(\zeta_2) - F(\zeta_1)}{\zeta_2 - \zeta_1}\right|^2.$$

This inequality is trivial when F has two boundary fixed points. In this case, equality holds if and only if F is a hyperbolic automorphism of Δ.

Without loss of generality, we can assume that ζ_2 is a fixed point of F. Suppose that $\zeta_2 = \tau$ is the Denjoy–Wolff point of F (see Sect. 2.5). We then get the following estimate for the angular derivative at a contact point:

$$|F'(\zeta_1)| \geq \left|\frac{F(\zeta_1) - \tau}{\zeta_1 - \tau}\right|^2.$$

Quoting M. Abate [1], we remark that Theorem 4.2 seems to have appeared for the first time (with different proofs) in Lewittes [100] and in Behan [23], but it was probably known earlier. We proceed with related results.

4.2 Cowen–Pommerenke Type Inequalities

The study of boundary regular fixed points and of mutual contact points along with relations between angular derivatives at these points is of certain intrinsic interest. For example, if $\zeta \in \partial\Delta$ is a regular contact point of an $F \in \mathrm{Hol}(\Delta)$, $z_0 \in \Delta$, and the values $F(\zeta) \in \partial\Delta$ and $F(z_0)$ are known, then the Julia–Carathéodory Theorem gives a simple lower bound on the derivative $F'(\zeta)$, namely

$$|F'(\zeta)| \geq \frac{|F(\zeta) - F(z_0)|^2}{1 - |F(z_0)|^2} \cdot \frac{1 - |z_0|^2}{|\zeta - z_0|^2}.$$

On the other hand, if $F(0) = 0$ and $F'(0)$ is given, Unkelbach's Theorem 4.1 asserts that

$$F'(\zeta) \ge 2\frac{1-\operatorname{Re} F'(0)}{1-|F'(0)|^2}.$$

Suppose we are given additional data of F, say, the angular derivatives of F at several contact points (or at several boundary fixed points). It turns out that the problem of obtaining sharp estimates in such a situation is much more delicate. A breakthrough in solving this problem is due to Carl C. Cowen and Christian Pommerenke [56].

For the sake of convenience, in what follows, we assume that functions under consideration are normalized such that the Denjoy–Wolff point τ is 0 or 1.

Theorem 4.3. *Let $F \in \operatorname{Hol}(\Delta)$ and $\tau, \zeta_1, \dots, \zeta_n$ be distinct fixed points of F in $\overline{\Delta}$.*

(i) If $\tau = 0$, then

$$\sum_{j=1}^{n} \frac{1}{F'(\zeta_j)-1} \le \operatorname{Re} \frac{1+F'(0)}{1-F'(0)}.$$

(ii) If $\tau = 1$ and $0 < F'(1) < 1$, then

$$\sum_{j=1}^{n} \frac{1}{F'(\zeta_j)-1} \le \frac{F'(1)}{1-F'(1)}. \tag{46}$$

(iii) If $\tau = 1$, then

$$\sum_{j=1}^{n} \frac{|1-\zeta_j|^2}{F'(\zeta_j)-1} \le 2\operatorname{Re}\left(\frac{1}{F(0)}-1\right).$$

Moreover, equality holds if and only if

(a) in case (i): F satisfies the functional equation

$$\frac{z+F(z)}{z-F(z)} = \sum_{j=1}^{n} \frac{1}{F'(\zeta_j)-1}\,\frac{1+z\overline{\zeta}_j}{1-z\overline{\zeta}_j} + i\gamma \quad \textit{for some} \quad \gamma \in \mathbb{R}, \tag{47}$$

hence F is a finite Blaschke product of order $n+1$;

(b) in cases (ii) and (iii) with $F'(1) < 1$: F has the form

$$F(z) = C^{-1}\left(C(z) + \frac{2}{(1-z)\displaystyle\sum_{j=1}^{n} \frac{1}{F'(\zeta_j)-1}\,\frac{1-\zeta_j}{z-\zeta_j}}\right), \tag{48}$$

where $C(z) = \dfrac{1+z}{1-z}$ is the Cayley transform of Δ;

(c) in cases (ii) and (iii): F is a finite Blaschke product of order n.

Note that a complete list of the fixed points of F need not be given, but the equality condition fails if we replace n with infinity. Assertion (i) generalizes inequality (44) to the case F has more than one boundary fixed points.

An essential part of this theorem was proved in [56] by Cowen and Pommerenke. Later, in [71], Mark Elin, David Shoikhet, and Nikolai Tarkhanov proved that the inequality in assertion (iii) holds for hyperbolic type mappings and found the explicit form (48) of extremal functions. Finally, in [36], Vladimir Bolotnikov, Elin and Shoikhet established that in case (i), extremal functions must satisfy functional equation (47).

Corollary 4.2 ([71]). *Suppose $F \in \mathrm{Hol}(\Delta)$ satisfies $F(1) = 1$ and $0 < F'(1) < 1$. Let $\zeta \neq 1$ be a boundary regular fixed point of F. Then F is an automorphism of Δ if and only if*

$$\frac{1 - \mathrm{Re}\,\zeta}{F'(\zeta) - 1} = \mathrm{Re}\left(\frac{1}{F(0)} - 1\right).$$

Theorem 4.3 was used in [56] to obtain an inequality relating the angular derivatives at mutual contact points of F. A geometrical approach to the study of mutual contact points was applied in [52] by Manuel D. Contreras, Santiago Díaz-Madrigal, and Christian Pommerenke. We combine their results as follows.

Theorem 4.4. *Let $F \in \mathrm{Hol}(\Delta)$ and $\{\zeta_1, \dots, \zeta_n\} \subset \partial\Delta$ be a collection of mutual regular contact points of F such that $F(\zeta_j) = w \in \partial\Delta$ for all $j = 1, 2, \dots, n$. Then*

$$\sum_{j=1}^{n} \frac{1}{|F'(\zeta_j)|} \le \mathrm{Re}\,\frac{w + F(0)}{w - F(0)} = \frac{1 - |F(0)|^2}{|w - F(0)|^2},$$

and equality holds if and only if

$$F(z) = wC^{-1}\left(\sum_{j=1}^{n} \frac{1}{|F'(\zeta_j)|}\,\frac{1 + z\overline{\zeta}_j}{1 - z\overline{\zeta}_j} + i\gamma\right) \quad \text{for some} \quad \gamma \in \mathbb{R}, \tag{49}$$

hence F is a Blaschke product of order n.

If, in addition, $\zeta \neq w$ is a boundary regular fixed point of F, then

$$\sum_{j=1}^{n} \frac{1}{|F'(\zeta_j)|} \left|\frac{\zeta - w}{\zeta - \zeta_j}\right|^2 \le F'(\zeta), \tag{50}$$

and equality in either inequality holds if and only if

$$F(z) = \frac{\zeta + (\zeta - w)A(z)}{1 - (1 - \zeta\overline{w})A(z)} \quad \text{with} \quad A(z) = \sum_{j=1}^{n} \frac{1}{|F'(\zeta_j)|}\,\frac{\zeta_j}{\zeta_j - \zeta}\,\frac{z - \zeta}{z - \zeta_j}.$$

The extremal functions for the first inequality are described in [36]. The second part of this theorem is an enhancement of Theorem 4.2 for the case that F has many contact points and gives the explicit form of extremal functions, namely, equality in Theorem 4.2 holds if and only if F is an automorphism of the unit disk Δ.

It is worth noticing that the first inequality of this theorem in the particular case $n = 1$ follows immediately from Yûsaku Komatu's 1961 result (see Theorem 4.17 below).

Osserman [114] considered the case that only one boundary fixed point of F is given and obtained a sharp estimate based on the values of the function and its hyperbolic derivative at zero.

Theorem 4.5. *Let $F \in \mathrm{Hol}(\Delta)$ and ζ be a boundary fixed point of F. Then*

$$|F'(\zeta)| \geq \frac{2}{1 + |F^*(0)|} \frac{1 - |F(0)|}{1 + |F(0)|} = \frac{2(1 - |F(0)|)^2}{1 - |F(0)|^2 + |F'(0)|}.$$

This theorem enhances assertion (i) of Theorem 4.1 and, for the case F has an only contact point, the first assertion of Theorem 4.4. Also, Osserman showed that if $F \in \mathrm{Hol}(\Delta)$ admits the power series expansion $F(z) = c_p z^p + c_{p+1} z^{p+1} + \ldots$, $c_p \neq 0$, then

$$|F(z)| \leq |z|^p \frac{|z| + |c_p|}{1 + |c_p z|}$$

(cf., Mercer's Theorem 3.2) and if, in addition, F has a regular contact point $\zeta \in \partial\Delta$, then

$$|F'(\zeta)| \geq p + \frac{1 - |c_p|}{1 + |c_p|}.$$

The last inequality is a particular case of a result due to Dubinin in [62], who strengthened the inequality $|F'(\zeta)| > 1$ by involving zeros of the function F. His result can be partially formulated as follows.

Theorem 4.6. *Suppose $F \in \mathrm{Hol}(\Delta)$ admits a power series expansion $F(z) = c_p z^p + c_{p+1} z^{p+1} + \ldots$, $c_p \neq 0$. Assume that F has a regular contact point $\zeta \in \partial\Delta$. Let $\{a_k\}_{k \in K}$ be a set of zeros of F in Δ that are different from $z = 0$, and p_k be the multiplicity of the zero a_k. Then*

$$|F'(\zeta)| \geq p + \sum_{k \in K} \text{ï£¡} n_k \frac{1 - |a_k|^2}{|\zeta - a_k|^2} + \frac{\prod_{k \in K} |a_k|^{n_k} - |c_p|}{\prod_{k \in K} |a_k|^{n_k} + |c_p|},$$

where n_k is a positive integer such that $n_k \leq p_k$, $k \in K$. If $\{a_k\}_{k \in K}$ is the set of all zeros of F in Δ that are different from zero, then

$$|F'(\zeta)| \geq p + \sum_{k\in K} \text{ï£¡} p_k \frac{1-|a_k|^2}{|\zeta - a_k|^2} - \frac{1}{2}\log\left|\frac{c_p}{\prod\limits_{k\in K} a_k^{p_k}}\right|.$$

In both formulas equality occurs for the Blaschke function

$$B(z) := z^p \prod_{k\in K}\left[\frac{|a_k|}{a_k}\frac{a_k - z}{1-\overline{a_k}z}\right]^{p_k},$$

where $\{a_k\}$ *is the set of points for which the product converges.*

In the same paper, Dubinin also proved that if 0 is the only zero of F in Δ, then the lower estimate of $F'(\zeta)$ can be strengthened to

$$|F'(\zeta)| \geq p - \frac{2|c_p|(\log|c_p|)^2}{2|c_p|\log|c_p| - |c_{p+1}|},$$

where $|c_{p+1}| \leq 2\left|c_p \log|c_p|\right|$. Here equality occurs for the function

$$F(z) = z^p \exp\left(\frac{1+z}{1-z}\log c_p\right).$$

As we shall see in the next sections, more can be said when F is univalent.

4.3 Estimates for Angular Derivatives of Univalent Functions

In this section, we consider strengthened inequalities for angular derivatives of univalent functions at their boundary contact (in particular, fixed) points.

We begin with the case that the Denjoy–Wolff point τ lies in Δ. The first theorem is due to Cowen and Pommerenke [56].

Theorem 4.7. *Let* $F \in \mathrm{Hol}(\Delta)$ *be univalent and satisfy* $F(0) = 0$. *If* $\zeta_1, \ldots, \zeta_n$ *are distinct fixed points of* F *on* $\partial\Delta$, *then*

$$\sum_{j=1}^{n} \frac{1}{\log F'(\zeta_j)} \leq 2\,\mathrm{Re}\left(B^{-1}\right),$$

where $B = \lim\limits_{r\to 1^-} \log \dfrac{F(r\zeta_1)}{F'(0)r\zeta_1}$. *Moreover, equality holds if and only if* $F(z) = \sigma^{-1}\left(F'(0)\sigma(z)\right)$, *where* $\sigma(z) = z\prod\limits_{j=1}^{n}\left(1-\overline{\zeta_j}z\right)^{-2\gamma_j B}$ *with* $\gamma_j = \dfrac{1}{\log F'(\zeta_j)}$.

In [13], James Milne Anderson and Alexander Vasil'ev considered the same situation and established the following weighted estimate.

Theorem 4.8. *Let* $F \in \mathrm{Hol}(\Delta)$, $F(0) = 0$, *be a univalent function which is conformal at its boundary fixed points* $\zeta_1 = e^{i\theta_1}, \dots, \zeta_n = e^{i\theta_n}$, $\theta_1 < \theta_2 < \dots < \theta_n$. *Then for every nonnegative vector* $t = (t_1, \dots, t_n)$ *such that* $\sum_{j=1}^{n} t_j = 1$,

$$|F'(0)| \cdot \prod_{j=1}^{n} \left(F'(\zeta_j)\right)^{2t_j^2} \geq 1. \tag{51}$$

Equality is attained only for the function $w = \hat{F}(z)$, *which satisfies the complex differential equation*

$$\frac{dw}{dz} = \frac{w \prod_{j=1}^{n}(z - e^{i\delta_j}) \prod_{k=1}^{n}(w - \zeta_k)}{z \prod_{j=1}^{n}(w - e^{i\delta_j}) \prod_{k=1}^{n}(\zeta - \zeta_k)},$$

where the numbers δ_k *are chosen such that* $\delta_k \in [\theta_k, \theta_{k+1})$, $\sum_{k=1}^{n}(\delta_k - \theta_k) = \pi$ *and satisfy the system* $2t_j = \dfrac{\prod_{k=1}^{n}(e^{i\theta_j} - e^{i\delta_k})}{e^{i\theta_j} \prod_{k \neq j}^{n}(e^{i\theta_j} - e^{i\theta_k})}$, $j = 1, \dots, n$.

Following [13], we remark that the image $\hat{F}(\Delta)$ is $\Delta \setminus S$, where S consists of at most n analytic arcs terminating at the points $e^{i\delta_k}$.

One can show that the minimum of the left-hand side of inequality (51) with respect to t is attained for

$$t_j = \left(\log F'(\zeta_j) \sum_{k=1}^{n} \frac{1}{\log F'(\zeta_k)}\right)^{-1}. \tag{52}$$

This leads to the following sharp inequality which resembles Theorem 4.7.

Corollary 4.3. *Under the assumptions of Theorem 4.8,*

$$\sum_{j=1}^{n} \frac{1}{\log F'(\zeta_j)} \leq \frac{-2}{\log |F'(0)|}.$$

Equality is attained only for the function $\hat{F}$ which satisfies the differential equation in Theorem 4.8 with t chosen as in (52).

In the case that a univalent function has contact (not necessarily fixed) points, the situation becomes more complicated. We present here a result by Pommerenke and Vasil'ev [119].

Proposition 4.1. *Let $F \in \mathrm{Hol}(\Delta)$ be univalent, $F(0) = 0$, such that at two points $\zeta_1, \zeta_2 \in \partial\Delta$ the angular limits satisfy $|F(\zeta_1)| = |F(\zeta_2)| = 1$. Suppose also that the angular derivatives $F'(\zeta_1)$ and $F'(\zeta_2)$ are finite. Then*

$$\left|\frac{F(\zeta_1) - F(\zeta_2)}{\zeta_1 - \zeta_2}\right|^{2t_1t_2} \geq \frac{1}{\sqrt{|F'(0)|}|F'(\zeta_1)|^{t_1^2}|F'(\zeta_2)|^{t_2^2}} \tag{53}$$

and this inequality is sharp for all $t_1, t_2 \in (0, \infty)$.

The extremal functions map Δ onto Δ with one or two analytic slits. In the case $t_1 + t_2 = 1$, the extremal map can be described explicitly (cf., Theorem 4.2).

We proceed with functions that have boundary Denjoy–Wolff points.

Theorem 4.9 ([56]). *Let $F \in \mathrm{Hol}(\Delta)$ be univalent and suppose that $\tau \in \partial\Delta$ is a boundary fixed point of F and $F'(\tau) < 1$. If $\zeta_1, \dots, \zeta_n$ are distinct fixed points of F on $\partial\Delta$, different from τ, then*

$$\sum_{j=1}^{n} \left(\log F'(\zeta_j)\right)^{-1} \leq -\left(\log F'(\tau)\right)^{-1}.$$

Moreover, equality holds if and only if

$$F(z) = s^{-1}\left(\sigma^{-1}\left(\sigma\left(s(z)\right) + 1\right)\right),$$

where

$$s(z) = \frac{\tau + z}{\tau - z} \quad \text{and} \quad \sigma(z) = \sum_{j=1}^{n} \frac{\log(s - s(\zeta_j))}{\log F'(\zeta_j)}.$$

In 2007, Contreras, Díaz-Madrigal, and Vasil'ev [53] provided a geometric proof of the last estimate which improves it in a certain sense. More precisely, they proved the following "weighted" inequality.

Theorem 4.10. *Let $F \in \mathrm{Hol}(\Delta)$ be univalent with an attractive Denjoy–Wolff point $\tau \in \partial\Delta$, and let $\zeta_1, \dots, \zeta_n$ be distinct boundary fixed points of F different from τ. Likewise, consider n nonnegative numbers $t_1, \dots, t_n$, such that $\sum_{k=1}^{n} t_k = 1$. Then*

$$F'(\tau) \cdot \prod_{k=1}^{n} F'(\zeta_k)^{t_k^2} \geq 1. \tag{54}$$

Moreover, this inequality is sharp.

This theorem is a "boundary" analog of Theorem 4.8.

It is natural to try to explain the exact relationship between Theorems 4.9 and 4.10. By a simple calculation, one can minimize the left-hand side of inequality (54). This recovers Theorem 4.9. At the same time, amongst all of the weighted inequalities, one is best, and this one is the original inequality of Cowen and Pommerenke. Nevertheless, all the weighted inequalities are sharp.

The next theorem treats the case $F'(\tau) = 1$. To formulate it, we need to define some parameters. If $F \in \mathrm{Hol}(\Delta)$ is univalent on Δ, the function $\log \dfrac{1-|F(0)|^2}{F'(0)z} \dfrac{F(z)-F(0)}{1-\overline{F(0)}F(z)} = \log(1+\ldots)$ is single valued on Δ. Let $\ell(z)$ be the branch of this function such that $\ell(0) = 0$. Suppose the Denjoy–Wolff point of F is $\tau = 1$ and $\zeta_1, \ldots, \zeta_n$ are fixed points of F on $\partial\Delta$. Let $b_j = \lim_{r\to 1^-} \mathrm{Im}\left(\ell(r\zeta_j) - \ell(r)\right)$. Thus, b_j is a value of $\arg \dfrac{(1-F(0)\overline{\zeta_j})(1-\overline{F(0)})}{(1-\overline{F(0)}\zeta_j)(1-F(0))}$.

Theorem 4.11 ([56]). *Let $F \in \mathrm{Hol}(\Delta)$ be univalent and suppose that 1 is the Denjoy–Wolff point of F and that $F'(1) = 1$. If $\zeta_1, \ldots, \zeta_n$ are distinct regular fixed points of F on $\partial\Delta$ different from 1, and b_j are defined as above, then*

$$\sum_{j=1}^{n} \frac{b_j^2}{\log F'(\zeta_j)} \leq 2 \log \frac{1-|F(0)|^2}{|F'(0)|}.$$

In 1990, Kin Y. Li [101] reproved the Cowen–Pommerenke results using the theory of contractively contained spaces and improved Theorem 4.11 as follows.

Theorem 4.12. *Suppose $F \in \mathrm{Hol}(\Delta)$ is univalent, 1 is the Denjoy–Wolff point of F, and $F'(1) = 1$. If $\zeta_1, \ldots, \zeta_n$ are distinct fixed points of F on $\partial\Delta$ different from 1, then*

$$\sum_{j=1}^{n} \frac{1}{\log F'(\zeta_j)} \left| \log \frac{1-F(0)}{1-F(0)\overline{\zeta_j}} \right|^2 \leq \log \frac{1-|F(0)|^2}{|1-F(0)|^2}.$$

Equality holds only if

$$\frac{F(z)-F(0)}{z} = (1-F(0)) \left(\frac{1-F(z)}{1-z}\right)^{\gamma} \cdot \prod_{j=1}^{n} \left(\frac{1-z\overline{\zeta_j}}{1-F(z)\overline{\zeta_j}}\right)^{\gamma_j},$$

where

$$\gamma_j = \frac{1}{\log F'(\zeta_j)} \log \frac{1 - F(0)}{1 - F(0)\overline{\zeta_j}} \quad \textit{and} \quad \gamma = 1 + \sum_{j=1}^{n} \gamma_j .$$

In the rest of this section, we turn to holomorphic self-mappings of the unit disk extended continuously to an arc $A \subset \partial\Delta$ that map A into $\partial\Delta$. This situation leads to the following question. What can be sad about the angular derivatives of such a mapping?

Pommerenke and Vasil'ev [119] considered the case that a univalent function not only maps a circle arc into $\partial\Delta$ but also has finite nonzero angular derivatives at its endpoints. To formulate their results we denote by p_c the classical conformal Pick map

$$p_c(z) = \frac{4cz}{1 - z + \sqrt{(1-z)^2 + 4cz}}.$$

Proposition 4.2. *Let $F \in \mathrm{Hol}(\Delta)$, $F(0) = 0$, be a univalent function which maps an arc $A \in \partial\Delta$ of length $s \le \pi$ onto the arc $F(A) \subsetneq \partial\Delta$ of length $\sigma < 2\pi$. Suppose F has finite nonzero angular derivatives $F'(\zeta_1)$, $F'(\zeta_2)$ at the endpoints of A. Then*

$$|F'(\zeta_1)F'(\zeta_2)| \ge \tan^2\frac{\sigma}{4}\cot^2\frac{s}{4} \tag{55}$$

and equality holds for the identity mapping as well as for the function $\xi p_c(\bar{\xi}z)$, where ξ is the midpoint of A and $c = \sin^2\frac{\sigma}{4}\sin^{-2}\frac{s}{4}$.

Löwner's Theorem 2.13 implies that inequality (55) is sharper than one in Theorem 4.2.

In applications of the Schwarz Lemma, one often requires a lower estimate of the derivative of a univalent function at its interior fixed point. It turns out that such an estimate can be obtained in the setting of the Löwner Theorem.

Proposition 4.3 ([119]). *Let $F \in \mathrm{Hol}(\Delta)$, $F(0) = 0$, be a univalent function that maps an arc $A \in \partial\Delta$ of length $s < 2\pi$ onto the arc $F(A) \subsetneq \partial\Delta$ of length σ. Then*

$$|F'(0)| \ge \frac{1 - \cos\frac{s}{2}}{1 - \cos\frac{\sigma}{2}}$$

and equality holds for the function $\xi p_c(\bar{\xi}z)$, where ξ is the midpoint of A.

Another question that often arises is whether there exists an upper estimate of the angular derivative at a contact point. Without knowledge of additional restrictions on the behavior of the function in the vicinity of the point, no such estimate can be found. In [62], Dubinin supplemented the Löwner Theorem 2.13 with the following fact.

Theorem 4.13. *Let $F \in \mathrm{Hol}(\Delta)$ be univalent, $F(0) = 0$, and suppose that $F'(0) = c$. Assume that there exists an arc $A \subset \partial\Delta$ of length s such that F maps A into $\partial\Delta$. Then for the midpoint ξ of A,*

$$|F'(\xi)| \leq |c|^{-1/(2\sin^2(s/4))}.$$

Equality occurs for $F(z) = cz$ with $|c| = 1$.

Taking into account inequality (51), we obtain, in the settings of the last theorem,

$$\frac{1}{\sqrt{c}} \leq F'(\xi) \leq \frac{1}{\sqrt{c^{1/\sin^2(s/4)}}}.$$

4.4 Improvements of Julia's Lemma

Julia's Lemma provides information concerning the location of the value $F(z)$ at a point $z \in \Delta$ under the assumption that for some boundary point ζ,

$$\liminf_{z\to\zeta} \frac{1-|F(z)|}{1-|z|} =: \alpha \tag{56}$$

exists finitely. In this section we formulate several results which give estimates sharper than Julia's Lemma.

To begin, we recall that Unkelbach's Theorem 4.1 deals with functions for which $z = 0$ is a fixed point. Weakening this condition creates a more complicated situation. More than 60 years after Unkelbach's paper, Mercer, in [108], proved a sharpened version of Julia's Lemma where the value of a function at an interior point is given.

To formulate his result, we set $\gamma_z := \dfrac{1-|z|^2}{|1-z|^2}$ for $z \in \Delta$.

Theorem 4.14. *Let $F \in \mathrm{Hol}(\Delta)$ and $z_0 \in \Delta$. Suppose that $\zeta = 1$ is a boundary fixed point of F at which the lower limit (56) exists finitely. Denote $w_0 = F(z_0)$. Then for each $z \in \Delta$ we have $F(z) \in \overline{B(c,r)}$, where*

$$c = m_{w_0}(\sigma m_{z_0}(z)) + \frac{m_{z_0}(z)}{\overline{\sigma}} \frac{\hat{\alpha}}{\hat{\alpha}\gamma_\mu + \gamma_z} \frac{1-|w_0|^2}{(1-\mu)^2},$$

$$r = |m_{z_0}(z)| \frac{\hat{\alpha}}{\hat{\alpha}\gamma_\mu + \gamma_z} \frac{1-|w_0|^2}{|1-\mu|^2},$$

with $\mu = \sigma\overline{w_0}m_{z_0}(z)$, $\sigma = \dfrac{(1-w_0)(1-\overline{z_0})}{(1-\overline{w_0})(1-z_0)}$, $\hat{\alpha} = \alpha\gamma_{w_0} - \gamma_{z_0}$.

Julia's Lemma implies that $\hat{\alpha} \geq 0$, and so $\alpha \geq \dfrac{|1 - F(0)|^2}{1 - |F(0)|^2}$. This improves the estimate $\alpha \geq \dfrac{1 - |F(0)|}{1 + |F(0)|}$ contained in Julia's Lemma.

Remark 4.1. In the particular case that F fixes zero, the substitution $z_0 = 0$ (thus, $\hat{\alpha} = \alpha - 1$, $\gamma_\mu = \sigma = 1$) in Theorem 4.14 leads to a boundary Schwarz Lemma with interior fixed point. In this case, $F(z) \in \overline{B(c,r)}$ with $c = \dfrac{z\gamma_z}{(\alpha - 1) + \gamma_z}$ and $r = \dfrac{|z|(\alpha - 1)}{(\alpha - 1) + \gamma_z}$, whereas by the Schwarz Lemma we only get $F(z) \in \overline{B(0, |z|)}$, while using Julia's Lemma we can show that $F(z) \in \overline{B\left(\dfrac{\gamma_z}{\alpha + \gamma_z}, \dfrac{\alpha}{\alpha + \gamma_z}\right)}$. Since $B(c,r) \subset B(0, |z|) \bigcap B\left(\dfrac{\gamma_z}{\alpha + \gamma_z}, \dfrac{\alpha}{\alpha + \gamma_z}\right)$, we conclude that Theorem 4.14 is sharper than that of Julia's Lemma. In addition, dividing the inequality $|F(z) - c| \leq r$ by z and then passing to the limit as $z \to 0$ leads to Unkelbach's inequality (44).

In [107], Mercer provided another improvement of the Julia Lemma. His result, based on the given value of the hyperbolic derivative $F^*(w)$ at some additional point w located "not farther" from the contact point than z, gives a sharper information about the variability region of $F(z)$.

Theorem 4.15. *Let $F \in \mathrm{Hol}(\Delta)$ satisfy* (56) *at some point $\zeta \in \partial\Delta$. Then there exists $\tau \in \partial\Delta$ such that for all $z \in \Delta$ and for each $w \in \overline{D(\zeta, \frac{1}{\gamma_z})}$,*

$$F(z) \in \overline{D\left(\tau, \frac{\alpha(1 + A)}{2\gamma_z}\right)}, \tag{57}$$

where $A = \dfrac{|F^*(w)| + |m_w(z)|}{1 + |F^*(w)||m_w(z)|}$ and $D(\zeta, \frac{1}{\gamma_z})$ is defined by (13).

One can see from the rigidity part of the Schwarz–Pick Lemma that if F is not an automorphism, then $A < 1$; thus, Theorem 4.15 improves Julia's Lemma (see Theorem 2.4). If, however, F is an automorphism, then $F(D(\zeta, k)) = D(\tau, \alpha k)$ for all $k > 0$.

If F has no fixed point in Δ, then by the Wolff Theorem 2.5, F has a boundary fixed point $\zeta \in \partial\Delta$. Consequently, and then by the Julia–Carathéodory Theorem 2.6, $F(D(\tau, k)) \subseteq D(\tau, \alpha k)$ for all $k > 0$. The Denjoy–Wolff Theorem asserts that the iterates of F converge locally uniformly to ζ. Although in the case $\alpha = 1$ this fact does not follow from Julia's Lemma, Theorem 4.15 makes it plausible (unless F is an automorphism).

Letting z_0 in (41) go out to the boundary along a suitable sequence, Mercer obtained the following Julia–Dieudonné type result.

Theorem 4.16. *Let $F \in \mathrm{Hol}(\Delta)$. Suppose that $\zeta = 1$ is a boundary fixed point of F at which the lower limit* (56) *exists finitely. Then*

$$\left| F'(z) - \frac{1}{\alpha}\left(\frac{1-F(z)}{1-z}\right)^2 \right| \leq \frac{1-|F(z)|^2}{1-|z|^2} - \frac{1}{\alpha}\left|\frac{1-F(z)}{1-z}\right|^2$$

for every $z \in \Delta$.

This last inequality improves the estimate of Julia's Lemma (Theorem 2.4), which only provides that its right-hand side is nonnegative. In the same sense, a similar estimate, established by Yûsaku Komatu in 1961, improves the first inequality of Theorem 4.4 for $n = 1$.

Theorem 4.17 (See [94]). *Let $F \in \mathrm{Hol}(\Delta)$ satisfy* (56) *at $\zeta = 1$. Then for all $z \in \Delta$,*

$$(1-|z|)^2 \left| \frac{F'(z)}{(1-F(z))^2} - \frac{1}{\alpha}\frac{1}{(1-z)^2} \right| \leq \frac{1-|F(0)|^2}{|1-F(0)|^2} - \frac{1}{\alpha}.$$

Equality at a point $z_0 \in \Delta \setminus \{0\}$ holds if and only if F is a rational function satisfying

$$\frac{1}{1-F(z)} = \frac{1}{\alpha}\frac{1}{1-z} + \left(\frac{1-|F(0)|^2}{|1-F(0)|^2} - \frac{1}{\alpha}\right)\frac{z_0}{z_0 - z|z_0|} - \frac{\overline{F(0)}}{1-\overline{F(0)}}.$$

In [20], Beardon and Minda used their Theorem 3.29 to prove a three-point Julia's Lemma. In its formulation it is assumed without loss of generality that $F \in \mathrm{Hol}(\Delta)$ has the boundary fixed point $\zeta = 1$.

Theorem 4.18. *Suppose that $F \in \mathrm{Hol}(\Delta)$ is not an automorphism of Δ and has angular derivative α at the point 1. Let $z, u, v \in \Delta$, $z \in \partial D(1, R)$, and $u \in \overline{D(1, R)}$. Then $F(z) \in \overline{D(1, \alpha R\mu)}$, where*

$$2\mu = 1 + \frac{|m_v(z)| + |\Delta_v F(u)|}{1 + |m_v(z)||\Delta_v F(u)|} < 2.$$

The case $u = v$ is due to Mercer [106].

In Theorems 4.14–4.18, Julia's type estimates were obtained under the assumption that the values of F at some interior points are given. In the case we have more data or data for boundary points, the situation seems to be more difficult. Nevertheless, the cases that such points are either mutual contact points or boundary regular fixed points were treated in [36].

Given angular derivative at contact points $\{\zeta_j\}$, we have, by the Julia–Wolff–Carathéodory Theorem, that for each j,

$$\frac{1-|F(z)|^2}{|w-F(z)|^2} \geq \frac{1}{|F'(\zeta_j)|}\frac{1-|z|^2}{|z-\zeta_j|^2}.$$

However, it is not clear how to combine these inequalities. The following result is an extension of the Julia–Wolff–Carathéodory theorem for the case of several mutual contact points (cf., [136]).

Theorem 4.19. *Let $F \in \mathrm{Hol}(\Delta)$ be such that $F(\zeta_j) = w \in \partial\Delta$ for a finite or countable collection $\{\zeta_j\} \subset \partial\Delta$. Then*

$$\frac{1-|F(z)|^2}{|w-F(z)|^2} \geq \sum_j \frac{1}{|F'(\zeta_j)|} \frac{1-|z|^2}{|z-\zeta_j|^2} \tag{58}$$

for all $z \in \Delta$. Moreover, equality in (58) *holds at some point $z \in \Delta$ if and only if F has the form* (49)*:*

$$F(z) = wC^{-1}\Big(\sum_j \frac{1}{|F'(\zeta_j)|} \frac{1+z\overline{\zeta}_j}{1-z\overline{\zeta}_j} + i\gamma\Big) \quad \textit{for some} \quad \gamma \in \mathbb{R}.$$

Choosing $z = 0$ in (58) gives an extension of the Cowen–Pommerenke inequalities for mutual contact points (see Theorem 4.4).

On the other hand, if a boundary point, say 1, is the Denjoy–Wolff point of F, then

$$\frac{|1-z|^2}{1-|z|^2} - \frac{|1-F(z)|^2}{1-|F(z)|^2} \geq 0.$$

Thus it is natural to look for sharper lower estimates of the difference in the left-hand side. One of such refinements was given in Theorem 4.16 above. Another approach is presented in the following Julia–Wolff–Carathéodory type theorem for the case of several boundary regular fixed points.

Theorem 4.20. *Let $F \in \mathrm{Hol}(\Delta)$, $F(1) = 1$, $F'(1) \leq 1$, and $F(\zeta_j) = \zeta_j$ for a finite or countable collection of boundary points $\{\zeta_j\}$ different from* 1*. Then*

$$\frac{|1-z|^2}{1-|z|^2} - \frac{|1-F(z)|^2}{1-|F(z)|^2} \geq \frac{|F(z)-z|^2}{1-|F(z)|^2} \sum_j \frac{|1-\zeta_j|^2}{F'(\zeta_j)-1} \frac{1}{|z-\zeta_j|^2}. \tag{59}$$

Moreover, if equality in (59) *holds at some point $z \in \Delta$, then F satisfies*

$$\frac{(1-F(z))(1-z)}{F(z)-z} = \frac{1}{2} \sum_j \frac{|1-\zeta_j|^2}{F'(\zeta_j)-1} \frac{1+z\overline{\zeta}_j}{1-z\overline{\zeta}_j} + i\gamma \tag{60}$$

for all $z \in \Delta$, with $\gamma \in \mathbb{R}$.

Letting z tend to 0 or to 1, we obtain generalizations of assertion (iii) or (ii) of Theorem 4.3, respectively. In turn, assertion (i) is a consequence of the following strengthened version of the Schwarz lemma.

Theorem 4.21. *Let $F \in \mathrm{Hol}(\Delta)$ satisfy $F(0) = 0$. Suppose there exists a (possibly, infinite) set of boundary regular fixed points $\{\zeta_j\} \subset \partial\Delta$ of F. Then*

$$|F(z)|^2 \leq |z|^2 - |F(z) - z|^2 \sum_j \frac{1}{F'(\zeta_j) - 1} \frac{1 - |z|^2}{|z - \zeta_j|^2} \tag{61}$$

for all $z \in \Delta$. Moreover, if the set of fixed points is finite and equality in (61) *holds at some point $z \in \Delta$, then F satisfies* (47).

We end this section with an improvement of the Julia–Wolff–Carathéodory Theorem for univalent functions proved by Pommerenke and Vasil'ev in [118].

Theorem 4.22. *Let $F \in \mathrm{Hol}(\Delta)$ be univalent with $F(0) = 0$ and $F'(0) = c \in (0, 1]$. Assume that for some point $\zeta \in \partial\Delta$ the lower limit* (56) *exists finitely. Then for all $z \in \Delta$,*

$$\frac{|F(\zeta) - F(z)|^2}{1 - |F(z)|^2} \cdot \frac{1 - |z|^2}{|\zeta - z|^2} \leq \frac{\sqrt{c}\alpha(1 + |z|)}{\sqrt{(1 - |z|)^2 + 4c|z|}},$$

$$\frac{1 - |F(z)|}{1 - |z|} \leq \alpha \sqrt{\left|\frac{F(z)}{z}\right|} \left(\frac{|\zeta - z|}{1 - |z|}\right)^2,$$

$$\left|z\frac{F'(z)}{F(z)}\right| \leq \sqrt{c}\alpha \frac{1 + |z|}{\sqrt{(1 - |z|)^2 + 4c|z|}} \left(\frac{|\zeta - z|}{1 - |z|}\right)^2.$$

All these inequalities are sharp in the sense that equality holds for $F(z) = p_c(z) = \dfrac{4cz}{\left(1 - z + \sqrt{(1 - z)^2 + 4cz}\right)^2}$, $\zeta = 1$, $0 \leq z < 1$.

4.5 Lower Schwarz–Pick Estimates and Angular Derivatives

Let $F \in \mathrm{Hol}(\Delta)$. Whence the Schwarz–Pick Lemma gives an upper estimate of $|F'(z)|$, $z \in \Delta$, Anderson and Vasil'ev, in [13], established a sharp lower estimate of $|F'(z)|$ for conformal homeomorphisms F of Δ such that the angular limit $\angle \lim\limits_{z \to 1} F(z) = 1$ exists and condition (56) holds at $\zeta = 1$. By the Julia–Carathéodory Theorem 2.6, this means that the angular limit $\angle \lim\limits_{z \to 1} F'(z) = \alpha$ exists finitely.

Theorem 4.23. *Let $F \in \mathrm{Hol}(\Delta)$ be a univalent function which is conformal at the boundary point $\zeta = 1$, $\angle \lim\limits_{z \to 1} F(z) = 1$, and condition* (56) *hold. Then*

$$|F'(z)| \geq \frac{1}{\alpha^2} \frac{(1 - |z|^2)^3}{|1 - z|^4} \frac{|1 - F(z)|^4}{(1 - |F(z)|^2)^3}.$$

With fixed $z \in \Delta$ *and* $F(z) = w$, *equality is attained only for the function* $\hat{F} = B_w^{-1} \circ p_c \circ B_z$, *where* $B_w(z) := \dfrac{1-\overline{w}}{1-w}\dfrac{z-w}{1-\overline{w}z}$ *and* $c = \dfrac{1}{\alpha^2}\dfrac{(1-|z|^2)^3}{|1-z|^4}\dfrac{(|1-\hat{F}(z)|^4}{(1-|\hat{F}(z)|^2)^3}$.

For the case of several boundary regular fixed points, Anderson and Vasil'ev also proved the following weighted estimate.

Theorem 4.24. *Let* $F \in \mathrm{Hol}(\Delta)$ *be a univalent function which is conformal at its boundary regular fixed points* $\zeta_1, \ldots, \zeta_n$. *Then for every nonnegative vector* $t = (t_1, \ldots, t_n)$ *such that* $\sum_{j=1}^{n} t_j = 1$,

$$|F'(z)| \geq \frac{1}{\prod_{j=1}^{n}(F'(\zeta_j))^{2t_j^2}} \cdot \prod_{j=1}^{n} \frac{\phi_j(z)}{\phi_j(F(z))},$$

where

$$\phi_j(z) = \frac{(1-|z|^2)^{t_j(2t_j+1)}}{|1-\zeta_j\overline{z}|^{4t_j^2}} \left(\prod_{k\neq j} \left| \frac{\zeta_j - z}{1-\zeta_j\overline{z}} - \frac{\zeta_k - z}{1-\zeta_k\overline{z}} \right|^{t_k} \right)^{2t_j}.$$

This inequality is sharp.

Equality in this theorem is attained only for the function $\hat{F}$, constructed analogously to that of Theorem 4.8.

In these theorems, the univalence hypothesis is essential. For example, $F(z) = z^2$ satisfies the conditions $F(1) = 1$ and $F'(1) = 2$. However, the inequality in Theorem 4.23 is not satisfied at $z = 0$ since $F'(0) = 0$. There do not seem to exist analogous results for non-univalent functions.

4.6 Higher Order Julia–Wolff–Carathéodory Theorems

In this section, we present several results involving higher order angular derivatives that enhance the Julia–Wolff–Carathéodory Theorem as well as the Cowen–Pommerenke inequalities.

We denote the class of functions which admit the representation

$$F(z) = \sum_{j=0}^{n} \frac{c_j}{j!}(z-\zeta)^j + \gamma_n(z)$$

in a neighborhood of a boundary point $\zeta \in \partial\Delta$ by $C^n(\zeta)$ if $\lim\limits_{z\to\zeta} \dfrac{\gamma_n(z)}{(z-\zeta)^n} = 0$ and by $C_A^n(\zeta)$ if $\angle\lim\limits_{z\to\zeta} \dfrac{\gamma_n(z)}{(z-\zeta)^n} = 0$. For $F \in C_A^n(\zeta)$, we denote $F^{(n)}(\zeta) := c_n$, and then $F^{(n)}(\zeta) = \angle\lim\limits_{z\to\zeta} F^{(n)}(z)$ (see, for instance, [69]).

We start with a result by Bolotnikov, Elin, and Shoikhet [36] which, in a sense, the size of the image of a self-mapping.

Proposition 4.4. *Let $\zeta \in \partial\Delta$ be a contact point of $F \in \mathrm{Hol}(\Delta)\bigcap C^2(\zeta)$ with $F(\zeta) = w \in \partial\Delta$. We denote*

$$\alpha := |F'(\zeta)| \quad \textit{and} \quad a := \frac{\mathrm{Re}\left(\overline{w}\zeta^2 F''(\zeta)\right) + \alpha(1-\alpha)}{\alpha^2}.$$

Then

$$\overline{F(\Delta)} \not\subset D\left(w, \frac{1}{a}\right) = \left\{\lambda \in \Delta : \frac{|w-\lambda|^2}{1-|\lambda|^2} < \frac{1}{a}\right\}. \tag{62}$$

Moreover, $F(\Delta) \subseteq D\left(w, \frac{1}{a}\right)$ if and only if

$$\frac{|w-F(z)|^2}{1-|F(z)|^2} \le \frac{\alpha|z-\zeta|^2}{\alpha|z-\zeta|^2 a + 1 - |z|^2}. \tag{63}$$

Note that inequality (63) is sharper than one in the Julia–Wolff–Carathéodory theorem. This inequality can be used to evaluate the rate of convergence of iterates of a parabolic type self-mapping F to its Denjoy–Wolff point.

We proceed with two higher order analogues of the Julia–Wolff–Carathéodory theorem. Recall that for a parabolic type mapping F, classical inequalities (12)–(15) imply that each horodisk is F-invariant. At the same time, if F is not automorphism, it should shrink horodisks.

Theorem 4.25. *Let $F \in \mathrm{Hol}(\Delta)$ satisfy the conditions*

$$F(1) = 0, \quad F''(1) = 0, \quad F'(1) = 1, \quad F'''(1) < 0. \tag{64}$$

Then

$$\frac{|1-z|^2}{1-|z|^2} - \frac{|1-F(z)|^2}{1-|F(z)|^2} \ge \frac{6}{|F'''(1)|}\,\frac{1}{|1-z|^2}\,\frac{|z-F(z)|^2}{1-|F(z)|^2} \tag{65}$$

for all $z \in \Delta$.

The next higher order analog of the Julia–Carathéodory Theorem 2.6 was obtained in [35] by Vladimir Bolotnikov and Alexander Kheifets.

Theorem 4.26. *Let $F : \Delta \mapsto \overline{\Delta}$, $\zeta \in \partial\Delta$, and $n = 0, 1, \ldots$. The following assertions are equivalent.*

(1) $d_1 := \liminf_{z\to\zeta} \frac{1}{(n!)^2} \frac{\partial^{2n}}{\partial z^n \partial \overline{z}^n} \frac{1-|F(z)|^2}{1-|z|^2} < \infty.$

(2) $d_2 := \angle\lim_{z\to\zeta} \frac{1}{(n!)^2} \frac{\partial^{2n}}{\partial z^n \partial \overline{z}^n} \frac{1-|F(z)|^2}{1-|z|^2} < \infty.$

(3) The boundary Schwarz–Pick matrix

$$\mathbf{P}_n^w(\zeta) := \angle\lim_{z\to\zeta} \left[\frac{1}{i!\,j!} \frac{\partial^{i+j}}{\partial z^i \partial \overline{z}^j} \frac{1-|F(z)|^2}{1-|z|^2} \right]_{i,j=0}^n \quad \textit{exists.}$$

(4) The non-tangential limits $w_j := \angle\lim_{z\to\zeta} \frac{F^{(j)}(z)}{j!}$, $(j = 0, 1, \ldots, 2n+1)$ *exist and satisfy* $|w_0| = 1$ *and* $\mathbb{P}_n^w(\zeta) \geq 0$, *where*

$$\mathbb{P}_n^w(\zeta) := \begin{bmatrix} w_1 & \ldots & w_{n+1} \\ \vdots & \ddots & \vdots \\ w_{n+1} & \ldots & w_{2n+1} \end{bmatrix} \boldsymbol{\Psi}_n(\zeta) \begin{bmatrix} \overline{w_0} & \ldots & \overline{w_n} \\ \vdots & \ddots & \vdots \\ 0 & \ldots & \overline{w_0} \end{bmatrix}$$

with $\boldsymbol{\Psi}_n(\zeta) = \left[\psi_{j\ell}\right]_{j,\ell=0}^n$ *is the upper triangular matrix with entries*

$$\psi_{j\ell} = \begin{cases} 0, & \textit{if } j > \ell, \\ (-1)^\ell \binom{\ell}{j} \zeta^{\ell+j+1}, & \textit{if } j \leq \ell. \end{cases}$$

Moreover, if these conditions hold, then $\mathbf{P}_n^w(\zeta) = \mathbb{P}_n^w(\zeta)$, *hence* $d_1 = d_2$.

This theorem motivates the investigation of functions that satisfy the equivalent conditions (1)–(4). More precisely, *given a point* $\zeta \in \partial\Delta$, *we say that a holomorphic self-mapping* $F \in \mathrm{Hol}(\Delta)$ *is of class* $\mathcal{S}^{(n)}(\zeta)$ *if* F *satisfies the Julia–Carathéodory type condition*

$$\liminf_{z\to\zeta} \left(\frac{\partial^2}{\partial z \partial \overline{z}} \right)^{n-1} \frac{1-|F(z)|^2}{1-|z|^2} < \infty.$$

The classes $\mathcal{S}^{(n)}$ have proved to be natural for boundary interpolation problems. Here, we present only meaningful replacements of Cowen–Pommerenke inequality (46) for parabolic type mappings belonging to classes $\mathcal{S}^{(2)}$ and $\mathcal{S}^{(3)}$.

Theorem 4.27. *Let* $F \in \mathcal{S}^{(2)}(1)$ *satisfy* $F(1) = 1 = F'(1)$ *and let* $\{\zeta_j\} \in \partial\Delta$ *be a finite or countable collection of repelling fixed points of* F. *Assume that* 1 *does not belong to the closure of* $\{\zeta_j\}$. *The angular boundary limits* $F''(1)$ *and* $F'''(1)$ *exist and satisfy*

$$\sum_{j=1}^{N} \frac{1}{F'(\zeta_j) - 1} \leq \frac{2}{3} \frac{F'''(1)}{(F''(1))^2} - 1. \tag{66}$$

Equality in (66) *holds if and only if* F *is a solution of equation* (60).

Theorem 4.28. *Let $F \in \mathcal{S}^{(3)}(1)$ satisfy (64) and $\{\zeta_j\} \subset \partial\Delta$ be a finite or countable collection of repelling fixed points of F. Suppose 1 lies away from the closure of $\{\zeta_j\}$. Then the angular boundary limits $F^{(4)}(1)$ and $F^{(5)}(1)$ exist and*

$$\sum_j \frac{1}{F'(\zeta_j) - 1} \leq \frac{3}{10} \frac{5F^{(4)}(1) + F^{(5)}(1)}{(F'''(1))^2} + \frac{3}{8} \frac{(F^{(4)}(1))^2}{(F'''(1))^3} - 1. \tag{67}$$

Moreover, equality in (67) holds if and only if F satisfies the equation

$$\frac{(1 - F(z))(1 - z)}{F(z) - z} = \frac{1}{2} \sum_{j=1}^{N} \frac{|1 - \zeta_j|^2}{F'(\zeta_j) - 1} \frac{1 + z\overline{\zeta}_j)}{1 - z\overline{\zeta}_j)} - \frac{3}{F'''(1)} \frac{1 + z}{1 - z} + i\gamma$$

for some $\gamma \in \mathbb{R}$ and all $z \in \Delta$.

5 The Uniqueness Part of the Boundary Schwarz Lemma

This section is devoted to boundary versions of the uniqueness part of the Schwarz Lemma. In our context, rigidity (or, uniqueness) principles state that if the expansion of a function $f \in \mathrm{Hol}(D, \mathbb{C})$, $D \subset \mathbb{C}$, in a neighborhood of a point $\zeta \in \overline{D}$ agrees up to a certain order with a specific rational map (e.g., a constant map, the identity, an automorphism of D, etc.), then f coincides with this map on D.

As has been mentioned above, the uniqueness part of the Schwarz Lemma implies that if $F \in \mathrm{Hol}(\Delta)$ agrees at 0 with the rotation $e^{i\theta}z$ up to first order, i.e., $F(0) = 0$ and $F'(0) = e^{i\theta}$, then $F(z) \equiv e^{i\theta}z$ in Δ; in particular, for $\theta = 0$, F is the identity map. Moreover, applying this fact to the function $m_\zeta \circ F \circ m_\zeta$, shows that if F agrees at a point $\zeta \in \Delta$ with the automorphism $G := m_\zeta \circ (\lambda m_\zeta)$, $|\lambda| = 1$, up to first order, i.e., $F(\zeta) = \zeta$ and $F'(\zeta) = \lambda$, then $F \equiv G$ on Δ.

These results do not hold for a boundary fixed point $\zeta \in \partial\Delta$ of F. Indeed, every parabolic automorphism agrees with the identity up to first order at its boundary Denjoy–Wolff point although it is not the identity map on Δ. Moreover, the Julia–Wolff–Carathéodory Theorem, which extends the Schwarz Lemma to the boundary, does not contain an analogous uniqueness part.

In this section, we consider a variety of boundary versions of the rigidity part of the Schwarz Lemma. We start from the famous Burns–Krantz Theorem and its various extensions for holomorphic self-mappings of Δ and other classes of holomorphic maps, and rigidity principles for commuting holomorphic mappings which provide sufficient conditions for their coincidence.

5.1 General Rigidity Principles for Holomorphic Self-Mappings of the Unit Disk

The uniqueness part of the boundary Schwarz Lemma was established in 1994 by Daniel M. Burns and Steven G. Krantz [42]:

Theorem 5.1 (Burns–Krantz' Theorem). *Let* $F \in \text{Hol}(\Delta)$ *satisfy*

$$F(z) = 1 + (z-1) + O((z-1)^4)$$

as $z \to 1$. *Then* $F(z) \equiv z$.

Similar results appeared earlier in the literature of conformal mappings (see, for instance, [138]) with the additional hypothesis that F be univalent (and often the function is assumed to be quite smooth—even analytic—in a neighborhood of 1). The theorem presented in [42] has no such hypothesis. The exponent 4 is sharp: simple geometric arguments show that the function

$$F(z) = z + \frac{1}{10}(z-1)^3$$

satisfies the conditions of the theorem with 4 replaced by 3. Note also that it follows from the proof that $O((z-1)^4)$ can be replaced by $o((z-1)^3)$.

The Burns–Krantz Theorem was improved in 1995 by Thomas L. Kriete and Barbara D. MacCluer [96], who replaced F with its real part and considered the radial limit in $o((z-1)^3)$ instead of the unrestricted limit. Here is a more precise statement of their result.

Theorem 5.2. *Let* $F \in \text{Hol}(\Delta)$ *with radial limit* $F(1) = 1$ *and angular derivative* $F'(1) = 1$. *If*

$$\liminf_{r\to 1^-} \frac{\text{Re}(F(r)-r)}{(1-r)^3} = 0,$$

then $F(z) \equiv z$.

Corollary 5.1. *Suppose* $F \in \text{Hol}(\Delta)$ *with angular derivative at some* $\zeta_0 \in \partial\Delta$ *satisfying* $|F'(\zeta_0)| = 1$. *If the maximum modulus function* $M(r) := \max_{\theta\in[0,2\pi]} |F(re^{i\theta})|$ *of* F *satisfies*

$$\liminf_{r\to 1^-} \frac{M(r)-r}{(1-r)^3} = 0,$$

then $M(r) \equiv r$ *and* F *is a rotation.*

They also considered the mappings

$$F(z) = z + t(1-z)^\beta, \quad t > 0,\ \beta > 1, \tag{68}$$

which generalize the mentioned example of Burns and Krantz (with $\beta = 3$ and $t = -\frac{1}{10}$) and showed that for each $1 < \beta \leq 3$ and $0 < t < 2^{1-\beta}$, F maps Δ to Δ and $|F(e^{i\theta})| < 1$ for all $e^{i\theta} \neq 1$. Moreover, $F(1) = 1$ and $F'(1) = 1$.

In 2006, Luca Baracco, Dmitri Zaitsev, and Giuseppe Zampieri [16] reformulated the Burns–Krantz Theorem 5.1 and gave a simple proof that its conclusion remains valid even if the theorem' assumption is satisfied only on a sequence in Δ that converges non-tangentially to 1.

Theorem 5.3. *Let $F \in \mathrm{Hol}(\Delta)$ and $z_k \in \Delta$ be a sequence converging non-tangentially to 1 such that $F(z_k) = z_k + o(|z_k - 1|^3)$ as $k \to \infty$. Then $F(z) \equiv z$.*

The next result, by Roberto Tauraso and Fabio Vlacci [136], gives sufficient conditions for a holomorphic self-mapping F of Δ to coincide with an arbitrary function $G \in \mathrm{Hol}(\Delta)$ if they agree up to the third order at a boundary contact point and one of the functions is "subordinated," in a sense, to the other one.

Theorem 5.4. *Let $F, G \in \mathrm{Hol}(\Delta)$ and $\zeta \in \partial\Delta$ be such that*

$$\lim_{r \to 1^-} \frac{F(r\zeta) - G(r\zeta)}{(r-1)^3} = \ell \tag{69}$$

for some $\ell \in \mathbb{C}$, and

$$\frac{1 - |F(z)|^2}{|w - F(z)|^2} \geq \frac{1 - |G(z)|^2}{|w - G(z)|^2}, \qquad z \in \Delta, \tag{70}$$

for some $w \in \partial\Delta$. Then $F \equiv G$ if and only if $\ell = 0$. Moreover, $\overline{w}\ell$ is a nonpositive real number.

The Burns–Krantz Theorem 5.1 is a particular case of this result for $\zeta = 1$ and $G = I$.

We close this section with a result established by Bolotnikov, Elin, and Shoikhet in [36] which deals with holomorphic mappings having infinitely many mutual contact points.

Proposition 5.1. *Suppose $F \in \mathrm{Hol}(\Delta, \overline{\Delta})$ has infinitely many regular mutual contact points $\{\zeta_j\}_{j=1}^{\infty}$ on the boundary, i.e., $F(\zeta_j) = w$ for some $w \in \partial\Delta$. If $\sup_{j \in \mathbb{N}} |F'(\zeta_j)| < \infty$, then $F(z) \equiv w$.*

5.2 Coincidence of a Self-Mapping with a Given Rational Map

In this section, we describe conditions under which a function $F \in \mathrm{Hol}(\Delta)$ coincides with a conformal automorphism of the disk, a linear fractional transformation, or a general rational map.

First, we note that the existence of extremal functions in the diverse theorems of Sects. 4.2 and 4.3 provides, in fact, rigidity results. For instance, Theorems 4.3 and 4.4 assert that if angular derivatives of a holomorphic self-mapping F of Δ at boundary fixed or mutual contact points satisfy a specific equality, then F coincides with a given rational map, in particular, F is a Blaschke product.

We now consider rigidity results involving angular derivatives of higher order. In [136], Tauraso and Vlacci investigated how rigid is the set of holomorphic self-mappings of the unit disk Δ after imposing some conditions on the boundary Schwarzian derivative of F defined by

$$\mathcal{S}_F(z) := \frac{F'''(z)}{F'(z)} - \frac{3}{2}\left(\frac{F''(z)}{F'(z)}\right)^2, \quad z \in \partial\Delta.$$

It is known that the Schwarzian derivative carries global information about F: it vanishes identically if and only if F is a Möbius transformation.

Initially, the original rigidity result of Burns and Krantz was extended in [136] from the identity to the parabolic automorphisms.

Theorem 5.5. *Let* $F \in \mathrm{Hol}(\Delta) \bigcap C_A^3(1)$. *If*

$$F(1) = 1, \quad F'(1) = 1, \quad \mathrm{Re}\, F''(1) = 0 \quad \textit{and} \quad \mathrm{Re}\, \mathcal{S}_F(1) = 0,$$

then F *is the parabolic automorphism of* Δ *defined by*

$$\frac{1 + F(z)}{1 - F(z)} = \frac{1 + z}{1 - z} + ib,$$

where $b = \mathrm{Im}\, F''(1)$.

In the particular case $F''(1) = F'''(1) = 0$, this reduces to the result of Burns and Krantz, i.e., $F(z) \equiv z$.

In 2010, Contreras, Díaz-Madrigal, and Pommerenke [54] supplemented Theorem 5.5 as follows.

Theorem 5.6. *(1) A nontrivial (i.e.,* $F \neq I$*) mapping* $F \in \mathrm{Hol}(\Delta)$ *is a parabolic automorphism if and only if there exists* $\zeta \in \partial\Delta$ *such that* $F \in C_A^3(\zeta)$ *and*

$$F(\zeta) = \zeta, \quad F'(\zeta) = 1, \quad \mathrm{Re}(\zeta F''(\zeta)) = 0 \quad \textit{and} \quad \mathcal{S}_F(\zeta) = 0.$$

(2) $F \in \mathrm{Hol}(\Delta)$ *is a hyperbolic automorphism if and only if there exist* $\zeta \in \partial\Delta$ *and* $\alpha \in (0, 1)$ *such that* $F \in C_A^3(\zeta)$ *and*

$$F(\zeta) = \zeta, \quad F'(\zeta) = \alpha, \quad \mathrm{Re}(\zeta F''(\zeta)) = \alpha(\alpha - 1) \quad \textit{and} \quad \mathcal{S}_F(\zeta) = 0.$$

To proceed, for a holomorphic self-mapping $F \in C_A^2(\zeta)$ of the disk with a boundary regular contact point $\zeta \in \partial\Delta$ such that $w = F(\zeta) \in \partial\Delta$ and $\alpha_\zeta := |F'(\zeta)|$, we define

$$a_\zeta = \frac{\operatorname{Re}\left(\overline{w}\zeta^2 F''(\zeta)\right) + \alpha(1-\alpha)}{\alpha^2} \tag{71}$$

and recall that by Proposition 4.4,

$$\inf_{z\in\Delta} \frac{1-|F(z)|^2}{|w-F(z)|^2} \le a_\zeta .$$

Using this fact, which was proved later than the original results of this section, we present them in slightly modified forms.

The following assertion provides boundary conditions under which F coincides identically with a given rational map.

Theorem 5.7 ([136]). *Let $F \in \operatorname{Hol}(\Delta)$, $w \in \partial\Delta$, and let $\{\zeta_1,\dots,\zeta_N\} \subset \partial\Delta$ be mutual distinct contact points of F such that $F(\zeta_k) = w$ and $0 < \alpha_k := |F'(\zeta_k)| < \infty$ for $k = 1,\dots,N$. Suppose that for some $j \in \{1,\dots,N\}$, $F \in C_A^3(\zeta_j)$ and $\inf_{z\in\Delta} \frac{1-|F(z)|^2}{|w-F(z)|^2} = a_{\zeta_j}$. Then*

$$\operatorname{Im}(\zeta_j^2 \mathcal{S}_F(\zeta_j)) = 0, \tag{72}$$

$$\operatorname{Re}(\zeta_j^2 \mathcal{S}_F(\zeta_j)) \le -6\alpha_j \sum_{k=1,k\neq j}^{N} \frac{1}{\alpha_k}\cdot\frac{1}{|\zeta_k-\zeta_j|^2}. \tag{73}$$

Moreover, equality in (73) holds if and only if F is the rational map

$$F(z) = \varphi_w^{-1}\left(\sum_{k=1}^{N} \frac{1}{\alpha_k}\varphi_{\zeta_k}(z) + a_{\zeta_j} + ib\right), \tag{74}$$

where $\varphi_\sigma(z) := \frac{\sigma+z}{\sigma-z}$ and

$$b := \frac{\operatorname{Im}(\zeta_j^2\overline{\tau}F''(\zeta_j))}{\alpha_j^2} + \sum_{k=1,k\neq j}^{N} \frac{1}{\alpha_k}\cdot\frac{2\operatorname{Im}(\zeta_k\overline{\zeta_j})}{|\zeta_k-\zeta_j|^2}. \tag{75}$$

In particular, if $\tau \in \partial\Delta$ is the Denjoy–Wolff point of $F \in \operatorname{Hol}(\Delta)\bigcap C_A^3(\tau)$ with $\alpha := F'(\tau)$ and $\operatorname{Re}(\tau F''(\tau)) = \alpha(\alpha-1)$, this result implies that $\mathcal{S}_F(\tau) = 0$ if and only if (cf. [54])

$$F(z) = \frac{(\tau F''(\tau) - 2\alpha^2)z + \overline{F''(\tau)}}{F''(\tau)z + \overline{\tau F''(\tau)} - 2\alpha^2}, \quad z\in\Delta. \tag{76}$$

The following boundary rigidity principles are due to Shoikhet [133]. He obtained a generalization of the Burns–Krantz rigidity theorem in the spirit of the classical Schwarz–Pick Lemma and established conditions on behavior of a holomorphic self-mapping F of Δ in a neighborhood of a boundary regular fixed point (not necessarily the Denjoy–Wolff point) under which F is a linear-fractional transformation.

It is known that if a mapping $F \in \mathrm{Hol}(\Delta)$ with the boundary regular fixed point $\tau = 1$ and $F'(1) =: \alpha$ is linear fractional, then for all $k > 0$,

$$F\left(D(1,k)\right) = D\left(1, \frac{\alpha k}{1+\alpha k a_1}\right),$$

where a_1 is defined by (71), i.e., for all $z \in \Delta$,

$$\frac{|1-F(z)|^2}{1-|F(z)|^2} = \frac{\alpha\,|1-z|^2}{(1-|z|^2)+\alpha a_1\,|1-z|^2}. \tag{77}$$

Moreover, F is an automorphism of Δ (either hyperbolic, $\alpha \neq 1$, or parabolic, $\alpha = 1$) if and only if $a_1 = 0$.

It turns out that, under some smoothness conditions, equality (77) (and even some weaker condition) is also sufficient for $F \in \mathrm{Hol}(\Delta)$ to be linear fractional. Namely (cf., Proposition 4.4 and Theorem 5.7):

Theorem 5.8. *Let $F \in \mathrm{Hol}(\Delta) \cap C_A^3(1)$, $F(1) = 1$ and $F'(1) = \alpha$. Then F is a linear fractional transformation if and only if the following conditions hold.*

(i) $F(\Delta) \subseteq D\left(1, \frac{1}{a}\right)$, *i.e.,* $\dfrac{|1-F(z)|^2}{1-|F(z)|^2} \le \dfrac{1}{a}, \quad z \in \Delta;$

(ii) the Schwarzian derivative $\mathcal{S}_F(1) = 0$.

So, if conditions (i) and (ii) are satisfied, then equality (77) holds for all $z \in \Delta$.

Theorem 5.9. *Let $F \in \mathrm{Hol}(\Delta) \cap C_A^3(1)$ and let $\tau = 1$ be the Denjoy–Wolff point of F with $\alpha = F'(1)$. For all $k \ge 0$, we denote $A_k = \frac{1}{\alpha^2}\left[\frac{1}{1+k}\operatorname{Re} F''(1) + \alpha(1-\alpha)\right]$. Then F is a linear fractional map if and only if $\operatorname{Re}\mathcal{S}_F(1) = 0$ and there exists $k > 0$ such that*

$$F\left(D\left(1,\frac{1}{k}\right)\right) \subseteq D\left(1, \frac{1}{k+(k+1)A_k}\right). \tag{78}$$

Corollary 5.2. *Let $F \in \mathrm{Hol}(\Delta) \cap C_A^3(1)$ be such that $F(1) = 1$ and $\alpha = F'(1) \in (0,1]$. Then F is an automorphism of Δ if and only if $\operatorname{Re}\mathcal{S}_F(1) = 0$ and $\operatorname{Re} F''(1) = (1+k)\alpha(\alpha-1)$ for some $k \ge 0$.*

Moreover, if $k > 0$, then F is either the identity mapping or a parabolic automorphism of Δ.

Another consequence of Theorem 5.9 provides conditions for a holomorphic self-mapping F of Δ to coincide with a given affine map.

Corollary 5.3. *Let* $F \in \mathrm{Hol}(\Delta) \cap C_A^3(1)$ *with* $\alpha = F'(1) \in (0,1]$. *The following are equivalent.*

(i) $F''(1) = F'''(1) = 0$ *and*

$$F(\Delta) \subseteq D\left(1, \frac{\alpha}{1-\alpha}\right),$$

or, equivalently,

$$F(D(1,k)) \subseteq D\left(1, \frac{k\alpha}{1+k(1-\alpha)}\right), \quad k > 0.$$

(ii) F *is an affine mapping of the form* $F(z) = \alpha z + 1 - \alpha$.

In 2001, Dov Chelst [49], in turn, established the following conditions on the local behavior of F near a finite set of boundary points which ensure that F is a finite Blaschke product.

Theorem 5.10. *Let* $F \in \mathrm{Hol}(\Delta)$ *and suppose* $G : \Delta \mapsto \Delta$ *is a finite Blaschke product which equals* $\tau \in \partial\Delta$ *on a finite set* $A_G \subset \partial\Delta$. *If*

(i) for a given $\zeta_0 \in A_G$, $F(z) = G(z) + O((z-\zeta_0)^4)$, *as* $z \to \zeta_0$, *and*
(ii) for all $\zeta \in A_G - \{\zeta_0\}$, $F(z) = G(z) + O((z-\zeta)^{k_\zeta})$, *for some* $k_\zeta \geq 2$ *as* $z \to \zeta$,

then $F(z) \equiv G(z)$.

The following rigidity result, in the spirit of Chelst's theorem, was established by Mustafa Arslan [14]. He proved it using an approach suggested by Baracco, Zaitsev, and Zampieri in [16].

Theorem 5.11. *Let* $G \in \mathrm{Hol}(\Delta, \mathbb{C})$ *have exactly* n *zeros, counting multiplicities, in* Δ. *Let* $F \in \mathrm{Hol}(\Delta, \mathbb{C})$ *satisfy* $|F| \leq |G|$ *on* $\partial\Delta$. *Suppose there exist* n *distinct points* ζ_j *on* $\partial\Delta$ *and positive integers* m_j, $j = 1, \ldots, n$, *with* $\sum m_j = n+1$, *such that* G *is bounded away from zero around* ζ_j *and*

$$F(z) = G(z) + o(|z-\zeta_j|^{2m_j-1}) \quad as \quad \Delta \ni z \to \zeta_j \tag{79}$$

for each j. *Then* $F \equiv G$.

The assumptions of this theorem are satisfied in the settings of Burns–Krantz' Theorem 5.1 and Chelst's Theorem 5.10 since all finite Blaschke products have modulus 1 on the boundary of Δ.

The conditions of Theorems 5.10 and 5.11 are sufficient to guarantee that $F \equiv G$ on Δ. The problem posed in [49] is to find conditions that are not only sufficient but also necessary (in some sense). This problem was solved by Bolotnikov in [34] as follows.

Theorem 5.12. *Let $F \in \mathrm{Hol}(\Delta)$, G be a finite Blaschke product of degree d, and $\zeta_1, \ldots, \zeta_n \in \partial\Delta$ be distinct points. Suppose that*

$$F(z) = G(z) + o((z - \zeta_j)^{m_j}) \quad \text{for} \quad j = 1, \ldots, n \tag{80}$$

as z tends to ζ_j non-tangentially, where $m_1, \ldots, m_n$ are positive integers. If

$$\left[\frac{m_1 + 1}{2}\right] + \ldots + \left[\frac{m_n + 1}{2}\right] > d = \deg G, \tag{81}$$

then $F \equiv G$ on Δ. Otherwise, the uniqueness result fails.

In other words, the points $\zeta_j \in \partial\Delta$ can be chosen arbitrarily (regardless of G) as well as the degrees of convergence. Thus, this result improves Theorems 5.10 and 5.11.

5.3 Rigidity Results for Commuting Self-Mappings of the Disk

In the previous section, we considered conditions under which a holomorphic self-mapping F of Δ coincides identically with a given function G. It is natural to assume that if F commutes with G, then F and G share some common properties and, consequently, coincide under weaker assumptions. So we begin this section with a brief discussion of joint characteristics of two commuting holomorphic self-mappings of the open unit disk. Such characteristics include common Denjoy–Wolff point, type (parabolic, hyperbolic, or with an interior fixed point), and membership in a common pseudo-iteration semigroup.

We begin by analyzing the fixed point sets of commuting holomorphic mappings. The first result in this direction is due to Allen Lowell Shields (see [130], 1964).

Theorem 5.13. *Let $\mathcal{A}$ be a commuting family of continuous functions mapping the closed unit disc $\overline{\Delta}$ into itself and analytic inside Δ. There exists a common fixed point $\tau \in \overline{\Delta}$ for all the functions of the family.*

In 1973, Shields' theorem was extended by Donald F. Behan [23] for holomorphic self-mappings of Δ which are not necessarily continuous on the boundary. Let $T(F)$ denote either the Denjoy–Wolff point of $F \in \mathrm{Hol}(\Delta)$ or, in the case F is an elliptic automorphism, the unique fixed point of F in Δ.

Theorem 5.14. *Let F and G be analytic self-mappings of Δ, neither of which is the identity mapping. If F is not a conformal hyperbolic automorphism of Δ and G commutes with F, then $T(F) = T(G)$.*

Note that *if F is a hyperbolic automorphism and $F \circ G = G \circ F$ $(G \neq I)$, then G is also a hyperbolic automorphism of Δ.*

This fact was established by Maurice H. Heins as a lemma in his 1941 paper [82]. In this case, F and G share two boundary fixed points; however, their Denjoy–Wolff points do not necessarily coincide.

What can be said about other fixed points (different from the Denjoy–Wolff points) of two commuting holomorphic self-mappings of Δ? The example $F(z) = z^2$ and $G(z) = z^3$ shows that the fixed point sets of two commuting holomorphic mappings are not necessarily the same.

This question was investigated by Filippo Bracci. An analysis of a variety of all the possible cases can be found in his paper [38]. He established, for example, the following result for univalent self-mappings. Let

$$\mathrm{Fix}(F) := \{p \in \partial\Delta : F(p) = p\} \cup \{\tau_F\},$$

where τ_F is the Denjoy–Wolff point of F and

$$\mathrm{Hol}_u^1(\Delta) := \{h \in \mathrm{Hol}(\Delta) : h \text{ is univalent and } h' \text{ is continuous on } \overline{\Delta}\}.$$

Theorem 5.15. *Suppose $F \in \mathrm{Hol}_u^1(\Delta)$ and τ_F is the Denjoy–Wolff point of F.*

(i) If $\tau_F \in \partial\Delta$, $F'(\tau_F) < 1$, and F commutes with $G \in \mathrm{Hol}_u^1(\Delta)$, then $\mathrm{Fix}(F) = \mathrm{Fix}(G)$.

(ii) If $\tau_F \in \Delta$, F commutes with $G \in \mathrm{Hol}_u^1(\Delta)$, and $\mathrm{Fix}(F) \cap \mathrm{Fix}(G)$ *contains two points, then* $\mathrm{Fix}(F) = \mathrm{Fix}(G)$.

(iii) If $\tau_F \in \Delta$, then there exists $m \in \mathbb{N}$ such that for any $G \in \mathrm{Hol}_u^1(\Delta) \setminus \mathrm{Aut}(\Delta)$ which commutes with F, $\mathrm{Fix}(F) = \mathrm{Fix}(G_m)$.

Another common characteristic of a pair of commuting holomorphic self-mappings F and G of Δ (neither of which is the identity) is their type: hyperbolic, parabolic or elliptic (having an interior fixed point). Indeed, it has been proved by Heins in 1941, that if F is a hyperbolic automorphism, then G is also a hyperbolic automorphism. The result in the case that one of the two mappings has an interior fixed point follows from Behan's Theorem 5.14. The result in the case F and G, neither of which is an automorphisms of Δ, share a common boundary fixed point is a consequence of the following result by Carl C. Cowen.

Theorem 5.16 ([55]). *Suppose $F, G \notin \mathrm{Aut}(\Delta)$ are commuting nonconstant analytic self-mappings of Δ. Denote the Denjoy–Wolff point of F by ζ.*

(1) If $F'(\zeta) = 0$, then $G'(\zeta) = 0$.
(2) If $0 < |F'(\zeta)| < 1$, then $0 < |G'(\zeta)| < 1$.
(3) If $F'(\zeta) = 1$, then $G'(\zeta) = 1$.

Thus, if F and G, neither of which is an automorphism of Δ, and share a common boundary Denjoy–Wolff point ζ, then they are of the same type, either hyperbolic or parabolic.

Finally, the fact that a parabolic automorphism of Δ cannot commute with a hyperbolic self-mapping of Δ was proved in 2008 by Elin, Levenshtein, Reich, and Shoikhet.

Proposition 5.2 ([69]). *Let F and G be two commuting holomorphic self-mappings of Δ and assume that G is not the identity. If F is of hyperbolic type, then G is also of hyperbolic type.*

The next characteristic of two commuting holomorphic self-mappings of Δ is that they belong to a common pseudo-iteration semigroup. This property was established by Cowen in [55] and developed in [72, 139], and [32].

Using these properties of commuting functions and the techniques based on building representative fractional models for holomorphic self-mappings of Δ, the following uniqueness principles were established.

In 2002, Bracci, Tauraso and Vlacci, in [39], presented a series of rigidity results for commuting holomorphic self-mappings of Δ. In particular, they proved that two such mappings which have the same expansion up to third order at their common boundary Denjoy–Wolff point coincide. They also showed that the order three is necessary only in a particular case (which contains the case studied by Burns and Krantz). In their main theorem [39, Theorem 2.4], a variety of possible cases are considered.

Theorem 5.17. *Suppose $F, G \in \mathrm{Hol}(\Delta)$ commute. If one of the following conditions holds, then $F \equiv G$.*

(1) If there exist $z_0 \in \Delta$ and $k \in \mathbb{N}$ such that $F(z_0) = G(z_0) = z_0$, $F^{(m)}(z_0) = G^{(m)}(z_0) = 0$ for $1 \le m < k$, and $F^{(k)}(z_0) = G^{(k)}(z_0) \neq 0$.
(2) If G is a hyperbolic automorphism of Δ with a fixed point $\tau \in \partial\Delta$ and $\lim_{r\to 1^-} F'(r\tau) = G'(\tau)$.
(3) If G is a parabolic automorphism of Δ with a fixed point $\tau \in \partial\Delta$ and $\lim_{r\to 1^-} F'(r\tau) = G'(\tau)$, $\lim_{r\to 1^-} F''(r\tau) = G''(\tau)$.
(4) If $G = I$ and there exists $\tau \in \partial\Delta$ such that $\lim_{r\to 1^-} F(r\tau) = \tau$, $\lim_{r\to 1^-} F'(r\tau) = 1$, $\lim_{r\to 1^-} F''(r\tau) = 0$, and $\lim_{r\to 1^-} F'''(r\tau) = 0$.
(5) If $F, G \notin \mathrm{Aut}(\Delta)$ have Denjoy–Wolff point $\tau \in \partial\Delta$ and $\lim_{r\to 1^-} F'(r\tau) = \lim_{r\to 1^-} G'(r\tau) < 1$.
(6) If $F, G \notin \mathrm{Aut}(\Delta)$ have Denjoy–Wolff point $\tau \in \partial\Delta$, $F \in C^{3+\varepsilon}(\tau)$, $G \in C^2(\tau)$, and $F'(\tau) = G'(\tau) = 1$, $F''(\tau) = G''(\tau) \neq 0$.
(7) If $F, G \notin \mathrm{Aut}(\Delta)$ have Denjoy–Wolff point $\tau \in \partial\Delta$, $F \in C^{5+\varepsilon}(\tau)$, $G \in C^4(\tau)$, and $F'(\tau) = G'(\tau) = 1$, $F''(\tau) = G''(\tau) = 0$, $F'''(\tau) = G'''(\tau)$.

Tauraso [135] improved this result in the case F is of parabolic type and supplemented it with the following assertion.

Theorem 5.18. *Let $F, G \in \mathrm{Hol}(\Delta)$. Suppose F is of parabolic type with Denjoy–Wolff point τ and commutes with G. If one of the following conditions holds, then $F \equiv G$.*

(1) $F \in C^2(\tau)$, $G \in C^2_K(\tau)$, $F''(\tau) = G''(\tau) \neq 0$ and $\mathrm{Re}(F''(\tau)\tau) > 0$;
(2) $F, G \in C^2(\tau)$, $F''(\tau) = G''(\tau) \neq 0$ and $\mathrm{Re}(F''(\tau)\tau) = 0$;
(3) $F \in C^3(\tau)$, $G \in C^3_K(\tau)$, $F''(\tau) = G''(\tau) = 0$ and $F'''(\tau) = G'''(\tau)$.

The following rigidity result was proved by Marina Levenshtein and Simeon Reich in [99].

Theorem 5.19. *Let* $F \in \mathrm{Hol}(\Delta)$ *be such that*

$$\lim_{z\to 1} \frac{F(z)-z}{(z-1)^2} = 0. \tag{82}$$

Suppose that $G \in \mathrm{Hol}(\Delta)$*,* $F \circ G = G \circ F$ *and*

$$\lim_{z\to 1} \frac{G(z)-z}{(z-1)^2} = \alpha \neq 0\,(\neq \infty). \tag{83}$$

Then $F \equiv I$*.*

In the case $\operatorname{Re}\alpha > 0$, the unrestricted limit in (82) can be replaced with the angular one.

5.4 Rigidity Principles for Other Classes of Holomorphic Mappings

In this section, for classes of holomorphic mappings different from $\mathrm{Hol}(\Delta)$, we are looking for boundary conditions which force a mapping F holomorphic in certain complex domain D to be identically constant. A number of such results with various smoothness hypotheses on the image of the boundary of D were established simultaneously under the assumption of infinite vanishing order of F at a boundary point $\zeta \in \partial D$ (see [7,9,11,24,85]).

We present several results by Serena Migliorini and Fabio Vlacci from their 2002 paper [110] in which they considered $F \in \mathrm{Hol}(D, \mathbb{C})$ for a generic domain $D \subset \mathbb{C}$ and determined up to what order F must vanish at a boundary point for F to be constant. Namely, they studied domains $D \subset \mathbb{C}$ with a local Dini-smooth corner of opening $\pi\alpha$ $(0 \leq \alpha \leq 2)$ at a boundary point $\zeta \in \partial D$ and considered the following three possible cases:

1. ∂D is a Dini-smooth Jordan arc in a neighborhood of ζ,
2. D has a local Dini-smooth corner of opening $\pi\alpha$ $(0 < \alpha \leq 2)$ at ζ, and
3. D has a local Dini-smooth corner of opening 0 at ζ, i.e., the case of an outward pointing cusp.

Regarding the first case, it was mentioned above that the Julia–Wolff–Carathéodory Theorem implies the following assertion.

If $F \in \mathrm{Hol}(\Delta, \overline{\Delta})$*, then the conditions* $\lim\limits_{r\to 1^-} F(r\zeta) = \zeta$ *and* $\lim\limits_{r\to 1^-} F'(r\zeta) = 0$ *at some* $\zeta \in \partial\Delta$ *imply that* $F \equiv \zeta$*.*

Applying this fact to the function $f \in \mathrm{Hol}(\Delta, \overline{\Pi^+})$ defined by $f(z) := \dfrac{\zeta - F(z)}{\zeta + F(z)}$, where $\Pi^+ := \{z \in \mathbb{C} : \operatorname{Re} z > 0\}$ is the right half-plane and $F \in \mathrm{Hol}(\Delta, \overline{\Delta})$, we get the following assertion (Lemma 3.1 in [110]).

Proposition 5.3. *Suppose* $f \in \mathrm{Hol}(\Delta, \overline{\Pi^+})$ *and there exists* $\zeta \in \partial\Delta$ *such that*

$$\angle \lim_{z \to \zeta} \frac{f(z)}{z - \zeta} = 0.$$

Then $f \equiv 0$.

Migliorini and Vlacci also stated the following form of this proposition.
Let $f, g \in \mathrm{Hol}(\Delta, \mathbb{C})$ *be such that*

$$\operatorname{Re} f(z) \geq \operatorname{Re} g(z), \quad z \in \Delta,$$

and there exists $\zeta \in \partial\Delta$ *such that*

$$\angle \lim_{z \to \zeta} \frac{f(z) - g(z)}{z - \zeta} = 0.$$

Then $f \equiv g$.

As a consequence of Proposition 5.3, they obtained the following result for holomorphic self-mappings of the right half-plane.

Corollary 5.4. *Let* $f \in \mathrm{Hol}(\Pi^+, \overline{\Pi^+})$ *and* $\zeta \in \partial\Pi^+$ *be such that*

$$\lim_{w \to \zeta} \frac{f(w)}{w - \zeta} = 0.$$

Then $f \equiv 0$.

Furthermore, they proved that for $f \in \mathrm{Hol}(\Delta, \mathbb{C})$ to be the zero mapping, it suffices that f vanish at $\zeta \in \partial\Delta$ up to first order and $f(\Delta)$ have a "good" support curve.

Theorem 5.20. *Let* $f \in \mathrm{Hol}(\Delta, \mathbb{C})$ *be a bounded function. Suppose that there is a Dini-smooth Jordan curve* C *passing through* $0 \in \partial f(\Delta)$ *such that* $f(\Delta)$ *is contained in the inner domain of* C*. If there exists* $\zeta \in \partial\Delta$ *such that*

$$\angle \lim_{z \to \zeta} \frac{f(z)}{z - \zeta} = 0,$$

then $f \equiv 0$.

The next result is more general.

Proposition 5.4. *Let $D \subset \mathbb{C}$ be a domain and $\zeta \in \partial D$ have a neighborhood U in $\mathbb{C}$ such that $\partial D \cap U$ is a Dini-smooth Jordan arc. Let $f \in \mathrm{Hol}(D, \Pi^+)$ be such that*

$$\lim_{w \to \zeta} \frac{f(w)}{w - \zeta} = 0.$$

Then $f \equiv 0$.

The next result provides suitable conditions under which a holomorphic map f defined on a domain with a local Dini-smooth corner vanishes identically.

Theorem 5.21. *Let $D \subset \mathbb{C}$ be a domain with a local Dini-smooth corner at $\zeta \in \partial D$ of opening $\pi\alpha$ $(0 < \alpha \leq 2)$ and $f \in \mathrm{Hol}(D, \Pi^+)$. If there exists $N \geq \frac{1}{\alpha}$ such that*

$$\lim_{w \to \zeta} \frac{f(w)}{(w - \zeta)^N} = 0,$$

then $f \equiv 0$.

For the case of holomorphic mappings from Δ into a domain Ω inner for a Jordan curve which has a Dini-smooth corner at $0 \in \partial\Delta$ of opening $\pi\alpha$ $(0 < \alpha \leq 2)$, the following assertion holds.

Theorem 5.22. *Let $f \in \mathrm{Hol}(\Delta, \mathbb{C})$ be a bounded function. Suppose that $f(\Delta)$ is contained in the inner domain Ω of a Jordan curve which has a Dini-smooth corner at $0 \in \partial\Omega$ of opening $\pi\alpha$ $(0 < \alpha \leq 2)$. If there exist $N \geq \alpha$ and $\zeta \in \partial\Delta$ such that*

$$\angle \lim_{z \to \zeta} \frac{f(z)}{(z - \zeta)^N} = 0,$$

then $f \equiv 0$.

This result immediately yields

Corollary 5.5. *Let $D \subset \mathbb{C}$ be a domain with a local Dini-smooth corner at $\zeta \in \partial D$ of opening $\pi\alpha$ $(0 < \alpha \leq 2)$ and let $f \in \mathrm{Hol}(D, \mathbb{C})$ be bounded. Suppose that $f(D)$ is contained in the inner domain Ω of a Jordan curve which has a Dini-smooth corner at $0 \in \partial\Omega$ of opening $\pi\beta$ $(0 < \beta \leq 2)$. If there exists $N \geq \frac{\beta}{\alpha}$ such that*

$$\lim_{w \to \zeta} \frac{f(w)}{(w - \zeta)^N} = 0,$$

then $f \equiv 0$.

However, in the case $\alpha = 0$, i.e., there is an outward pointing cusp, Theorems 5.21 and 5.22 no longer hold. To formulate rigidity results for such domains, Migliorini and Vlacci defined the domain G_a with an outward pointing cusp at 0 by

$$G_a := \left\{\rho e^{i\vartheta} : 0 < \rho < c,\ \vartheta_-(\rho) < \vartheta < \vartheta_+(\rho)\right\},$$

where $\vartheta_{\pm}$ are locally absolutely continuous functions such that

$$\vartheta_{+}(\rho)-\vartheta_{-}(\rho)\sim a\rho \quad \text{as} \quad \rho\to 0, \qquad \int_0^c \vartheta_{\pm}'(\rho)d\rho<\infty \tag{84}$$

with $a>0$. From the geometrical point of view, the first of conditions (84) means that the domain G_a has an outward pointing cusp whose "opening" is somehow measured by the parameter a.

Theorem 5.23. *Let $D\subset\mathbb{C}$ be a domain with a local Dini-smooth corner at $0\in\partial D$ of opening $\pi\alpha$ $(0<\alpha\le 2)$ and let $G_b\in\mathbb{C}$ be a domain with an outward pointing cusp at 0. Let $f\in \operatorname{Hol}(D,G_b)$. If there exists $N\ge\frac{1}{\alpha}$ such that*

$$\lim_{w\to 0}\left|\frac{f(w)}{w^N}\right|=L<+\infty,$$

then $f\equiv 0$.

Theorem 5.24. *Let G_a and G_b be two domains of $\mathbb{C}$ with outward pointing cusps at 0. If $f\in\operatorname{Hol}(G_a,G_b)$ is such that*

$$\lim_{w\to 0}\left|\frac{f(w)}{w}\right|=L<\frac{a}{b},$$

then $f\equiv 0$.

6 Infinitesimal Versions of the Schwarz Lemma and the Julia–Wolff–Carathéodory Theorem

We exhibit here notable applications of old and recent developments of one-parameter continuous semigroups and properties of their generators in the spirit of the Schwarz Lemma and the Julia–Wolff–Carathéodory Theorem.

6.1 One-Parameter Continuous Semigroups and Their Generators

We begin this section by recalling a formal definition of a one-parameter continuous semigroup.

Let $D\subset\mathbb{C}$ be a domain. A family $\mathcal{S}=\{F_t\}_{t\ge 0}\subset\operatorname{Hol}(D)$ is called a *one-parameter continuous semigroup on D* if

(i) $F_{t+s}(z)=F_t(F_s(z))$ *for all* $t,s\in[0,\infty)$ *and* $z\in D$;
(ii) $\lim_{t\to 0^+}F_t(z)=z$ *for all* $z\in D$.

In the case that conditions (i) and (ii) are satisfied for all $t, s \in \mathbb{R}$, the family $\mathcal{S}$ is, in fact, *a one-parameter continuous group of automorphisms* of D. The set of all automorphisms of D is usually denoted by $\mathrm{Aut}(D)$.

It is a well-known fact (see, for example, [2, 27] and [131]) that condition (ii) implies that for all $s > 0$ and $z \in D$,

$$\lim_{t \to s} F_t(z) = F_s(z).$$

Moreover, it follows from a result of Earl Berkson and Horacio Porta [27] that every continuous semigroup $\mathcal{S}$ on a simply connected domain D is differentiable in $t \in \mathbb{R}^+ = [0, \infty)$. So, for each $z \in D$, there exists the limit

$$\lim_{t \to 0+} \frac{z - F_t(z)}{t} =: f(z) \tag{85}$$

which defines a holomorphic mapping $f \in \mathrm{Hol}(D, \mathbb{C})$. The convergence in (85) is uniform on compacta in D. For the finite-dimensional case, see Marco Abate [2] (see also [121, 124] for Banach spaces).

Moreover, the function $u : \mathbb{R}^+ \times D \mapsto D$ defined by $u(t, z) = F_t(z)$ is the unique solution of the Cauchy problem

$$\begin{cases} \dfrac{\partial u(t, z)}{\partial t} + f(u(t, z)) = 0, & t \geq 0, \\ u(0, z) = z, & z \in D. \end{cases} \tag{86}$$

It follows from the uniqueness of the solution of Cauchy problem (86) that each element of a continuous semigroup is a univalent function on D. Furthermore, one can show [121, 131] that the function $u(t, z)$ satisfies the partial differential equation

$$\frac{\partial u(t, z)}{\partial t} + \frac{\partial u(t, z)}{\partial z} \cdot f(z) = 0, \quad z \in D.$$

The function $f \in \mathrm{Hol}(D, \mathbb{C})$ defined by (85) is called *the (infinitesimal) generator of the one-parameter continuous semigroup* $\mathcal{S} = \{F_t\}_{t \geq 0}$. We let $\mathcal{G}(D)$ denote the set of all holomorphic generators on D. If D is a convex domain, the set $\mathcal{G}(D)$ is a real cone in $\mathrm{Hol}(D, \mathbb{C})$, while the set of all group generators on D is a real Banach algebra. The set of group generators on D is usually denoted by $\mathrm{aut}(D)$ (see [121]).

Different parametric representation of the class $\mathcal{G}(\Delta)$ has been established.

Theorem 6.1. *Let* $f \in \mathrm{Hol}(\Delta, \mathbb{C})$. *The following are equivalent.*

(i) $f \in \mathcal{G}(\Delta)$;
(ii) there exists a unique point $\tau \in \overline{\Delta}$ *such that*

$$f(z) = (z - \tau)(1 - \overline{\tau} z) p(z), \quad z \in \Delta, \tag{87}$$

where $p \in \mathrm{Hol}(\Delta, \mathbb{C})$ *with* $\mathrm{Re}\, p(z) \geq 0$.

(iii) f *admits the representation*

$$f(z) = a - \overline{a}z^2 + zq(z), \quad z \in \Delta, \tag{88}$$

where $a \in \mathbb{C}$ *and* $q \in \mathrm{Hol}(\Delta, \mathbb{C})$ *with* $\mathrm{Re}\, q(z) \geq 0$. *Moreover,* $f \in \mathrm{Aut}(\Delta)$ *if and only if* $\mathrm{Re}\, q(z) = 0$, *i.e.,* $q(z) = ib$, $b \in \mathbb{R}$.

Thus, the equivalence of (i) and (iii) implies that the set $\mathcal{G}(\Delta)$ and the cone $\mathbb{C} \times \mathsf{P}$ (here P is the set of all holomorphic functions on Δ with the nonnegative real part) are isometric. Furthermore, all $f \in \mathrm{aut}(\Delta)$ are, in fact, polynomials of at most order 2 and have the form $f(z) = a - \overline{a}z^2 + ibz$, $a \in \mathbb{C}$, $b \in \mathbb{R}$.

Note that representation (87) due to Berkson and Porta [27] is unique. Representation (88) was established by Dov Aharonov, Mark Elin, Simeon Reich, and David Shoikhet in [3].

A natural question is whether, for $f \in \mathrm{Hol}(\Delta, \mathbb{C})$, there exist conditions which insure that $f \in \mathcal{G}(\Delta)$. Reich and Shoikhet in [121] provided a simple criterion for a special case. Namely, they proved that *if* $f \in \mathrm{Hol}(\Delta, \mathbb{C})$ *has a continuous extension to* $\overline{\Delta}$, *then* $f \in \mathcal{G}(\Delta)$ *if and only if* $\mathrm{Re}(f(z)\overline{z}) \geq 0$ *for all* $z \in \partial\Delta$.

However, there are holomorphic functions on Δ which have no continuous extension to $\overline{\Delta}$. The following assertion combines criteria that were established by Aharonov, Reich, and Shoikhet in [4], by Aharonov, Elin, Reich, and Shoikhet in [3], and by Reich and Shoikhet in [122].

Theorem 6.2. *Let* $f \in \mathrm{Hol}(\Delta, \mathbb{C})$. *The following are equivalent.*

(i) $f \in \mathcal{G}(\Delta)$.
(ii) $\mathrm{Re}(f(z)\overline{z}) \geq \mathrm{Re}(f(0)\overline{z})\left(1 - |z|^2\right)$ *for all* $z \in \Delta$. *Moreover, equality holds if and only if* $f \in \mathrm{aut}(\Delta)$.
(iii) $\mathrm{Re}\, f'(0) \geq 0$ *and*

$$\mathrm{Re}\Big(f(0)\overline{z}\Big)(1 - |z|^2) + \mathrm{Re}\, f'(0)\, \frac{1 + |z|}{1 - |z|}\, |z|^2 \geq \mathrm{Re}\Big(f(z)\overline{z}\Big) \geq$$

$$\geq \mathrm{Re}\Big(f(0)\overline{z}\Big)(1 - |z|^2) + \mathrm{Re}\, f'(0)\, \frac{1 - |z|}{1 + |z|}\, |z|^2$$

for all $z \in \Delta$. *Moreover, equality holds if and only if* $\mathrm{Re}\, f'(0) = 0$. *In this case* $f \in \mathrm{aut}(\Delta)$.
(iv) $\mathrm{Re}\left[\dfrac{f(z)\overline{z}}{1 - |z|^2} + \dfrac{f(w)\overline{w}}{1 - |w|^2}\right] \geq \mathrm{Re}\, \dfrac{\overline{z}f(w) + w\overline{f(z)}}{1 - \overline{z}w}$ *for all* $z, w \in \Delta$.

Assertion (iii) can be considered a distortion theorem for the class $\mathcal{G}(\Delta)$. It can be obtained using assertion (ii) theorem and Harnack's inequality for holomorphic functions with positive real part:

$$\mathrm{Re}\Big(p(0)\Big)\frac{1 - |z|}{1 + |z|} \leq \mathrm{Re}\Big(p(z)\Big) \leq \mathrm{Re}\Big(p(0)\Big)\frac{1 + |z|}{1 - |z|}, \quad z \in \Delta.$$

Harnack's inequality itself can be obtained directly from the Schwarz Lemma. Indeed, let $p \in \mathrm{Hol}(\Delta, \Pi^+)$. If $\mathrm{Re}\, p(0) = 0$, the inequality is obvious. Otherwise, the function F defined by $F(z) = \frac{p(z)-p(0)}{p(z)+\overline{p(0)}}$ belongs to $\mathrm{Hol}(\Delta)$ and satisfies $F(0) = 0$. Since $\mathrm{Re}\, p(z) = \mathrm{Re}\, p(0)\frac{1-|F(z)|^2}{|1-F(z)|^2}$, we have $\mathrm{Re}\, p(0)\frac{1-|F(z)|}{1+|F(z)|} \le \mathrm{Re}\, p(z) \le \mathrm{Re}\, p(0)\frac{1+|F(z)|}{1-|F(z)|}$. By the Schwarz Lemma, $|F(z)| \le |z|$, and we done.

In our context, assertion (iv) can be considered an infinitesimal version of the Schwarz–Pick Lemma.

We now review some fundamental properties of continuous semigroups and their generators that follow from the Berkson–Porta representation (87). We consider a semigroup $\mathcal{S} = \{F_t\}_{t\ge 0} \subset \mathrm{Hol}(\Delta)$ generated by $f(z) = (z-\tau)(1-\overline{\tau}z)p(z)$, $\mathrm{Re}(p(z)) \ge 0$ for all $z \in \Delta$ and make the following observations.

- If τ in (87) is an interior point of Δ and f does not vanish identically on Δ, then τ is the unique null point of f in Δ. Moreover, (due to the uniqueness of the solution to the Cauchy problem (86)) τ is a common fixed point of $\mathcal{S}$, i.e.,

$$F_t(\tau) = \tau \quad \text{for all} \quad t \ge 0.$$

- If $\tau \in \partial\Delta$, then f has no fixed point in Δ. In this case, τ is a boundary fixed point of F_t for each $t \ge 0$ in the sense that

$$\lim_{r\to 1^-} F_t(r\tau) = \tau.$$

- If for at least one t_0, the function F_{t_0} is neither the identity nor an elliptic automorphism of Δ, then there is a unique point $\tau \in \overline{\Delta}$ such that the semigroup $\{F_t\}_{t\ge 0}$ converges to τ as $t \to \infty$ uniformly on compact subsets of Δ., i.e., the point $\tau \in \overline{\Delta}$ in (87) is an attractive fixed point of the semigroup $\mathcal{S}$, namely,

$$\lim_{t\to\infty} F_t(z) = \tau \quad \text{for all} \quad z \in \Delta.$$

The last assertion is a continuous analog of the Denjoy–Wolff theorem. The first general continuous analog of the Denjoy–Wolff Theorem was given by Berkson and Porta in terms of semigroup generators [27] (see also [123] and [131]). The point τ in (87) is called the *Denjoy–Wolff point* of the semigroup $\mathcal{S} = \{F_t\}_{t\ge 0}$. In case $\tau \in \partial\Delta$, it is also often called Wolff's point, or the sink point of $\mathcal{S}$.

6.2 *Upper and Lower Schwarz–Pick Type Estimates*

Let f be the generator of a one-parameter continuous semigroup $\mathcal{S} = \{F_t(z)\}_{t\ge 0}$ on Δ. Suppose that $\mathcal{S}$ is not trivial, does not contain elliptic automorphisms, and that τ is the unique null point of f in Δ. In this case, τ is the attractive fixed point of the semigroup $\mathcal{S}$ and $F_t'(\tau) = e^{-t\,\mathrm{Re}\, f'(\tau)}$ (see [131]). By the Schwarz Lemma,

$e^{-t\,\mathrm{Re}\,f'(\tau)} \leq 1$ (see Sect. 2.1); therefore $\mathrm{Re}\,f'(\tau) \geq 0$. In addition, assuming $f(0) = 0$, we have, by the Schwarz Lemma, the invariance result: $|F_t(z)| \leq |z|$ for all $z \in \Delta$ and $t \geq 0$. It turns out that the fact that $\mathcal{S}$ is a continuous semigroup generated by f leads to a more qualified estimate (see, for example, [131]).

Theorem 6.3. *Let $f \in \mathcal{G}(\Delta)$ and $\mathcal{S} = \{F_t\}_{t\geq 0}$ be the semigroup generated by f. Assume that $f(0) = 0$ and $\lambda := \mathrm{Re}\,f'(0) > 0$. There exists $c \in [0,1]$ such that for all $z \in \Delta$ and $t \geq 0$,*

(i) $|z| \cdot \exp\left(-\lambda t \dfrac{1+c|z|}{1-c|z|}\right) \leq |F_t(z)| \leq |z| \cdot \exp\left(-\lambda t \dfrac{1-c|z|}{1+c|z|}\right),$

(ii) $\exp(-\lambda t)\dfrac{|z|}{(1+c|z|)^2} \leq \dfrac{|F_t(z)|}{(1-c\,|F_t(z)|)^2} \leq \exp(-\lambda t)\dfrac{|z|}{(1-c|z|)^2}.$

Here, we have not only refined the upper bound estimate for $|F_t(z)|$ in Δ, but the lower bound as well. In fact, inequality (i) implies that for each $z \in \Delta$, the rate of convergence of the semigroup to its interior Denjoy–Wolff point is exponential.

Estimate (i) with $c = 1$ is due to Kenneth R. Gurganus [80], while estimate (ii) was established by Tadeusz Poreda [120].

Applying Theorem 6.3 to an appropriate Möbius transform gives the following result.

Corollary 6.1. *Let $\tau \in \Delta$, $f \in \mathcal{G}(\Delta)$ and $\mathcal{S} = \{F_t\}_{t\geq 0}$ be the semigroup generated by f. Assume that $f(\tau) = 0$ and $\lambda := \mathrm{Re}\,f'(\tau) > 0$. There exists $c \in [0,1]$ such that for all $z \in \Delta$ and $t \geq 0$, the following estimates hold*

$$\delta(z,\tau)\cdot\exp\left(-\lambda t\frac{1+c\delta(z,\tau)}{1-c\delta(z,\tau)}\right) \leq \delta\left(F_t(z),\tau\right) \leq \delta(z,\tau)\cdot\exp\left(-\lambda t\frac{1-c\delta(z,\tau)}{1+c\delta(z,\tau)}\right),$$

where

$$\delta(z,\tau) = |m_\tau(z)| = \left|\frac{z-\tau}{1-\overline{\tau}z}\right|$$

is the pseudo-hyperbolic distance on Δ.

6.3 Semigroup Generators with Boundary Null Points

In this section, we exhibit various representations for semigroup generators having boundary regular null points. Most estimates for semigroups and their generators in the spirit of the Julia–Charathéodory Theorem 2.6 are based on these formulas.

Suppose that the semigroup $\mathcal{S}$ generated by $f \in \mathcal{G}(\Delta)$ has a boundary Denjoy–Wolff point $\tau \in \partial\Delta$. By the Berkson–Porta formula (87), τ is a boundary null point of the semigroup generator f. The Riesz–Herglotz representation of functions with positive real part was used in [65] to show that the angular derivative $f'(\tau)$ exists finitely and is a real nonnegative number.

Note that, in general, the generator $f \in \mathcal{G}(\Delta)$ may have more than one boundary null point, and for each such point $\zeta \in \partial\Delta$, $\zeta \neq \tau$, the angular derivative $f'(\zeta)$ exists and is either a real negative number or infinity (see the discussion in Sect. 6.4 below). A point $\zeta \in \partial\Delta$ such that $f(\zeta) = 0$ and the angular derivative $f'(\zeta)$ is finite is called a *boundary regular null point* of $f \in \mathrm{Hol}(\Delta, \mathbb{C})$. If $\zeta \in \partial\Delta$ is a boundary regular null point of $f \in \mathcal{G}(\Delta)$ and $f'(\zeta) < 0$, then ζ is a repelling fixed point for the semigroup generated by f.

For a point $\zeta \in \overline{\Delta}$, we define the class

$$\mathcal{G}[\zeta] := \{f \in \mathcal{G}(\Delta) : \ f(\zeta) = 0 \quad \text{and} \quad f'(\zeta) \quad \text{exists finitely}\}. \tag{89}$$

For the boundary Denjoy–Wolff point, the equivalence of the analysis of conditions (87) and (88) leads to the following decomposition theorem which was established in [132].

Theorem 6.4. *Let $\zeta \in \partial\Delta$. Then $f \in \mathcal{G}[\zeta]$ admits the representation*

$$f(z) = (z - \zeta)\left(1 - z\overline{\zeta}\right) p(z) + \frac{\lambda}{2}\left(\overline{\zeta} z^2 - \zeta\right), \tag{90}$$

where

$$\lambda = f'(\zeta), \quad \operatorname{Re} p\,(z) \geq 0 \quad \text{and} \quad \angle \lim_{z \to \zeta} \left(1 - z\overline{\zeta}\right) p\,(z) = 0. \tag{91}$$

The first term in (90) is a parabolic type generator, while the second term generates a group of hyperbolic type. It is clear that the point ζ in (90) is the Denjoy–Wolff point of the corresponding semigroup if and only if $\lambda \geq 0$.

Victor Goryainov gave a representation of semigroup generators more precise than (87) in the case a semigroup generator has many boundary null points.

Theorem 6.5 (See [76]). *Let $\mathcal{S} = \{F_t\}_{t \geq 0}$ be a semigroup generated by $f \in \mathcal{G}(\Delta)$ and $\tau = 1$ be the Denjoy–Wolff point of $\mathcal{S}$. If $-1, \zeta_2, \zeta_3, \ldots, \zeta_n$ are distinct repelling fixed points of $\mathcal{S}$ with finite angular derivatives $F_t'(\zeta_k)$ for all $t \geq 0$ and $k = 0, 1, \ldots, n$, then f admits the representation*

$$f(z) = \alpha(1 + z)(1 - z)^2 \frac{1 + \varphi(z)}{1 - z\varphi(z)} \tag{92}$$

with $\alpha \leq 0$, $\varphi \in \mathrm{Hol}(\Delta)$ and $\sup \left\{ \dfrac{|1 + \psi(z)|^2}{1 - |\psi(z)|^2} \dfrac{1 - |z|^2}{|z - \zeta_k|^2} : z \in \Delta \right\} < \infty$ *for $k = 2, 3, \ldots, n$, where* $\psi(z) = \dfrac{(1 + z)\varphi(z) + z(1 + \varphi(z))}{(1 + z) + (1 + \varphi(z))}$.

In the particular case that only one repelling fixed point $\zeta = -1$ with a finite angular derivative is given, the semigroup generator f admits the representation

$$f(z) = \alpha(1+z)(1-z)^2 \frac{1+\varphi(z)}{1-z\varphi(z)}$$

with $\alpha \le 0$, $\varphi \in \mathrm{Hol}(\Delta)$.

Notice that null points of a generator are singular points of the corresponding dynamical system. From this point of view, formula (92) has two disadvantages: the null points of the generator do not appear explicitly in its representation and the distinction between parabolic and hyperbolic type generators is not demonstrated.

On the other hand, consider the group $\mathcal{S} = \{F_t\}_{t\ge 0}$ of hyperbolic automorphisms having attractive point $\tau = 1$ and repelling fixed point at $\zeta = -1$. By (87) the generator of $\mathcal{S}$ is of the form $f = \frac{\beta}{2}(z-1)(z+1)$, and its singularities can be separated as

$$\frac{1}{f(z)} = \frac{1}{\beta}\left(\frac{1}{z-1} - \frac{1}{z+1}\right).$$

It turns out that, in general, each rational function f described in the form

$$\frac{1}{f(z)} = \sum_{k=1}^{n} \frac{1}{|f'(\zeta_k)|}\left(\frac{1}{z-1} - \frac{1}{z-\zeta_k}\right),$$

with $\zeta_1, \zeta_2, \ldots, \zeta_n \in \partial\Delta$ generates a semigroup which has Denjoy–Wolff point 1 and repelling null points $\zeta_1, \zeta_2, \ldots, \zeta_n$.

In the opposite direction, Elin et al. [71] recently established the following theorem on separation of singularities.

Theorem 6.6. *Let $f \in \mathcal{G}(\Delta)$ generate a semigroup with the Denjoy–Wolff point $\tau \in \overline{\Delta}$. Suppose that $\zeta_1, \zeta_2, \ldots, \zeta_n$ are boundary regular null points of f distinct from τ (each $f'(\zeta_k)$ being negative). There exist a number $r \ge 0$ and a function $h \in \mathcal{G}[\tau]$ satisfying $\angle \lim\limits_{z\to\zeta_k} \dfrac{h(z)}{z-\zeta_k} = \infty$ for all $k = 1, 2, \ldots, n$ such that*

(i) if $\tau = 0$, then

$$\frac{1}{f(z)} = \sum_{k=1}^{n} \frac{1}{|f'(\zeta_k)|}\left(\frac{1}{2z} - \frac{1}{z-\zeta_k}\right) + \frac{r}{h(z)};$$

(ii) if $\tau = 1$, then

$$\frac{1}{f(z)} = \sum_{k=1}^{n} \frac{1}{|f'(\zeta_k)|}\left(\frac{1}{z-1} - \frac{1}{z-\zeta_k}\right) + \frac{r}{h(z)}$$

with $h'(1) = f'(1)$.

6.4 Rates of Convergence

The sharp Schwarz type inequalities pointed out in Sect. 6.2 actually give rates of convergence of a semigroup to its interior Denjoy–Wolff point. Hence, it is natural to consider an analog of the Julia–Wolff–Carathéodory Theorem (see Sect. 2.5) for a study of the asymptotic behavior of a semigroup in terms of its generator. The following result, established by Elin and Shoikhet, is a continuous version of the Julia–Wolff–Carathéodory Theorem.

Theorem 6.7 ([65]). *Let $\mathcal{S} = \{F_t\}_{t\geq 0}$ be a one-parameter continuous semigroup generated by $f \in \mathcal{G}$. The following are equivalent.*

(i) *f has no null point in Δ.*
(ii) *f admits the Berkson–Porta representation*

$$f(z) = (1 - z\bar{\tau})(z - \tau)p(z)$$

for some $\tau \in \partial\Delta$ and $\operatorname{Re} p(z) \geq 0$ *everywhere.*
(iii) *There exists a point $\tau \in \partial\Delta$ such that* $\operatorname{Re} f'(\tau) \geq 0$.
(iv) *There exists a point $\tau \in \partial\Delta$ such that*

$$\angle \lim_{z\to\tau} \frac{f(z)}{z - \tau} = \beta$$

with $\operatorname{Re} \beta \geq 0$.
(v) *There exist a point $\tau \in \partial\Delta$ and a real number $\gamma > 0$, such that*

$$\frac{|F_t(z) - \tau|^2}{1 - |F_t(z)|^2} \leq e^{-t\gamma} \frac{|z - \tau|^2}{1 - |z|^2}. \tag{93}$$

Moreover,

(a) *the points $\tau \in \partial\Delta$ in (ii)–(v) are the same;*
(b) *the number β in (iv) is, in fact, a nonnegative real number, which is the maximum of all $\gamma \geq 0$ that satisfy (93).*

Remark 6.1. Note that in the above theorem the implication $(iv) \Rightarrow (v)$ follows from the relation $\frac{\partial F_t}{\partial z}(\tau) = e^{-t\beta}$ for all elements F_t of $\mathcal{S}$ by the Julia–Carathéodory Theorem 2.6. Later, Shoikhet [132] showed that a point $\zeta \in \partial\Delta$ is a boundary regular null point of f if and only if ζ is a repelling fixed point for all elements F_t of $\mathcal{S}$ and $\frac{\partial F_t}{\partial z}(\zeta) = e^{-tf'(\zeta)}$. Thus

$$\frac{|F_t(z) - \zeta|^2}{1 - |F_t(z)|^2} \leq e^{-t\gamma} \frac{|z - \zeta|^2}{1 - |z|^2}$$

at each boundary regular null point ζ of f. The same fact was proved later by Contreras and Díaz-Madrigal in [51].

Observe that, in contrast to the case of interior null points, a boundary null point ζ of a generator f may be not a fixed point of the generated semigroup if $f'(\zeta)$ does not exist finitely (see Example 1, p. 104 in [131]).

It follows that if $\zeta = \tau \in \partial\Delta$ is the Denjoy–Wolff point of the semigroup ($f'(\tau) \geq 0$), then the horocycles

$$D(\tau, k) = \left\{ z \in \Delta : \frac{|z-\tau|^2}{1-|z|^2} < k \right\}, \ k > 0, \tag{94}$$

internally tangent to $\partial\Delta$ at τ are F_t-invariant for all $t \geq 0$. For hyperbolic type semigroups the inequality

$$\frac{|F_t(z) - \tau|^2}{1 - |F_t(z)|^2} \leq e^{-tf'(\tau)} \frac{|z-\tau|^2}{1-|z|^2} \tag{95}$$

establishes the rate of convergence in terms of the non-Euclidean "distance" $d(w, z)$ from $w \in \overline{\Delta}$ to $z \in \Delta$ defined by

$$d(w, z) := \frac{|z-w|^2}{1-|z|^2}. \tag{96}$$

In this case the rate of convergence is at least exponential.

For semigroups of parabolic type ($f'(\tau) = 0$), inequality (95) does not provide any rate of convergence.

Thus, there remain two yet unanswered questions.

- In the hyperbolic case, is the rate of convergence indeed exponential? Equivalently, does there exist a lower bound for $d(\zeta, F_t(z))$ in (95), and if so, is it of exponential type?
- What is the rate of convergence for the parabolic type semigroups?

It turns out that for both parabolic and hyperbolic cases, estimate (95) can be improved in cases that the generator f of a semigroup $\mathcal{S}$ satisfies some additional natural conditions. We present here several results by Elin, Reich, Shoikhet, and Fiana Yacobzon in [70]. Without losing any of the generality, we set $\tau = 1$. The function f then admits the representation

$$f(z) = -(1-z)^2 q(z), \tag{97}$$

where $\operatorname{Re} q(z) \geq 0, z \in \Delta$.

Theorem 6.8. *Let $\mathcal{S} = \{F_t\}_{t \geq 0} \subset \operatorname{Hol}(\Delta)$ be the semigroup generated by f of form (97) with $f'(1) = \beta \geq 0$. There exists a real constant $m \geq 0$ such that $\operatorname{Re} q(z) \geq m$ for all $z \in \Delta$ if and only if*

$$d\left(1, F_t(z)\right) \leq \begin{cases} \dfrac{d(1,z)e^{-\beta t}}{1 + \dfrac{1-e^{-\beta t}}{\beta} 2md(1,z)}, & \textit{if } \beta \neq 0, \\ \dfrac{d(1,z)}{1 + 2mtd(1,z)}, & \textit{if } \beta = 0. \end{cases} \tag{98}$$

Moreover, if for some $z_0 \in \Delta$ and $t_0 > 0$, equality in (98) holds for $z = z_0$ and all $t \in [0, t_0]$, then the semigroup $\mathcal{S}$ consists of linear fractional mappings (LFM's).

For hyperbolic type semigroups, the first inequality improves estimate (95). In the particular case $m = 0$, estimates (95) and (98) coincide. For parabolic type semigroups ($\beta = 0$), the second inequality establishes a rate of convergence in terms of the non-Euclidean "distance" d. The rate of convergence is at least $O\left(\frac{1}{t}\right)$ except in the case $m = 0$; in that case, estimate (98) coincides with (95).

Note that by the Julia–Carathéodory Theorem applied to the positive real part function q defined by (97), the inequality $\operatorname{Re} q(z) \geq m$ implies that

$$\operatorname{Re} q(z) \geq \frac{\beta}{2} \frac{1-|z|^2}{|1-z|^2} + m. \tag{99}$$

If $m > 0$, then $\tau = 1$ is the only null point in $\overline{\Delta}$ of the function f in (97). In this case, condition (99) means that the semigroup generated by f has no repelling points on $\partial\Delta$.

We are able to improve the estimates (95) even more for a narrow class of generators (which is, nevertheless, often used in different applications of semigroup theory, for example, in the theory of branching processes).

Theorem 6.9. *Let $\mathcal{S} = \{F_t\}_{t\geq 0} \subset \operatorname{Hol}(\Delta)$ be the semigroup generated by f of form (97) with $f'(1) = \beta \geq 0$. Fix $0 \leq \alpha < 1$ and $m \geq 0$. Then*

$$\operatorname{Re} q(z) \geq \frac{\beta}{2} \frac{1-|z|^2}{|1-z|^2} + m \left(\frac{1-|z|^2}{|1-z|^2}\right)^{\alpha}, \quad z \in \Delta, \tag{100}$$

if and only if

$$d\left(1, F_t(z)\right) \leq \begin{cases} \dfrac{d(1,z)e^{-\beta t}}{\left[1 + \dfrac{1-e^{-\beta t(1-\alpha)}}{\beta} 2m(d(1,z))^{1-\alpha}\right]^{\frac{1}{1-\alpha}}}, & \textit{if } \beta \neq 0, \\ \dfrac{d(1,z)}{\left[1 + 2mt(d(1,z))^{1-\alpha}\right]^{\frac{1}{1-\alpha}}}, & \textit{if } \beta = 0. \end{cases} \tag{101}$$

For $\alpha = 0$, the first part of Theorem 6.8 is just a particular case of Theorem 6.9. For $0 < \alpha < 1$, inequality (101) provides a better estimate for the rate of convergence than (98). The example of the function

$$f(z) = -(1-z)^2\left(\frac{\beta}{2}\frac{1+z}{1-z} + m\left(\frac{1+z}{1-z}\right)^{\alpha}\right)$$

shows that the class of generators which satisfy condition (100) is not trivial (see [70] for details).

Theorems 6.8 and 6.9 do not give a lower bound for the non-Euclidean "distance" $d(1, F_t(z))$; hence, they do not give the precise rate of convergence of semigroups to their boundary Denjoy–Wolff point τ. So we still need to consider the following question.

- *What conditions on the generator f of a semigroup $\mathcal{S} = \{F_t\}_{t\geq 0}$ ensure the existence of a function $g(t)$ such that the limit*

$$\varepsilon(z) := \lim_{z\to\tau} g(t)d(\tau, F_t(z)) \tag{102}$$

exists and is different from zero?

The next theorem establishes a criterion for exponential convergence of hyperbolic type semigroups. Recall that each generator which belongs to $\mathcal{G}[1]$ with $f'(1) \geq 0$ can be represented as

$$f(z) = -(1-z)^2 p(z) + \frac{\beta}{2}\left(z^2 - 1\right), \tag{103}$$

where $\operatorname{Re} p(z) \geq 0$, $\angle\lim_{z\to 1}(1-z)p(z) = 0$ and $\beta = f'(1)$ (see Theorem 6.4).

Theorem 6.10. *Let $f \in \mathcal{G}[1]$ be of form (103) and $\mathcal{S} = \{F_t\}_{t\geq 0}$ be a semigroup on Δ generated by f.*

(i) The limit

$$\lim_{t\to\infty} e^{\beta t} d(1, F_t(z)) := \varepsilon(z) \tag{104}$$

exists.

(ii) The function $\varepsilon(z)$ is either identically zero or $\varepsilon(z) > 0$ for all $z \in \Delta$.

(iii) $\varepsilon(z) > 0$ for all $z \in \Delta$ if and only if the integral

$$\int_0^z p(\xi)\, d\xi$$

converges non-tangentially as $z \to 1$.

The following example shows that the convergence might be faster than exponential.

Example 6.1. Let $f \in \mathcal{G}[1]$ be of form (103) with

$$p(z) = \left((1-z) \log \frac{2}{1-z} \right)^{-1}.$$

Here $\angle \lim_{z \to 1} (1-z)\, p(z) = 0$, $\beta = f'(1) > 0$ and

$$\operatorname{Re} \frac{1}{p(z)} = \operatorname{Re} (1-z) \ln \left| \frac{2}{1-z} \right| + \arg (1-z) \operatorname{Im} (1-z).$$

Since $\arg (1-z) \operatorname{Im} (1-z) \geq 0$, we see that $\operatorname{Re} p(z) \geq 0$ for all $z \in \Delta$. It is easy to verify that the integral $\int_0^z p(\xi)\, d\xi$ is divergent.

Now let $\mathcal{S} = \{F_t\}_{t \geq 0}$ be the semigroup on Δ generated by f. It follows from assertion (iii) of Theorem 6.10 that $\lim_{t \to \infty} e^{\beta t} d(1, F_t(z)) = 0$. Thus, the convergence of the semigroup $\mathcal{S}$ to its Denjoy–Wolff point $\tau = 1$ is faster than exponential.

The problem of finding the precise rate of convergence of a hyperbolic type semigroup to its Denjoy–Wolff point in the case the function ε in (104) is identically zero is still open. This question is closely related to the problem of finding a lower bound for the function $d(\tau, F_t(z))$, where $d(\tau, z)$ is defined by (96). In the parabolic case, the problem of finding a function $g(t)$ that establishes a criterion similar to that of the hyperbolic case (102) is also still open.

It is natural to suppose that under stronger assumptions on the smoothness of a generator f at the boundary Denjoy–Wolff point $\tau = 1$, one can derive better estimates of the rate of convergence of the corresponding semigroup. In fact, the following theorem by Elin and Shoikhet [67] gives complete quantitative characteristics of the rate of convergence of parabolic type semigroups under additional hypothesis on the smoothness of their generators f at τ.

Theorem 6.11. *Let $\mathcal{S} = \{F_t\}_{t \geq 0}$ be a continuous semigroup of holomorphic self-mappings of the open unit disk Δ and f be its generator.*

(i) Suppose that f admits the representation

$$f(z) = b(z-1)^2 + R(z), \tag{105}$$

where $R \in \operatorname{Hol}(\Delta, \mathbb{C})$ and $\lim_{z \to 1} \frac{R(z)}{(z-1)^2} = 0$. Then

$$\frac{1}{1 - F_t(z)} = -bt + G(z,t), \quad \text{where } \lim_{t \to \infty} \frac{G(z,t)}{t} = 0. \tag{106}$$

(ii) If f admits the representation

$$f(z) = b(z-1)^2 + c(z-1)^3 + R_1(z) \tag{107}$$

with $b \neq 0$ and $\lim_{z\to 1} \frac{R_1(z)}{(z-1)^3} = 0$, then

$$\frac{1}{1-F_t(z)} = -bt - \frac{c}{b}\log(t+1) + G_1(z,t), \tag{108}$$

where $\lim_{t\to\infty} \frac{G_1(z,t)}{\log(t+1)} = 0$.

(iii) If the function R_1 in (107) satisfies the condition $\lim_{z\to 1} \frac{R_1(z)}{(z-1)^{3+\varepsilon}} = 0$ for some $\varepsilon > 0$, then

$$\frac{1}{1-F_t(z)} = -bt - \frac{c}{b}\log(t+1) + A(z) + G_2(z,t), \tag{109}$$

where $\lim_{t\to\infty} G_2(z,t) = 0$.

In particular, if the generator of a parabolic type semigroup with the boundary Denjoy–Wolff point $\tau = 1$ is at least twice differentiable at τ, then the rate of convergence of the semigroup to τ is $O\left(\frac{1}{t}\right)$.

If a semigroup consists of LFMs, the rate of convergence to its attractive boundary fixed point can be computed exactly for both the hyperbolic and parabolic cases.

Theorem 6.12 (See [87]). *Let $\mathcal{S} = \{F_t\}_{t\geq 0}$ be a semigroup with $F_1(z) = \frac{az+b}{cz+d}$, $F_1(1) = 1$, and $F_1'(1) > 0$. Then*

$$d\,(1, F_t(z)) = \frac{\lambda^t\,(1-\lambda)\,|1-z|^2}{\left(1-|z|^2\right)(1-\lambda) + \lambda\,(1-\lambda^t)\,\mathrm{Re}\,\alpha\,|1-z|^2}, \tag{110}$$

where $\lambda := F_1'(1) = \frac{a-c}{c+d}$ and $\alpha := \frac{b-c}{a-c}$.

Note that $\lambda = e^{-\beta}$, where β is the angular derivative of the generator f of $\mathcal{S}$ at 1 (see [65]). If $\beta \geq 0$, then formula (110) gives the rate of convergence of the semigroup to its Denjoy–Wolff point $\tau = 1$ in terms of the non-Euclidean "distance" (cf., (98))

$$d\left(1, F_t(z)\right) = \begin{cases} \dfrac{d(1,z)e^{-\beta t}}{1 + \dfrac{1-e^{-\beta t}}{\beta} 2m_\beta d(1,z)}, & \text{if } \beta \neq 0, \\ \dfrac{d(1,z)}{1 + 2m_\beta t d(1,z)}, & \text{if } \beta = 0. \end{cases} \tag{111}$$

with

$$m_\beta = \begin{cases} \dfrac{\operatorname{Re}\alpha}{2} \dfrac{\beta e^{-\beta}}{1-e^{-\beta}}, & \text{if } \beta \neq 0, \\ \dfrac{\operatorname{Re}\alpha}{2}, & \text{if } \beta = 0. \end{cases} \tag{112}$$

Thus, for a hyperbolic semigroup consisting of LFMs with the boundary Denjoy–Wolff point $\tau = 1$, the rate of convergence is exponential. The rate of convergence for a parabolic type semigroup consisting of LFMs is $O\left(\frac{1}{t}\right)$.

6.5 *Rigidity of Holomorphic Generators*

We have already mentioned at the beginning of Sect. 5 that by the Schwarz–Pick Lemma, if a self-mapping F agrees with an elliptic automorphism up to first order at its interior fixed point $\tau \in \Delta$, then the two mappings coincide on Δ. The infinitesimal version of this fact follows, for instance, by Corollary 6.1: *if for some $\tau \in \Delta$, a semigroup generator $f \in \mathcal{G}[\tau]$ satisfies* $\operatorname{Re} f'(\tau) = 0$, *then f generates a group of elliptic automorphisms.* These results do not hold if τ is replaced with a boundary point. Whence various rigidity principles for a single self-mapping of the disk were considered in Sect. 5, here we examine their infinitesimal counterparts. We note that since for each $F \in \operatorname{Hol}(\Delta)$ the function $f := I - F$ belongs to the class $\mathcal{G}(\Delta)$, any uniqueness result for the class of semigroup generators automatically implies a uniqueness result for holomorphic self-mappings of the unit disk.

From this point of view, one expects that the Burns–Krantz Theorem 5.1 can be generalized as follows. If a holomorphic generator vanishes up to third order at a boundary point, then it vanishes identically in Δ and, consequently, generates the trivial semigroup of the identity mappings. This result was proved by Elin, Levenshtein, Shoikhet, and Tauraso in [68]. More precisely:

Theorem 6.13. *Let $f \in \mathcal{G}(\Delta)$. Suppose that for some $\tau \in \partial\Delta$,*

$$f(z) = a(z-\tau)^3 + o(|z-\tau|^3)$$

as $z \to \tau$ in each non-tangential approach region at τ. Then $a\tau^2 \geq 0$. Moreover, $a = 0$ if and only if $f \equiv 0$.

Now let us consider class of functions $f \in \mathrm{Hol}(\Delta, \mathbb{C})$ consisting of functions continuous on $\overline{\Delta}$ that satisfy the boundary condition

$$\operatorname{Re} f(z)\overline{z} \geq |f(z)| \cos \frac{\alpha\pi}{2}, \quad z \in \partial\Delta, \tag{113}$$

for some $\alpha \in (0, 2]$. If $\alpha \leq 1$, then condition (113) implies that $\operatorname{Re}(f(z)\overline{z}) \geq 0$, $z \in \partial\Delta$, and consequently, f is a generator on Δ. Conversely, if a generator f on Δ is continuous on $\overline{\Delta}$, then (113) holds with $\alpha = 1$. So in a sense, this class generalizes the class of holomorphic generators which are continuous on $\overline{\Delta}$.

Theorem 6.14. *Let $f \in \mathrm{Hol}(\Delta, \mathbb{C})$ be continuous on $\overline{\Delta}$ and satisfy the condition (113). The condition*

$$\lim_{z \to \tau, z \in \overline{\Delta}} \frac{f(z)}{(z - \tau)^{2+\alpha}} = 0 \quad \textit{for some} \quad \tau \in \partial\Delta \tag{114}$$

implies that $f \equiv 0$.

The following assertion is an immediate consequence of this theorem.

Corollary 6.2. *Let $F \in \mathrm{Hol}(\Delta, \mathbb{C})$ be continuous on $\overline{\Delta}$ and satisfies the boundary condition*

$$\operatorname{Re} F(z)\overline{z} \leq 1 - |F(z) - z| \cos \frac{\alpha\pi}{2}, \quad z \in \partial\Delta, \tag{115}$$

for some $\alpha \in (0, 2]$. If there exists $\tau \in \partial\Delta$ such that

$$F(z) = \tau + (z - \tau) + o(|z - \tau|^{2+\alpha})$$

as $z \to \tau$, then $F \equiv I$.

As we have already seen in Sect. 5.3, if holomorphic self-mappings F and G of the open unit disk commute, then they share some common properties; in particular, they coincide under some weak assumptions. It is natural to consider holomorphic generators f and g of commuting semigroups $\{F_t\}_{t \geq 0}$ and $\{G_s\}_{s \geq 0}$ on Δ (i.e., such that $F_t \circ G_s = G_s \circ F_t$ for all $s, t \geq 0$). It turns out, as expected, that f and g coincide identically if they agree up to some order at a boundary point.

Theorem 6.15 ([68]). *Let f and g be generators of one-parameter commuting semigroups $\{F_t\}_{t \geq 0}$ and $\{G_t\}_{t \geq 0}$, respectively, and suppose that $f(\tau) = 0$ at some point $\tau \in \overline{\Delta}$.*

(i) If $\tau \in \Delta$ and $f'(\tau) = g'(\tau)$, then $f \equiv g$.
(ii) If $\tau \in \partial\Delta$ and f and g admit the representations

$$f(z) = f'(\tau)(z - \tau) + \ldots + \frac{f^{(m)}(\tau)}{m!}(z - \tau)^m + o(|z - \tau|^m) \tag{116}$$

and

$$g(z) = g(\tau) + g'(\tau)(z-\tau) + \ldots + \frac{g^{(m)}(\tau)}{m!}(z-\tau)^m + o(|z-\tau|^m) \quad (117)$$

as $z \to \tau$ *along some curve lying in* Δ *and ending at* τ *and if* $f^{(m)}(\tau) = g^{(m)}(\tau) \neq 0$, *then* $f \equiv g$.

Remark 6.2. If $\tau \in \partial\Delta$ is the Denjoy–Wolff point of a semigroup generated by a mapping $h \in \mathcal{G}(\Delta)$, then h admits the expansion

$$h(z) = h'(\tau)(z-\tau) + o(z-\tau)$$

as $z \to \tau$ in each non-tangential approach region at τ and $h'(\tau) = \angle\lim_{z\to\tau} h'(z)$. Moreover, in this case $h'(\tau) \geq 0$ and $h'(0) = 0$ if and only if h generates a semigroup of parabolic type (see [65]). Therefore, if f (or g) in Theorem 6.15 generates a semigroup of hyperbolic type with the Denjoy–Wolff point $\tau \in \partial\Delta$, then the condition $f'(\tau) = g'(\tau)$ suffices to ensure that $f \equiv g$.

Remark 6.3. As a matter of fact, if f and g have expansions (116) and (117), respectively, as $z \to \tau$ in each non-tangential approach region at $\tau \in \partial\Delta$ up to order $m = 3$, and $f'(\tau) = g'(\tau)$, $f''(\tau) = g''(\tau)$ and $f'''(\tau) = g'''(\tau)$, then $f \equiv g$.

If, in particular, $f^{(i)}(\tau) = g^{(i)}(\tau) = 0$, $i = 1, 2, 3$ then both f and g are equal zero identically on Δ.

The question as to what conditions ensure that $f \in \mathcal{G}(\Delta)$ generates a group of automorphisms, or, more generally, a semigroup of linear-fractional transformations, is of special interest. Shoikhet in [133] established such conditions in the class $\mathcal{G}[1]$.

Theorem 6.16. *Let* $f \in \mathrm{Hol}(\Delta, \mathbb{C}) \bigcap C_A^3(1)$ *satisfy the Berkson–Porta representation* $f(z) = -(1-z)^2 p(z)$ *with*

$$\inf_{z\in\Delta}\left[\mathrm{Re}\, p(z) - \frac{1}{2}\beta\frac{1-|z|^2}{|1-z|^2}\right] =: m \geq 0. \quad (118)$$

Then f *generates a semigroup* $S = \{F_t\}_{t\geq 0}$ *of linear-fractional transformations if and only if the following two conditions hold:*

(i) $f'(1) - \mathrm{Re}\, f''(1) \leq 2m$;
(ii) $f'''(1) = 0$.

Moreover, in this case, $m = 0$ *if and only if* f *is a generator of a group of automorphisms of* Δ.

Corollary 6.3. *Let $f \in \mathcal{G}(\Delta) \cap C_A^3(1)$ be of the form $f(z) = -(1-z)^2 p(z)$. Then f is a generator of a semigroup S of affine self-mappings of Δ if and only if the following two conditions hold:*

(i) $\operatorname{Re} p(z) \geq \frac{1}{2} f'(1)$, $z \in \Delta$;
(ii) $f''(1) = f'''(1) = 0$.

In this case, $\beta := f'(1) \geq 0$ *and* $f(z) = \beta(z-1)$.

Note that under conditions of Theorem 6.16, the generator f should be a polynomial of order two. In the case that a semigroup generator has additional boundary regular null points (so, is at least of higher order), it is natural to look for conditions that ensure that f is a rational function of minimal possible order. It turns out that the answer may be considered an infinitesimal analog of Cowen–Pommerenke inequalities (cf., Sect. 4.2). More precisely, Theorem 6.6 leads to the following relation between angular derivatives of a generator at its boundary regular null points. The case of equality is a rigidity result.

Theorem 6.17 ([71]). *Let $f \in \mathcal{G}(\Delta)$ generate a semigroup with the Denjoy–Wolff point $\tau \in \overline{\Delta}$. Suppose that $\zeta_1, \zeta_2, \ldots, \zeta_n$ are boundary regular null points of f different from τ (each $f'(\zeta_k)$ being negative).*

(i) If $\tau = 0$ then

$$\sum_{k=1}^{n} \frac{1}{|f'(\zeta_k)|} \leq 2 \operatorname{Re} \frac{1}{f'(0)}$$

and equality holds if and only if

$$\frac{1}{f(z)} = \sum_{k=1}^{n} \frac{1}{|f'(\zeta_k)|} \left(\frac{1}{2z} - \frac{1}{z-\zeta_k} \right) + \frac{ic}{z},$$

for some $c \in \mathbb{R}$.

(ii) If $\tau = 1$ then

$$\sum_{k=1}^{n} \frac{1}{|f'(\zeta_k)|} \leq 2 \operatorname{Re} \frac{1}{f'(0)}.$$

In addition,

$$\sum_{k=1}^{n} \frac{1 - \operatorname{Re} \zeta_k}{|f'(\zeta_k)|} \leq -\operatorname{Re} \frac{1}{f(0)} \tag{119}$$

and equality holds if and only if

$$\frac{1}{f(z)} = \sum_{k=1}^{m} \frac{1}{|f'(\zeta_k)|} \left(\frac{1}{z-1} - \frac{1}{z-\zeta_k} \right) + \frac{ic}{(z-1)^2},$$

for some $c \in \mathbb{R}$. Moreover, if $c = 0$, then f is of hyperbolic type. If $c \neq 0$, then f is of parabolic type.

These estimates are sharp and can be considered an infinitesimal version of the Cowen–Pommerenke inequalities for a single holomorphic self-mapping of the unit disk (see Sect. 4.2, Theorem 4.3). Note that the inequality (119) was obtained earlier by Contreras, Díaz-Madrigal, and Pommerenke in [52]. Moreover, they proved that the equality holds in (119) if and only if $F_t(z) = h^{-1}(h(z) + t)$, where

$$h(z) = \frac{z}{(1-\overline{\tau}z)f(0)}$$
$$+2\sum_{k=1}^{n} \frac{1-\mathrm{Re}(\overline{\tau}\zeta_k)}{\beta_k} \left[\frac{\overline{\tau}\zeta_k}{(1-\overline{\tau}\zeta_k)^2} \log\left(\frac{1-\overline{\tau}z}{1-\overline{\zeta_k}z} \right) + \frac{\overline{\zeta_k}}{1-\overline{\zeta_k}z} \frac{z}{1-\overline{\tau}z} \right].$$

We complete our overview with a special rigidity principle for hyperbolic type generators.

Theorem 6.18. *Let f be the generator of a hyperbolic type semigroup with the Denjoy–Wolff point $\tau = 1$, i.e. $f(1) = 0$ and $m := f'(1) > 0$. Suppose $\zeta_1, \ldots, \zeta_n$ are boundary regular null points of f different from τ. Then*

$$\frac{1}{f'(\tau)} \geq \sum_{j=1}^{n} \frac{1}{|f'(\zeta_j)|}.$$

Moreover, equality holds if and only if

$$\frac{1}{f(z)} = \sum_{j=1}^{n} \frac{1}{|f'(\zeta_j)|} \left(\frac{1}{z-1} - \frac{1}{z-\zeta_j} \right). \tag{120}$$

References

1. M. Abate, *Iteration Theory of Holomorphic Maps on Taut Manifolds* (Mediterranean, Rende, 1989)
2. M. Abate, The infinitesimal generators of semigroups of holomorphic maps. Ann. Math. Pura Appl. **161**, 167–180 (1992)
3. D. Aharonov, M. Elin, S. Reich, D. Shoikhet, Parametric representations of semi-complete vector fields on the unit balls in $\mathbb{C}^n$ and in Hilbert space. Atti Accad. Naz. Lincei Cl. Sci. Fis. Mat. Natur. Rend. Lincei **10**, 229–253 (1999)

4. D. Aharonov, S. Reich, D. Shoikhet, Flow invariance conditions for holomorphic mappings in Banach spaces. Math. Proc. R. Ir. Acad. **99A**, 93–104 (1999)
5. L.V. Ahlfors, An extension of Schwarz's lemma. Trans. Am. Math. Soc. **43**, 359–364 (1938)
6. L.V. Ahlfors, *Collected Papers* (Birkhäuser, Boston, 1982)
7. H. Alexander, Boundary behavior of certain holomorphic maps. Mich. Math. J. **38**, 117–128 (1991)
8. D.S. Alexander, *A History of Complex Dynamics: From Schröder to Fatou and Julia* (Vieweg, Braunschweig, 1994)
9. H. Alexander, A weak Hopf lemma for holomorphic mappings. Indag. Math. (N.S.) **6**, 1–5 (1995)
10. D.S. Alexander, F. Iavernaro, A. Rosa, *Early Days in Complex Dynamics: A History of Complex Dynamics in One Variable During 1906–1942* (American Mathematical Society, Providence, 2011)
11. S. Alinhac, M.S. Baouendi, L. Rothschild, Unique continuatin and regularity at the boundary for holomorphic functions. Duke Math. J. **61**, 635–653 (1990)
12. J.M. Anderson, J. Rovnyak, On generalized Schwarz–Pick estimates. Mathematika **53**, 161–168 (2006)
13. J.M. Anderson, A. Vasil'ev, Lower Schwarz–Pick estimates and angular derivatives. Ann. Acad. Sci. Fenn. Math. **33**, 101–110 (2008)
14. M. Arslan, Rigidity of analytic functions at the boundary. arXiv:math.CV/0605026 v1 (2006)
15. F.G. Avkhadiev, K.-J. Wirths, *Schwarz–Pick Type Inequalities* (Birkhäuser, Basel, 2009)
16. L. Baracco, D. Zaitsev, G. Zampieri, A Burns–Krantz type theorem for domains with corners. Math. Ann. **336**, 491–504 (2006)
17. L. Baribeau, P. Rivard, E. Wegert, On hyperbolic divided differences and the Nevanlinna–Pick problem. Comput. Methods Funct. Theory **9**, 391–405 (2009)
18. A.F. Beardon, The Schwarz–Pick Lemma for derivatives. Proc. Am. Math. Soc. **125**, 3255–3256 (1997)
19. A.F. Beardon, T.K. Carne, A strengthening of the Schwarz–Pick inequality. Am. Math. Mon. **99**, 216–217 (1992)
20. A.F. Beardon, D. Minda, A multi-point Schwarz–Pick lemma. J. Anal. Math. **92**, 81–104 (2004)
21. A.F. Beardon, D. Minda, Dieudonné points of holomorphic self-maps of regions. Comput. Methods Funct. Theory **8**, 409–432 (2008)
22. E.F. Beckenbach, A relative of the lemma of Schwarz. Bull. Am. Math. Soc. **44**, 698–707 (1938)
23. D.F. Behan, Commuting analytic functions without fixed points. Proc. Am. Math. Soc. **37**, 114–120 (1973)
24. S. Bell, L. Lempert, A C^∞ Schwarz reflection principle in one and several complex variables. J. Differ. Geom. **32**, 889–915 (1990)
25. E. Beltrami, Teoria fondamentale degli spazii di curvatura costante. Ann. Math. Pura Appl. **2**, 232–255 (1868)
26. C. Bénéteau, A. Dahlner, D. Khavinson, Remarks on the Bohr phenomenon. Comput. Methods Funct. Theory **4**, 1–19 (2004)
27. E. Berkson, H. Porta, Semigroups of analytic functions and composition operators. Mich. Math. J. **25**, 101–115 (1978)
28. L. Bernal-González, M.C. Calderón-Moreno, Two hyperbolic Schwarz lemmas. Bull. Aust. Math. Soc. **66**, 17–24 (2002)
29. D. Betsakos, Geometric versions of Schwarz's lemma for quasiregular mappings. Proc. Am. Math. Soc. **139**, 1397–1407 (2011)
30. D. Betsakos, Multi-point variations of the Schwarz lemma with diameter and width conditions. Proc. Am. Math. Soc. **139**, 4041–4052 (2011)
31. D. Betsakos, S. Pouliasis, Versions of Schwarz's lemma for condenser capacity and inner radius. Can. Math. Bull. **56**, 241–250 (2013)

32. C. Bisi, G. Gentili, Commuting holomorphic maps and linear fractional models. Complex Variables Theory Appl. **45**, 47–71 (2001)
33. H.P. Boas, Julius and Julia: mastering the art of the Schwarz lemma. Am. Math. Mon. **117**, 770–785 (2010)
34. V. Bolotnikov, A uniqueness result on boundary interpolation. Proc. Am. Math. Soc. **136**, 1705–1715 (2008)
35. V. Bolotnikov, A. Kheifets, A higher order analogue of the Carathéodory–Julia theorem. J. Funct. Anal. **237**, 350–371 (2006)
36. V. Bolotnikov, M. Elin, D. Shoikhet, Inequalities for angular derivatives and boundary interpolation. Anal. Math. Phys. **3**, 63–96 (2013)
37. P.S. Bourdon, J.H. Shapiro, Cyclic phenomena for composition operators. Mem. Am. Math. Soc. **125** (1997)
38. F. Bracci, Fixed points of commuting holomorphic mappings other than the Wolff point. Trans. Am. Math. Soc. **355**, 2569–2584 (2003)
39. F. Bracci, R. Tauraso, F. Vlacci, Identity principles for commuting holomorphic self-maps of the unit disc. J. Math. Anal. Appl. **270**, 451–473 (2002)
40. R.B. Burckel, D.E. Marshall, P. Poggi-Corradini, On a theorem of Landau and Toeplitz, preprint arXiv:math/0603579 (2006)
41. R.B. Burckel, D.E. Marshall, D. Minda, P. Poggi-Corradini, T.J. Ransford, Area, capacity and diameter versions of Schwarz's lemma. Conform. Geom. Dyn. **12**, 133–152 (2008)
42. D.M. Burns, S.G. Krantz, Rigidity of holomorphic mappings and a new Schwarz lemma at the boundary. J. Am. Math. Soc. **7**, 661–676 (1994)
43. C. Carathéodory, Sur quelques applications du théorème de Landau-Picard. C. R. Acad. Sci. Paris **144**, 1203–1206 (1907)
44. C. Carathéodory, Untersuchungen über die konformen Abbildungen von festen und veränderlichen Gebieten. Math. Ann. **72**, 107–114 (1912)
45. C. Carathéodory, *Über die Winkelderivierten von beschränkten analytischen Funktionen* (Sitzungsber. Preuss. Akad. Wiss, Berlin, 1929), pp. 39–54
46. C. Carathéodory, *Theory of functions of a complex variable*, Vol. 1. (Math. Mag., **28**, 1954)
47. T. Carroll, J. Ratzkin, Isoperimetric inequalities and variations on Schwarz's lemma, preprint, arXiv:1006.2310v1[math.SP] (2010)
48. T. Carroll, J. Ratzkin, Two isoperimetric inequalities for the Sobolev constant. Z. Angew. Math. Phys. **63**, 855–863 (2012)
49. D. Chelst, A generalized Schwarz Lemma at the boundary. Proc. Am. Math. Soc. **129**, 3275–3278 (2001)
50. K.H. Cho, S. Kim, T. Sugawa, On a multi-point Schwarz–Pick Lemma. Comput. Methods Funct. Theory **12**, 483–499 (2012)
51. M.D. Contreras, S. Díaz-Madrigal, Analytic flows on the unit disk: angular derivatives and boundary fixed points. Pac. J. Math. **222**, 253–286 (2005)
52. M.D. Contreras, S. Díaz-Madrigal, Ch. Pommerenke, On boundary critical points for semigroups of analytic functions. Math. Scand. **98**, 125–142 (2006)
53. M.D. Contreras, S. Díaz-Madrigal, A. Vasil'ev, Digons and angular derivatives of analytic self-maps of the unit disk. Complex Variables Elliptic Equ. **52**, 685–691 (2007)
54. M.D. Contreras, S. Díaz-Madrigal, C. Pommerenke, Second angular derivatives and parabolic iteration in the unit disk. Trans. Am. Math. Soc. **362**, 357–388 (2010)
55. C.C. Cowen, Commuting analytic functions. Trans. Am. Math. Soc. **283**, 685–695 (1984)
56. C.C. Cowen, Ch. Pommerenke, Inequalities for the angular derivative of an analytic function in the unit disk. J. Lond. Math. Soc. **26**, 271–289 (1982)
57. C.C. Cowen, B.D. MacCluer, *Composition Operators on Spaces of Analytic Functions* (CRC, Boca Raton, 1995)
58. S. Dai, Y. Pan, Note on Schwarz–Pick estimates for bounded and positive real part analytic functions. Proc. Am. Math. Soc. **136**, 635–640 (2008)
59. J. Dieudonné, Recherches sur quelques problèmes relatifs aux polynômes et aux fonctions bornées d'une variable complexe. Ann. Sci. École Norm. Sup. **48**, 247–358 (1931)

60. S. Dineen, *The Schwarz Lemma* (Oxford Math. Monogr., Oxford, 1989)
61. T. Dubejko, K. Stephenson, The branched Schwarz Lemma: a classical result via circle packing. Mich. Math. J. **42**, 211–234 (1995)
62. V.N. Dubinin, On the Scwarz inequality on the boundary for functions regular in the disk. J. Math. Sci. **122**, 3623–3629 (2004)
63. V.N. Dubinin, The Schwarz lemma and estimates for the coefficients of regular functions with a free domain (Russian). Mat. Sb. **196**, 1605–1625 (2005)
64. P. Duren, *Univalent Functions* (Springer, New York, 1983)
65. M. Elin, D. Shoikhet, Dynamic extension of the Julia–Wolff–Carathéodory theorem. Dyn. Syst. Appl. **10**, 421–437 (2001)
66. M. Elin, D. Shoikhet, *Linearization Models for Complex Dynamical Systems. Topics in Univalent Functions, Functions Equations and Semigroup Theory* (Birkhäuser, Basel, 2010)
67. M. Elin, D. Shoikhet, Boundary behavior and rigidity of semigroups of holomorphic mappings. Anal. Math. Phys. **1**, 241–258 (2011)
68. M. Elin, M. Levenshtein, D. Shoikhet, R. Tauraso, Rigidity of holomorphic generators and one-parameter semigroups, *Dyn. Syst. Appl.* **16**, 251–266 (2007)
69. M. Elin, M. Levenshtein, S. Reich, D. Shoikhet, Commuting semigroups of holomorphic mappings. Math. Scand. **103**, 295–319 (2008)
70. M. Elin, S. Reich, D. Shoikhet, F. Yacobzon, Rates of convergence of one-parameter semigroups with boundary Denjoy–Wolff fixed points, in *Fixed Point Theory and Its Applications* (Yokohama Publishers, Yokohama, 2008), pp. 43–58
71. M. Elin, D. Shoikhet, N. Tarkhanov, Separation of boundary singularities for holomorphic generators. Ann. Math. Pura Appl. **190**, 595–618 (2011)
72. G. Gentili, F. Vlacci, Pseudo-iteration semigroups and commuting holomorphic maps. Atti Accad. Naz. Lincei Cl. Sci. Fis. Mat. Nat. Rend. Lincei **5**, 33–42 (1994)
73. G. Gentili, F. Vlacci, Rigidity for regular functions over Hamilton and Cayley numbers and a boundary Schwarz Lemma. Indag. Math. (N.S.) **19**, 535–545 (2008)
74. P. Ghatage, D. Zheng, Hyperbolic derivatives and generalized Schwarz–Pick estimates. Proc. Am. Math. Soc. **132**, 3309–3318 (2004)
75. K. Goebel, S. Reich, *Uniform Convexity, Hyperbolic Geometry and Nonexpansive Mappings* (Marcel Dekker, New York, 1984)
76. V.V. Goryainov, Fractional iterates of functions that are analytic in the unit disk with given fixed points. Math. USSR-Sbs. **74**, 29–46 (1993)
77. G.M. Goluzin, Some estimations of derivatives of bounded functions. Rec. Math. [Mat. Sbs.] N.S. **16**, 295–306 (1945)
78. G.M. Goluzin, *Geometric Theory of Functions of a Complex Variable*. Translations of Mathematical Monographs, vol. 26 (American Mathematical Society, Providence, 1969)
79. I. Graham, D. Minda, A Schwarz Lemma for multivalued functions and distortion theorems for Bloch functions with branch points. Trans. Am. Math. Soc. **351**, 4741–4752 (1999)
80. K.R. Gurganus, Φ-like holomorphic functions in $\mathbb{C}^n$ and Banach spaces. Trans. Am. Math. Soc. **205**, 389–406 (1975)
81. L.A. Harris, On the size of balls covered by analytic transformations. Monatsh. Math. **83**, 9–23 (1977)
82. M.H. Heins, A generalization of the Aumann–Carathéodory "Starrheitssatz". Duke Math. J. **8**, 312–316 (1941)
83. A. Herzig, Die Winkelderivierte und das Poisson-Stieltjes-Integral. Math. Z. **46**, 129–156 (1940)
84. X. Huang, L. Chen, Generalized Schwarz lemmas for meromorphic functions. Bull. Korean Math. Soc. **49**, 417–422 (2012)
85. X.J. Huang, S.G. Krantz, A unique continuation problem for holomorphic mappings. Comm. Partial Differ. Equ. **18**, 241–263 (1993)
86. S. Ivashkovich, J.P. Rosay, Schwarz-type lemmas for solutions of $\bar{\partial}$-inequalities and complete hyperbolicity of almost complex manifolds. Ann. Inst. Fourier (Grenoble) **54**, 2387–2435 (2004)

87. F. Jacobzon, S. Reich, D. Shoikhet, Linear fractional mappings: invariant sets, semigroups and commutativity. J. Fixed Point Theory Appl. **5**, 63–91 (2009)
88. G. Julia, Mémoire sur l'itération des fonctions rationnelles. J. Math. Pure Appl. **8**, 47–245 (1918)
89. G. Julia, Extension nouvelle d'un lemme de Schwarz. Acta Math. **42**, 349–355 (1920)
90. G. Julia, Sur les moyennes des modules de fonctions analytiques. Bull. Sci. Math. **51**, 198–214 (1927)
91. H.T. Kaptanoğlu, Some refined Schwarz–Pick lemmas. Mich. Math. J. **50**, 649–664 (2002)
92. L. Keen, N. Lakic, *Hyperbolic Geometry from a Local Viewpoint* (Cambridge University Press, Cambridge, 2007)
93. K.-T. Kim, H. Lee, *Schwarz's Lemma from a Differential Geometric Viewpoint* (IISc Press, Bangalore, 2011)
94. Y. Komatu, On angular derivative. Kōdai Math. Semin. Rep. **13**, 167–179 (1961)
95. S.G. Krantz, The Schwarz Lemma at the boundary. Complex Variables Elliptic Equ. **56**, 455–468 (2011)
96. T.L. Kriete, B.D. MacCluer, A rigidity theorem for composition operators on certain Bergman spaces. Mich. Math. J. **42**, 379–386 (1995)
97. E. Landau, O. Toeplitz, Über die grösste Schwankung einer analytischen Funktion in einem Kreise. Arch. Math. Phys. **11**, 302–307 (1907)
98. E. Landau, G. Valiron, A deduction from Schwarz's lemma. J. Lond. Math. Soc. **4**, 162–163 (1929)
99. M. Levenshtein, S. Reich, A rigidity theorem for commuting holomorphic functions. J. Nonlinear Convex Anal. **11**, 65–70 (2010)
100. J. Lewittes, An extension of hyperbolic geometry and Julia's theorem. Duke Math. J. **35**, 777–782 (1968)
101. K.Y. Li, Inequalities for fixed points of holomorphic functions. Bull. Lond. Math. Soc. **22**, 446–452 (1990)
102. E. Lindelöf, Mémoire sur sertaines inégalités dans la théorie des fonctions monogénes et sur quelques properiétés nouvelles de ces fonctions dans le voisinage d'un point singulier essentiel (Acta Soc. Sci. Fenn. **35** 1909)
103. J.E. Littlewood, *Lectures on the Theory of Functions* (Oxford University Press, Oxford, 1944)
104. K. Löwner, Untersuchungen über schlichte konforme Abbildungen des Einheitskreises. I. Math. Ann. **89**, 103–121 (1923)
105. B.D. Maccluer, K. Stroethoff, R. Zhao, Generalized Schwarz–Pick estimates. Proc. Am. Math. Soc. **131**, 593–599 (2003)
106. P.R. Mercer, Sharpened versions of the Schwarz lemma. J. Math. Anal. Appl. **205**, 508–511 (1997)
107. P.R. Mercer, On a strengthened Schwarz–Pick inequality. J. Math. Anal. Appl. **234**, 735–739 (1999)
108. P.R. Mercer, Another look at Julia's lemma. Complex Variables Theory Appl. **43**, 129–138 (2000)
109. P.R. Mercer, Schwarz–Pick–type estimates for the hyperbolic derivative. J. Inequal. Appl. **2006**, 1–7 (2006)
110. S. Migliorini, F. Vlacci, A new rigidity result for holomorphic maps. Indag. Math. **13**, 537–549 (2002)
111. Z. Nehari, A generalization of Schwarz' Lemma. Duke Math. J. **14**, 1035–1049 (1947)
112. R. Osserman, From Schwarz to Pick to Ahlfors and beyond. Not. Am. Math. Soc. **46**, 868–873 (1999)
113. R. Osserman, A new variant of the Schwarz–Pick–Ahlfors lemma. Manuscr. Math. **100**, 123–129 (1999)
114. R. Osserman, A sharp Schwarz inequality on the boundary. Proc. Am. Math. Soc. **128**, 3513–3517 (2000)
115. G. Pick, Über eine Eigenschaft der konformen Abbildung kreisförmiger Bereiche. Math. Ann. **77**, 1–6 (1915)

116. G. Pólya, G. Szegö, *Aufgaben und Lehrsätze aus der Analysis, I* (Springer, Berlin, 1925)
117. Ch. Pommerenke, *Boundary Behavior of Conformal Maps* (Springer, New York, 1992)
118. Ch. Pommerenke, A. Vasil'ev, On bounded univalent functions and the angular derivative. Ann. Univ. Mariae Curie Skłodowska Sect. A **54**, 79–106 (2000)
119. Ch. Pommerenke, A. Vasil'ev, Angular derivatives of bounded univalent functions and extremal partitions of the unit disk. Pac. J. Math. **206**, 425–450 (2002)
120. T. Poreda, On generalized differential equations in Banach spaces. Diss. Math. (Rozprawy Mat.) **310**, 1–50 (1991)
121. S. Reich, D. Shoikhet, Generation theory for semigroups of holomorphic mappings in Banach spaces. Abstr. Appl. Anal. **1**, 1–44 (1996)
122. S. Reich, D. Shoikhet, Semigroups and generators on convex domains with the hyperbolic metric. Atti Accad. Naz. Lincei Cl. Sci. Fis. Mat. Nat. Rend. Lincei **8**, 231–250 (1997)
123. S. Reich, D. Shoikhet, The Denjoy–Wolff theorem. Ann. Univ. Mariae Curie Skłodowska **51**, 219–240 (1997)
124. S. Reich, D. Shoikhet, *Nonlinear Semigroups, Fixed Points, and Geometry of Domains in Banach Spaces* (Imperial College Press, London, 2005)
125. P. Rivard, A Schwarz–Pick theorem for higher order hyperbolic derivatives. Proc. Am. Math. Soc. **139**, 209–217 (2011)
126. W. Rogosinski, Zum Schwarzschen Lemma. Jber. Dtsch. Math.-Verein. **44**, 258–261 (1934)
127. W. Rudin, *Function Theory in the Unit Ball of* $\mathbb{C}^n$ (Springer, Berlin, 1980)
128. St. Ruscheweyh, Two remarks on bounded analytic functions. Serdica **11**, 200–202 (1985)
129. J.H. Shapiro, *Composition Operators and Classical Function Theory* (Springer, Berlin, 1993)
130. A.L. Shields, On fixed points of commuting analytic functions. Proc. Am. Math. Soc. **15**, 703–706 (1964)
131. D. Shoikhet, *Semigroups in Geometrical Function Theory* (Kluwer, Dordrecht, 2001)
132. D. Shoikhet, Representations of holomorphic generators and distortion theorems for spirallike functions with respect to a boundary point. Int. J. Pure Appl. Math. **5**, 335–361 (2003)
133. D. Shoikhet, Another look at the Burns-Krantz theorem. J. Anal. Math. **105**, 19–42 (2008)
134. A. Solynin, A Schwarz lemma for meromorphic functions and estimates for the hyperbolic metric. Proc. Am. Math. Soc. **136**, 3133–3143 (2008)
135. R. Tauraso, Commuting holomorphic maps of the unit disc. Ergodic Theory Dyn. Syst. **24**, 945–953 (2004)
136. R. Tauraso, F. Vlacci, Rigidity at the boundary for holomorphic self-maps of the unit disk. Complex Variables Theory Appl. **45**, 151–165 (2001)
137. H. Unkelbach, Über die Randverzerrung bei konformer Abbildung. Math. Z. **43**, 739–742 (1938)
138. J.A. Velling, Spherical geometry and the Schwarzian differential equation. Thesis (Ph.D.), Stanford University, 1985
139. F. Vlacci, On commuting holomorphic maps in the unit disc of $\mathbb{C}$. Complex Variables Theory Appl. **30**, 301–313 (1996)
140. J. Wolff, Sur une généralisation d'un théorème de Schwarz. C. R. Acad. Sci. Paris **182**, 918–920 (1926)
141. S. Yamashita, The Pick version of the Schwarz lemma and comparision of the Poincaré densities. Ann. Acad. Sci. Fenn. Ser. A I Math. **19**, 291–322 (1994)
142. S. Yamashita, Goluzin's extension of the Schwarz–Pick inequality. J. Inequal. Appl. **1**, 345–356 (1997)
143. S.T. Yau, A general Schwarz lemma for Kähler manifolds. Am. J. Math. **100**, 197–203 (1978)
144. W. Yuan, D. Chen, P. Wang, A strengthened Schwarz–Pick inequality for derivatives of the hyperbolic metric. Tamkang J. Math. **37**, 131–134 (2006)
145. M.Z. Zhang, Generalized Schwarz–Pick lemma. Acta Math. Sin. (China Ser.) **49**, 613–616 (2006)

Coorbit Theory and Bergman Spaces

H. G. Feichtinger and M. Pap

Abstract *Coorbit theory* arose as an attempt to describe in a unified fashion the properties of the continuous wavelet transform and the STFT (Short-time Fourier transform) by taking a group theoretical viewpoint. As a consequence H.G. Feichtinger and K.H. Gröchenig have established a rather general approach to atomic decomposition for families of Banach spaces (of functions, distributions, analytic functions, etc.) through integrable group representations (see Feichtinger and Gröchenig (Lect. Notes Math. 1302:52–73, 1988; J. Funct. Anal. 86(2):307–340, 1989; Monatsh. Math. 108(2–3):129–148, 1989), Gröchenig (Monatsh. Math. 112(3):1–41, 1991)), now known as *coorbit theory*.

They gave also examples for the abstract theory and until now this approach gives new insights on atomic decompositions, even for cases where concrete examples can be obtained by other methods. Due to the flexibility of this theory the class of possible atoms is much larger than it was supposed to be in concrete cases. It is a remarkable fact that almost all classical function spaces in real and complex variable theory occur naturally as coorbit spaces related to certain integrable representations.

In the present paper we present an overview of the general theory and applications for the case of the weighted Bergman spaces over the unit disc, indicating the benefits of the group theoretic perspective (more flexibility, at least at a qualitative level, more general atoms).

Keywords Atomic decomposition • Bergman spaces • Coorbit spaces

---

H.G. Feichtinger (✉)
Faculty of Mathematics, NuHAG, University Vienna, Nordbergstraße 15, 1090 Wien, Austria
e-mail: hans.feichtinger@univie.ac.at

M. Pap
Institute of Mathematics and Informatics,University of Pécs, Ifjúság útja 6,7634 Pécs, Hungary
e-mail: papm@gamma.ttk.pte.hu

A. Vasil'ev (ed.), *Harmonic and Complex Analysis and its Applications*, Trends in Mathematics, DOI 10.1007/978-3-319-01806-5_4,

1 Coorbit Theory and Atomic Decomposition Results

In signal and image processing both the wavelet and the Gábor transform play an important role. In both cases a certain group action on the Hilbert space $\mathcal{H} = L^2(\mathbb{R})$ is giving rise to a continuous transform, which then can be discretized and provides good (coherent) frames. Through the theory of *coorbit spaces* a unified theory for the Gábor and the wavelet transforms has been created. The common generalization of these transforms is called *voice-transform* (see [19, 21, 25]), following earlier work.

In this section we summarize the basic notions used in the definition of voice-transform and coorbit spaces and present a short description of the Feichtinger and Gröchenig theory which produces atomic decomposition for these spaces (see [19, 21, 25, 29, 41]).

In the construction of voice-transform the starting point will be a locally compact topological group $(\mathcal{G}, \cdot)$, with Haar measure m. In the definition of voice-transform a *unitary and irreducible representation* π of the group $(\mathcal{G}, \cdot)$ on a Hilbert space $\mathcal{H}$ is used, which is supposed to be strongly continuous. The *voice transform* of $f \in \mathcal{H}$ generated by the representation π and by the *analyzing window* or atom $g \in \mathcal{H}$ is the (possibly complex-valued) function on $\mathcal{G}$ defined by

$$(V_g f)(x) := \langle f, \pi(x)g \rangle \quad (x \in \mathcal{G},\ f, g \in \mathcal{H}). \tag{1}$$

For any such representation π and for each $f, g \in \mathcal{H}$ the voice transform $V_g f$ is a continuous and bounded function on $\mathcal{G}$ and $V_g : \mathcal{H} \to C^b(\mathcal{G})$ is a bounded linear operator satisfying the following inequalities:

$$|(V_g f)(x)| = |\langle f, \pi(x)g \rangle| \leq \|f\| \|g\|,$$

consequently $\|V_g f\|_\infty \leq \|f\|$ if $\|g\| = 1$.

Taking as starting point not necessarily commutative locally compact groups we can construct in this way important transformations in signal processing and control theory. For example, the affine wavelet transform and the Gábor-transform are all special voice transforms (see [19, 29, 41]).

A voice transform V_g generated by an unitary representation π is one-to-one for all $g \in \mathcal{H} \setminus \{0\}$ if π is irreducible.

The function $V_g f$ is continuous on $\mathcal{G}$ but in general is not square integrable. If there exist $g \in \mathcal{H}, g \neq 0$ such that $V_g g \in L^2_m(\mathcal{G})$, then the representation π is *square integrable* and the g is called admissible for π. For a fixed square integrable π the collection of admissible elements of $\mathcal{H}$ will be denoted by $\mathcal{H}^2$. Normalizing the vector g if necessary the voice transform $V_g : \mathcal{H} \to L^2_m(\mathcal{G})$ will be isometric. This is a consequence of the following theorem (see [29, 41]):

Theorem 1. *Let π be an irreducible square integrable representation of $\mathcal{G}$ in $\mathcal{H}$. Then the collection of admissible elements $\mathcal{H}^2$ is a linear subspace of $\mathcal{H}$ and for every $g \in \mathcal{H}^2$ the voice transform of the function f is square integrable on $\mathcal{G}$,* *namely* $V_g f \in L^2_m(\mathcal{G})$ *if* $f \in \mathcal{H}$. *Moreover there is a unique positive, self-adjoint, densely defined operator* A *on* $\mathcal{H}$ *such that* $V_g g \in L^2_m(\mathcal{G})$ *if* $g \in domA$ *and the orthogonality relation hold:*

$$[V_{g_1} f, V_{g_2} g] = \langle Ag_1, Ag_2\rangle\langle f, g\rangle, \tag{2}$$

for all $f, g \in \mathcal{H}$ *and* $g_1, g_2 \in domA$, *where* $[\cdot,\cdot]$ *is the usual inner product in* $L^2_m(\mathcal{G})$. *If the group* $\mathcal{G}$ *is unimodular, then* A *is just a scalar multiple the identity operator and every element of* $\mathcal{H}^2$ *is admissible. By normalizing* g *such that* $\langle Ag, Ag\rangle = 1$ *we have*

$$[V_g f, V_g g] = \langle f, g\rangle\,, \quad (f, g \in \mathcal{H}). \tag{3}$$

An important consequence of this theorem is the following reproducing formula: For $g \in \mathcal{H}^2$with $\|Ag\|^2 = 1$ we have the following convolution relation (on $\mathcal{G}$):

$$V_g f = V_g f * V_g g, \quad f \in \mathcal{H}. \tag{4}$$

By a specific choice of a group and a suitable group representations this formula and its extensions permit non-orthogonal wavelet expansion for Besov–Triebel–Lizorkin spaces on $\mathbb{R}^n$, the Gabor-type expansions for modulation spaces and atomic decomposition results in Banach spaces of analytic functions. The atoms for all these spaces are transforms of a single function, where the transformations are given by a certain unitary group representation. Formula (4) and its extensions are the very reasons for the unification of all different examples mentioned before.

In [20] stronger integrability condition on π are imposed in order to handle other spaces than Hilbert spaces. Thus it covers the family of Besov spaces in the wavelet case, or weighted Bergman spaces. Let us consider a positive, continuous *submultiplicative weight* w on $\mathcal{G}$, i.e., $w(xy) \leq w(x)w(y)$, $w(x) \geq 1$, $\forall x, y \in \mathcal{G}$. Assume that the representation is integrable i.e., the set of analyzing vectors is not trivial:

$$\mathcal{A}_w = \{g \in \mathcal{H} : V_g g \in L^1_w(\mathcal{G})\} \neq \{0\}. \tag{5}$$

With this assumption the reproducing formula given by the convolution (4) can be analyzed. Let us define a space of possible atoms:

$$\mathcal{H}^1_w := \{f \in \mathcal{H} : V_g f \in L^1_w(\mathcal{G})\}. \tag{6}$$

with norm

$$\|f\|_{\mathcal{H}^1_w} = \|V_g f\|_{L^1_w(\mathcal{G})}. \tag{7}$$

The definition of $\mathcal{H}_w^1$ is independent of the choice of $g \in \mathcal{A}_w$, and it is the minimal π invariant Banach space in $\mathcal{H}$ with the property $\|\pi(x)(f)\|_{\mathcal{H}_w^1} \leq C_f w(x) \|f\|_{\mathcal{H}_w^1}$ for all $f \in B, x \in \mathcal{G}$. Furthermore it is dense in $\mathcal{H}$ (see [19] Corollary 4.7).

The papers [19–21, 25] describe a unified approach to atomic decomposition through integrable group representations. In what follows we will outline first how it can be obtained atomic decomposition results in the minimal π invariant Banach space $\mathcal{H}_w^1$ following the exposition published in [19]. For the Schrödinger representation of the Heisenberg group this reduces to the Segal algebra $S_0(\mathcal{G})$ (see [18]). We set $\mathbf{S} = \{F : F = V_g f \text{ for some } f \in \mathcal{H}_w^1\}$.

The convolution operator given by (4), which is the identity on $\mathbf{S}$, can be approximated by series of translation operator, similar to a Riemanian sum. This is achieved using the so-called BUPUs (*bounded uniform partitions of unity*). Let us fix some notions.

Definition 1. Given a compact neighborhood of the identity Q a countable family $X = (x_i)$ in $\mathcal{G}$ is said to be *Q-dense* if $\bigcup x_i Q = \mathcal{G}$. It is *separated*, if one has for some compact neighborhood V of the identity $x_j V \cap x_j V = \emptyset,\ j \neq i$.

Definition 2. We call a family of (continuous) function $\Psi = \{\psi_i\}_{i \in I}$ on $\mathcal{G}$ a *bounded uniform partition of unity* of size Q (in short: Q-BUPU) if there exists a family of points in $(x_i)_{i \in I}$ in $\mathcal{G}$ such that

- $0 \leq \psi_i(x) \leq 1$,
- $\operatorname{supp} \psi_i \subseteq x_i Q$,
- $\sum_i \psi_i(x) \equiv 1$,
- $\sup_{z \in G} \#\{i \in I : z \in x_i Q\} < \infty$.

In order to approximate $V_g f$ by a discrete sum of translates of $G := V_g g$ let us interpret the reproducing formula (4) as convolution on $\mathcal{G}$, i.e.

$$\int_{\mathcal{G}} V_g f(x) V_g g(x^{-1} y)\, dm(x) = V_g f(y),$$

becomes $F = F * G$ for $F = V_g f$. Define the operators $TF = F * G$ and T_Ψ associated with a particular bounded uniform partition of unity Ψ, by

$$T_\Psi(y) = \sum_i \langle F, \psi_i \rangle G(x_i{}^{-1} y). \tag{8}$$

Lemma 4.3 of [19] shows that if $F \in L_w^1(\mathcal{G})$ the sequence of coefficients $\Lambda = (\lambda_i)_{i \in I}$, given by $\lambda_i = \langle F, \psi_i \rangle$ belongs to ℓ_w^1, more precisely, given a fixed compact neighborhood Q of unity there exists a constant C_0 such that the norms of the linear operators $F \to \Lambda$ are uniformly bounded by C_0 for all Q-BUPUs. Conversely, if $g \in \mathcal{A}_w$ and $\Lambda = (\lambda_i)_{i \in I} \in \ell_w^1$, then $F := \sum_i \lambda_i G(x_i^{-1} y) \in L_w^1(\mathcal{G})$, the sum being absolutely convergent in $L_w^1(\mathcal{G})$ and there is a universal constant C_1 such that $\|F\|_{L_w^1(\mathcal{G})} \leq C_1 \|\Lambda\|_{\ell_w^1}$. As a consequence the set of operators $\{T_\Psi\}$, where Ψ runs through the family of Q-BUPUs acts uniformly bounded on $L_w^1(\mathcal{G})$.

According to Lemma 4.5 of [19] the net $\{T_\Psi\}$ of Q-BUPUs, directed according to inclusions of the neighborhoods Q of unity, is norm convergent to T as operators on $L^1_w(\mathcal{G})$. This implies that $\{T_\Psi\}$ acts as identity on $L^1_w(\mathcal{G}) * G$ and is invertible as operator on $L^1_w(\mathcal{G}) * G$ for a sufficiently small neighborhood Q of the identity. We thus obtain for $f \in \mathcal{H}^1_w$

$$F(y) = V_g(f)(y) = T_\Psi(T_\Psi^{-1}F)(y) = \sum_i \langle \psi_i, T_\Psi^{-1}F\rangle G(x_i^{-1}y),$$

or equivalently

$$\langle f, \pi(y)g\rangle = \sum_i \langle \psi_i, T_\Psi^{-1}F\rangle \langle \pi(x_i)g, \pi(x_i)g\rangle = \left\langle \sum_i \langle \psi_i, T_\Psi^{-1}F\rangle \pi(x_i)g, \pi(y)g \right\rangle.$$

This implies the following atomic decomposition result for $\mathcal{H}^1_w$:

Theorem 2 ([19]). *For any $g \in \mathcal{A}_w \setminus \{0\}$ there exists a neighborhood Q of identity, and a constant C_0 (both only dependent of g), such that for any collection of points $\{x_i\} \subset \mathcal{G}$ which is Q-dense any $f \in \mathcal{H}^1_w$ can be written*

$$f = \sum \lambda_i(f)\pi(x_i)g, \quad \textit{with} \quad \sum_i |\lambda_i(f)| w(x_i) \le C_0 \|f\|_{\mathcal{H}^1_w} \tag{9}$$

where the sum is absolutely convergent in $\mathcal{H}^1_w$. In fact, for any Q-BUPU Ψ associated with $(x_i)_{i\in I}$ there is a linear mapping of the form $f \mapsto (\lambda_i(f)) = (\langle T_\Psi^{-1}V_g f, \psi_i\rangle)_{i\in I}$ providing suitable coefficients. The family (x_i) can even be replaced by a V-separated subfamily, for a suitably chosen open subset $V \subset Q$.[1]

We call such representations an *atomic decomposition* of $f \in \mathcal{H}^1_w$ with respect to a coherent family of atoms of the form $(\pi(x_i)g)$. Such atomic decompositions can be found for a much larger class of Banach spaces, related to $\mathcal{H}$ and characterized by the membership of the voice transform in some solid and translation invariant Banach space of functions (let us call it $\boldsymbol{Y}$) on $\mathcal{G}$.

Denote by $\mathcal{H}^{1*}_w$ the dual of the minimal π-invariant Banach space $\mathcal{H}^1_w$ and by $\mathcal{H}^{1\sim}_w$ the antidual-space of all continuous conjugate-linear functionals on $\mathcal{H}^1_w$. The use of the antidual as a reservoir space in the definition of the coorbit spaces is more convenient because to view the pairing between test functions (in $\mathcal{H}^1_w$) and distributions as an extension of the sesquilinear-form derived from the Hilbert space setting has various advantages. Above all it allows to carry over the notations and formulas from the Hilbert spaces without modifications, among them *reproducing formula*. Since the antidual $\mathcal{H}^{1\sim}_w$ can always be identified with $\mathcal{H}^{1*}_w$ (isometrically and additively) using the correspondence $f \to \overline{f}$ this technical

[1]This option will become important for the L^p-theory, with $p > 1$.

detail is of no real downside. The antidual space $\mathcal{H}_w^{1\sim}$ will be a sufficiently big Banach space, invariant under the group action. For its elements (view them as distributions in concrete cases) the voice transform is well defined and satisfies the growth condition $V_g(f)(x) = O(1/w(x))$. Thus it makes sense to use it for the definition of coorbit spaces with respect to most general function spaces on $\mathcal{G}$ (which need *not* be contained in $L^2(\mathcal{G})$).

For the description of general *coorbit-spaces* we need a solid Banach space of functions $(\boldsymbol{Y}, \|\cdot\|_Y)$ on $\mathcal{G}$, which is a left and right translation invariant and continuously included in $L^1_{\mathrm{loc}}(\mathcal{G})$. Furthermore certain convolution operators are supposed to be well defined and bounded on $\boldsymbol{Y}$. From now on let us consider the weight function

$$u(x) = \sup_{\|F\|_Y=1} \frac{\|L_x F\|_Y}{\|F\|_Y}, \quad v(x) = \sup_{\|F\|_Y=1} \frac{\|R_{x^{-1}} F\|_Y}{\|F\|_Y} \Delta(x^{-1})$$

the operator norms of the left translation operator $L_x F(y) = F(x^{-1}y)$ and right translation operator $R_x f(y) = F(yx)$ over the group $\mathcal{G}$.

We consider always pairs $(\boldsymbol{Y}, w)$ where w is a weight on $\mathcal{G}$ such that for a constant C and all $x \in \mathcal{G}$ one has:

$$\max\{u(x), u(x^{-1}), v(x), v(x^{-1})\Delta(x^{-1})\} \leq Cw(x),$$

If one w satisfies the above inequality, we say that w is a *canonical weight* for $\boldsymbol{Y}$, and we will call $(\boldsymbol{Y}, w)$ an admissible pair. Given such a pair, and any nonzero $g \in \mathcal{A}_w$ the coorbit space can be defined in the following way:

$$\boldsymbol{Co}(\boldsymbol{Y}) = \{f \in \mathcal{H}_w^{1\sim} : V_g f \in \boldsymbol{Y}\}. \tag{10}$$

This set is independent of the choice of $g \in \mathcal{A}_w$ and the canonical weight used. In fact, if w_1 is another weight with $w(x) \leq Cw_1(x)$ choosing as reservoir space the larger space $\mathcal{H}_{w_1}^{1\sim}$ one still obtains the same space $\boldsymbol{Co}(\boldsymbol{Y})$ (see Theorem 4.2 of [20]).

$\boldsymbol{Co}(\boldsymbol{Y})$ can be turned into Banach space by introducing the norm $\|f\|_{\boldsymbol{Co}(\boldsymbol{Y})} = \|V_g f\|_Y$. Furthermore one has the continuous embeddings

$$\mathcal{H}_w^1 \hookrightarrow \boldsymbol{Co}(\boldsymbol{Y}) \hookrightarrow \mathcal{H}_w^{1\sim}.$$

For example

$$\mathcal{H} = \boldsymbol{Co}(L^2(\mathcal{G})), \quad \mathcal{H}_w^1 = \boldsymbol{Co}(L_w^1(\mathcal{G})), \quad \text{and} \quad \mathcal{H}_w^{1\sim} = \boldsymbol{Co}(L_{1/w}^\infty).$$

In many concrete cases the minimal isometrically π-invariant space corresponds to $\boldsymbol{Co}(L^1(\mathcal{G})$ and correspondingly the maximal space is $\boldsymbol{Co}(L^\infty)$. For the Schrödinger representation of the (reduced) Heisenberg group these are the Segal algebra $S_0(\mathbb{R}^d)$ and its dual space (see [18] for details).

According to Theorem 4.1 and Proposition 4.3 of [20] the spaces $\boldsymbol{Co}(\boldsymbol{Y})$ are π-invariant Banach spaces of distributions which are isometrically isomorphic to a reproducing kernel Banach subspace of $\boldsymbol{Y}$, more exactly with $\boldsymbol{Y} * V_g g$ with $V_g g \in L^1_w(\mathcal{G})$ and the reproducing formula (4) can be extended, namely if $g \in \mathcal{A}_w$ such that $\|Ag\|^2 = 1$, then

$$V_g f = V_g f * V_g g, \quad \forall f \in \mathcal{H}^{1\sim}_w. \tag{11}$$

$\boldsymbol{Y} = L^p_w(\mathcal{G})$ defines the coorbit space with the simplest Banach structure. In this case we denote them by

$$\mathcal{H}^p_w := \{f \in \mathcal{H}^{1\sim}_w : V_g f \in L^p_w(\mathcal{G})\}. \tag{12}$$

From the point of view of the applications these coorbit spaces are the most important cases.

For the characterization of the coefficients of an atomic decomposition in $\boldsymbol{Co}(\boldsymbol{Y})$ we need an associated sequence space. In general any given solid Banach function space $\boldsymbol{Y}$ may be naturally associated with a sequence space $\boldsymbol{Y}_d(X)$ (see Definition 3.4 of [20]). The identification of this sequence space with the well-known sequence spaces is the easiest for $\boldsymbol{Y} = \mathcal{H}^p_w$. Then the associated sequence space is $\boldsymbol{Y}_d(X) = \ell^p_w(X)$, the discrete weight $w = (w_i)$ (logically different from the continuous weight) is given coordinatewise by the choice $w_i = w(x_i)$. For this reason in what follows we will present the general atomic decomposition results restricted for $\boldsymbol{Y} = \mathcal{H}^p_w$ spaces. The most general setting is described in [20] and [25].

To give atomic decomposition results for $\mathcal{H}^p_w$ we need better analyzing vectors. We need a control of the local behavior of the extended representation coefficients $V_g(f)$. To this end we have to restrict the set of analyzing vectors.

Choose $Q \subset \mathcal{G}$ compact with nonvoid interior and $e \in Q$. The maximal function Mf of f is defined as $Mf(x) = \sup_{y \in xQ} |f(y)|$. The *Wiener amalgam space* $\mathcal{M}(L^1(\mathcal{G}))$ is the space of all f such that $Mf \in L^1(\mathcal{G})$. Now define the set of basic atoms

$$\mathcal{B}_w = \{g \in \mathcal{H} : \langle \pi(.)g, g\rangle \in \mathcal{M}(L^1(\mathcal{G}))\}. \tag{13}$$

It can be proved that $\mathcal{B}_w \subseteq \mathcal{H}^1_w$ and is still dense in $\mathcal{H}^1_w$. If $\mathcal{G}$ has compact invariant neighborhood of unity (i.e., $\mathcal{G}$ is [IN]-group), then $\mathcal{B}_w = \mathcal{H}^1_w$. In general $\mathcal{B}_w$ and $\mathcal{H}^1_w$ are different.

We will need another maximal function for the description of the *local oscillations*. Let us write G for $V_g(g)$. The function defined by

$$G^\#_Q(x) = \sup_{u \in Q} |G(ux) - G(x)|$$

is the Q-oscillation of G.

Using the invertibility of the discretization operator T_ψ the atomic characterization of the simplest coorbit spaces $\mathcal{H}^p_w$ was given in [19] Theorem 6.2.

Theorem 3 ([19]).

i) *Given any $g \in \mathcal{B}_w$ and any relatively separated family $X = (x_i)_{i \in I}$ on $\mathcal{G}$ one has: the synthesis mapping*

$$(\lambda_i)_{i \in I} \to \sum_{i \in I} \lambda_i \pi(x_i) g$$

defines a bounded linear operator from the sequence space ℓ_w^p into $\mathcal{H}_w^p$ and the sum is unconditionally norm convergent for $1 \leq p < \infty$, w^*-convergent in the case $p = \infty$.

ii) *Conversely, for every nonzero $g \in \mathcal{B}_w$ a neighborhood Q of the unity can be found such that for any Q-dense and relatively separated set $X = (x_i)_{i \in I}$ there exists a constant C and a linear mapping $\Lambda : f \mapsto \Lambda(f)$ of norm at most C which is a right inverse to the synthesis mapping above, i.e. for each $f \in \mathcal{H}_w^p$ $(1 \leq p < \infty)$ there exists a family where the sequence of coefficients $\Lambda(f) = \lambda_i(f)$ depending linearly on f with*

$$\|\Lambda(f)\|_{\ell_w^p} < C \, \|f\|_{\mathcal{H}_w^p} \quad \text{and} \quad f = \sum_{i \in I} \lambda_i \pi(x_i) g. \tag{14}$$

Later in the paper [20] (Theorem 6.1) this result was generalized for coorbit spaces. In the paper [25] (Theorem T) Gröchenig gave an explicit estimate on the size of Q in terms of the local oscillation $G_Q^{\#}$ which permitted also a lucid argument for the invertibility of T_Ψ.

Theorem 4 ([25]). *Assume that $(\boldsymbol{Y}, \|\cdot\|_{\boldsymbol{Y}})$ is a left and right translation invariant solid Banach function space continuously included in $L^1_{\mathrm{loc}}(\mathcal{G})$. Let w be a canonical weight, and suppose that the irreducible, unitary representation π is w-integrable and choose with $g \in \mathcal{B}_w$ such that $\|Ag\| = 1$. Choose a neighborhood Q so small that*

$$\left\| G_Q^{\#} \right\|_{L^1_w} < 1.$$

Then for any Q-dense and relatively separated set $X = (x_i)_{i \in I}$, $\boldsymbol{Co}(\boldsymbol{Y})$ has the following atomic decomposition: If $f \in \boldsymbol{Co}(\boldsymbol{Y})$, then

$$f = \sum_{i \in I} \lambda_i \pi(x_i) g, \tag{15}$$

where the coefficients $\lambda_i(f) = \langle T_\Psi^{-1} V_g f, \psi_i \rangle$ depend linearly on f and satisfy

$$\|(\lambda_i(f))_{i \in I}\|_{\boldsymbol{Y}_d(X)} < C \, \|f\|_{\boldsymbol{Co}_\pi(\boldsymbol{Y})} \, .$$

with a constant C depending on g.

Conversely, if $(\lambda_i(f))_{i\in I} \in \boldsymbol{Y}_d(X)$, *then*

$$f = \sum_{i\in I} \lambda_i \pi(x_i)g.$$

is in $\mathcal{C}o(\boldsymbol{Y})$, *the convergence of the sum is in the norm of* $\mathcal{C}o(\boldsymbol{Y})$, *if the bounded functions of compact support are dense in* $\boldsymbol{Y}$, *and in* w^**-sense within* $\mathcal{H}_w^{1\,*}$ *otherwise.*

In order to approximate the convolution operator $T : F \to F * V_g(g)$ by discrete sums Gröchenig in [25] used two more discretization operators:

$S_\Psi F = \sum_i F(x_i)\psi_i * V_g g$,

$U_\Psi F = \sum_i c_i F(x_i) L_{x_i} V_g g$, with $c_i = \int \psi_i$.

As a consequence of Theorems 4.11 and 4.13 of [25] we get that if $g \in \mathcal{B}_w$ the discretization operators S_Ψ, U_Ψ have a bounded inverse and thus it is possible to give other atomic decompositions and frames for $\mathcal{C}o(\boldsymbol{Y})$. From the invertibility of S_ψ the following result can be obtained:

Theorem 5 ([25]). *Under the general conditions of the previous theorem let us choose the neighborhood* Q *small enough to ensure the validity of the estimate*

$$\left\| G_U^\# \right\|_{L_w^1} < 1/ \left\| G \right\|_{L_w^1}.$$

Then for any Q*-dense and relatively separated set* $X = (x_i)_{i\in I}$, *the set* $\{\pi(x_i)g : i \in I\}$ *is a Banach frame for* $\mathcal{C}o(\boldsymbol{Y})$, *i.e.,*

i) $f \in \mathcal{C}o(\boldsymbol{Y})$ *if and only if* $\langle \pi(x_i)g, f\rangle_{i\in I} \in \boldsymbol{Y}_d(X)$.

ii) There exist two constants $C_1, C_2 > 0$ *depending only on* $g \in \mathcal{B}_w$ *such that*

$$C_1 \left\| f \right\|_{\mathcal{C}o(\boldsymbol{Y})} \le \left\| \langle \pi(x_i)g, f\rangle_{i\in I} \right\|_{\boldsymbol{Y}_d(X)} \le C_2 \left\| f \right\|_{\mathcal{C}o(\boldsymbol{Y})}.$$

iii) $f \in \mathcal{C}o(\boldsymbol{Y})$ *can be unambiguously reconstructed from the coefficients* $\langle \pi(x_i)g, f\rangle_{i\in I}$. *If the space of bounded functions with compact support is dense in* $\boldsymbol{Y}$, *this reconstruction may be achieved as follows: there exists a system* $e_i \in \mathcal{H}_w^1$, $i \in I$ *such that*

$$f = \sum_{i\in I} \langle \pi(x_i)g, f\rangle\, e_i \qquad (16)$$

with convergence in $\mathcal{C}o(\boldsymbol{Y})$.

Essentially, these results say that if $g \in \mathcal{B}_w$ then every $f \in \mathcal{C}o(\boldsymbol{Y})$ admits a decomposition into elementary pieces (atoms) if $\{x_i\}_{i\in I}$ is a sufficiently dense set, and the atoms arise from a single element under the group action. Several well-known decomposition theories are contained as special examples and are unified under the aspect of group theory in these results. By specific choices of group and a representation non-orthogonal wavelet expansions for Besov–Triebel–Lizorkin space on $\mathbb{R}^n$, the Gabor-type expansions for modulation spaces can be obtained.

To see how these decomposition results arise as special case of the general theory, see [19]. In the last years atomic decomposition results connected to the Bergman spaces on the unit disc were obtained using the described technique as we will present in the next section.

2 Results Connected to the Weighted Bergman Spaces over the Unit Disc

In the paper [19] the following was formulated: the Möbius invariant function spaces on the unit disc cannot be handled with the coorbit theory, because they arise with the representation of $\mathbb{S}U(1,1)$ on the Bergman space on the unit disc which is not integrable. The minimal Möbius invariant Banach space can definitely not be identified with the coorbit space of the representation of the Bergman space. It was suggested to take into consideration also projective representations to explain the Bergman spaces for the unit disc for the whole range as coorbit spaces, or equivalently, representations which are integrable only modulo the center of the group as in the case for universal covering group of the Heisenberg group.

In the last years results concerning the weighted Bergman spaces on the unit disc were obtained in two different ways independently.

M. Pap studying the properties of a special voice transform of the Blaschke group (which is related to the $\mathbb{S}U(1,1)$) outlined by the coorbit theory proved that not only $g = 1_{\mathbb{D}}$ will generate atomic decomposition, but also every function from the minimal Möbius invariant space will generate an atomic decomposition in some weighted Bergman spaces.

J.G. Christensen and G. Olafsson gave first a generalization of the coorbit spaces for dual pairs. They presented a generalized coorbit theory which is able to account for the examples where the integrability condition is not satisfied. As an example, the smooth vectors of the discrete series representation of $\mathbb{S}U(1,1)$ are used to described some weighted Bergman spaces of holomorphic functions on the unit disc as generalized coorbit spaces and they give atomic decomposition results corresponding to the cyclic vector $g = 1_{\mathbb{D}}$.

In what follows we give an overview of these results.

2.1 *The Weighted Bergman Spaces A^p_γ over the Unit Disc*

Let us denote by $\mathbb{D}$ the unit disc and $\mathcal{A}$ the set of functions $f : \mathbb{D} \to \mathbb{C}$ which are analytic in $\mathbb{D}$. Denote by

$$dA_\gamma(z) := \frac{\gamma+1}{\pi}(1-|z|^2)^\gamma \, dxdy\,, \quad z = x+iy$$

the weighted area measure on $\mathbb{D}$. Let us denote by $dA(z) := dA_0(z) := \frac{1}{\pi}\,dxdy$. For all $\gamma > -1$ let us consider the following subset of analytic functions:

$$A^p_\gamma := \left\{ f \in \mathcal{A} : \int_{\mathbb{D}} |f(z)|^p dA_\gamma(z) < \infty \right\}.$$

The set $\mathcal{H} = A^2_\gamma$ is a Hilbert space with the scalar product

$$\langle f, g\rangle_\gamma := \int_{\mathbb{D}} f(z)\overline{g(z)} dA_\gamma(z).$$

In the special case when $\gamma = 0$, $A^2 = A^2_0$ is the so-called *Bergman space* (see [17, 30]).

For $-1 < \gamma < +\infty$ the weighted Bergman projection

$$P_\gamma : L^2(\mathbb{D}, dA_\gamma) \to A^2_\gamma$$

is defined by

$$P_\gamma f(z) = \int_{\mathbb{D}} f(\xi) \frac{1}{(1-\bar{\xi}z)^{\gamma+2}} dA_\gamma(\xi).$$

P_γ is an orthogonal projection operator, which satisfies $P_\gamma f = f$ for $f \in A^2_\gamma$ and is a pointwise formula. The projection operator can be extended to $L^1(\mathbb{D}, dA_\gamma)$ by mapping each $f \in L^1(\mathbb{D}, dA_\gamma)$ to an analytic function in $\mathbb{D}$, and

$$f(z) = \frac{\gamma+1}{\pi} \int_{\mathbb{D}} f(\xi) \frac{1}{(1-\bar{\xi}z)^{\gamma+2}} (1-|\xi|^2)^\gamma d\xi_1 d\xi_2,$$

$$(f \in A^1_\gamma \quad z,\ \xi \in \mathbb{D},\ \xi = \xi_1 + i\xi_2)$$

and the integral converges uniformly in z in every compact subset of $\mathbb{D}$ (see [30] p. 6). We will need the following two theorems:

Theorem A (see [30]). *For any $-1 < \gamma < +\infty$ and any real δ, let*

$$I_{\gamma,\delta}(z) = \int_{\mathbb{D}} \frac{(1-|w|^2)^\gamma}{|1-z\overline{w}|^{2+\gamma+\delta}} dA(w), \quad z \in \mathbb{D}.$$

Then we have the following estimates for $|z| \to 1^-$*:*

$$I_{\gamma,\delta}(z) \sim \begin{cases} 1, & \delta < 0, \\ \log \frac{1}{1-|z|^2}, & \delta = 0, \\ \frac{1}{(1-|z|^2)^\delta}, & \delta > 0 \end{cases} .$$

Theorem B (see [30]). *Suppose* $-1 < \gamma$, $\delta < +\infty$ *and* $1 \le p < +\infty$. *Then* P_γ *is a bounded projection from* $L^p(\mathbb{D}, dA_\delta)$ *onto* A^p_δ *if and only if* $(\delta + 1) < (\gamma + 1)p$.

2.2 *The Blaschke Group* $\mathbb{B}$

Let us denote by

$$B_a(z) := \epsilon \frac{z-b}{1-\bar{b}z} \quad (z \in \mathbb{C}, a = (b,\epsilon) \in \mathbb{B} := \mathbb{D} \times \mathbb{T}, \bar{b}z \neq 1),$$

the so-called *Blaschke functions*, where

$$\mathbb{D} := \{z \in \mathbb{C} : |z| < 1\}, \quad \mathbb{T} := \{z \in \mathbb{C} : |z| = 1\}.$$

If $a \in \mathbb{B}$, then B_a is a one-to-one map on $\mathbb{T}$, $\mathbb{D}$, respectively. The restrictions of the Blaschke functions on the set $\mathbb{D}$ or on $\mathbb{T}$ with the operation $(B_{a_1} \circ B_{a_2})(z) := B_{a_1}(B_{a_2}(z))$ form a group. In the set of the parameters $\mathbb{B} := \mathbb{D} \times \mathbb{T}$ let us define the operation induced by the function composition in the following way $B_{a_1} \circ B_{a_2} = B_{a_1 \circ a_2}$. The group $(\mathbb{B}, \circ)$ will be isomorphic with the group $(\{B_a, a \in \mathbb{B}\}, \circ)$. If we use the notations $a_j := (b_j, \epsilon_j)$, $j \in \{1,2\}$ and $a := (b,\epsilon) =: a_1 \circ a_2$, then

$$b = \frac{b_1\bar{\epsilon}_2 + b_2}{1 + b_1\bar{b}_2\bar{\epsilon}_2} = B_{(-b_2\epsilon_2, \bar{\epsilon}_2)}(b_1), \quad \epsilon = \epsilon_1 \frac{\epsilon_2 + b_1\bar{b}_2}{1 + \epsilon_2\bar{b}_1 b_2} = B_{(-b_1\bar{b}_2, \epsilon_1)}(\epsilon_2).$$

The neutral element of the group $(\mathbb{B}, \circ)$ is $e := (0,1) \in \mathbb{B}$ and the inverse element of $a = (b,\epsilon) \in \mathbb{B}$ is $a^{-1} = (-b\epsilon, \bar{\epsilon})$.

The integral of the function $f : \mathbb{B} \to \mathbb{C}$, with respect to this left invariant Haar-measure m of the group $(\mathbb{B}, \circ)$, is given by

$$\int_{\mathbb{B}} f(a)\, dm(a) = \frac{1}{2\pi} \int_{-\pi}^{\pi} \int_{\mathbb{D}} \frac{f(b, e^{it})}{(1-|b|^2)^2}\, db_1 db_2 dt,$$

where $a = (b, e^{it}) = (b_1 + ib_2, e^{it}) \in \mathbb{D} \times \mathbb{T}$.

It can be shown that this integral is invariant with respect to the inversion transformation $a \to a^{-1}$, so this group is unimodular.

2.3 The Relation of the Blaschke Group with $\mathbb{S}U(1,1)$ and the Möbius Group

Let us denote by

$$\mathbb{S}U(1,1) = \left\{ \mathbf{g} = \begin{pmatrix} \alpha & \beta \\ \overline{\beta} & \overline{\alpha} \end{pmatrix} : |\alpha|^2 - |\beta|^2 = 1 \right\}.$$

The group $\mathbb{S}U(1,1)$ acts on the unit disc by

$$\tilde{\varphi}_{\mathbf{g}}(z) = \frac{\alpha z + \beta}{\overline{\alpha} z + \overline{\beta}}.$$

The topological group $\mathbb{S}U(1,1)$ is homeomorphic to the space $\mathbb{B} = \mathbb{D} \times \mathbb{T}$. If we set $b = -\frac{\beta}{\alpha}$ and $\theta = \arg\alpha(mod 2\pi)$, then the map $\mathbf{g} \to (b,\theta) \to (b, e^{i\theta})$, $\theta \in [-\pi,\pi)$ maps $\mathbf{g} \in \mathbb{S}U(1,1)$ into the product of the unit disc and the circle. So the Blaschke group can be viewed as another parametrization of the $\mathbb{S}U(1,1)$. This above map is even a diffeomorphism and the inverse of this map is obtained by setting $\alpha = e^{i\theta}(1-|b|^2)^{-1/2}$ and $\beta = -be^{i\theta}(1-|b|^2)^{-1/2}$. Consequently $\mathbb{S}U(1,1)$ can be parameterized by (b,θ), namely

$$\mathbb{S}U(1,1) = \{(b,\theta) : |b| < 1, \theta \in [-\pi,\pi)\}.$$

The real line is the universal covering of the circle. Thus the universal covering group of the $\mathbb{S}U(1,1)$ is topologically equivalent to the product of the same disc and the real line. It cannot be realized in matrix form.

The same parameters serve to describe the universal covering of $\mathbb{S}U(1,1)$ which is $\widetilde{\mathbb{S}U(1,1)} = \{(b,\theta) : |b| < 1, \theta \in \mathbb{R}\}$.

So using the parametrization of the Blaschke group reflects better in the same time the properties of the covering group and the action of the representation.

Let us write Möb($\mathbb{D}$) for the group of biholomorphic automorphism of the unit disc. Each $\varphi_{2\theta,b}$ from Möb($\mathbb{D}$) has the form

$$\varphi_{2\theta,b} = e^{2i\theta}\frac{z-b}{1-\overline{b}z} = B_a(z),\ a = (b,\epsilon^2),\ \epsilon = e^{i\theta}.$$

The map $\tilde{\varphi}_g$ can be written as

$$\tilde{\varphi}_{\mathbf{g}}(z) = e^{2i\theta}\frac{z-b}{1-\overline{b}z} = B_a(z),\ a = (b,\epsilon^2),\ \epsilon = e^{i\theta}.$$

This relation exhibits the Blaschke group as two fold covering group of Möb($\mathbb{D}$).

2.4 The Representations of the $\mathbb{S}U(1,1)$ and of the Blaschke Group on the Hilbert Space A_γ^2

The representation of $\mathbb{S}U(1,1)$ on the weighted Bergman space is given by

$$\Pi^{\gamma+2}\left(\begin{pmatrix}\alpha & \beta\\ \overline{\beta} & \overline{\alpha}\end{pmatrix} f\right)(z) := \frac{1}{(-\overline{\beta}z+\alpha)^{\gamma+2}} f\left(\frac{\overline{\alpha}z-\beta}{-\overline{\beta}z+\alpha}\right).$$

In the papers [33,34] the voice transform induced by the following representation of the Blaschke group on the weighted Bergman spaces was studied:

$$\pi^{\gamma}(a^{-1})f(z) := e^{i\frac{\gamma+2}{2}\psi}\frac{(1-|b|^2)^{\frac{\gamma+2}{2}}}{(1-\overline{b}z)^{\gamma+2}} f\left(e^{i\psi}\frac{z-b}{1-\overline{b}z}\right) \quad (a=(b,e^{i\psi})\in\mathbb{B}). \tag{17}$$

One can show that for all $\gamma \geq 0$ π^{γ} is a unitary, irreducible, square integrable representation of the group $\mathbb{B}$ on the Hilbert space A_γ^2.

It is simpler to take the expression of the representation for $a^{-1} \in \mathbb{B}$, correspondingly it is easier to study the voice transform in $a^{-1} \in \mathbb{B}$, $(a=(b,e^{i\psi}) \in \mathbb{B},\ f,g \in A_\gamma^2)$:

$$(V_g f)(a^{-1}) = (V_g f)(-b\epsilon,\overline{\epsilon}) := \langle f, \pi^{\gamma}(a^{-1})g\rangle_{\gamma}. \tag{18}$$

The function $V_g f$ is continuous and bounded on $\mathbb{B}$. It can be shown that every element from A_γ^2 is admissible. Taking into consideration that the Blaschke group is unimodular Theorem 1 implies that for $f,g \in A_\gamma^2$, with $g \neq 0$ and $\|Ag\| = 1$ the following reproducing formula is valid:

$$V_g f = V_g f * V_g g, \quad \text{i.e.,} \quad V_g f(y^{-1}) = \int_{\mathbb{B}} V_g f(x^{-1}) V_g g(x \circ y^{-1})\, dm(x). \tag{19}$$

2.5 Bounded Uniform Partitions of Unity for the Blaschke-Group

As we have seen before in the unified approach of the atomic decomposition the Q density, the V-separated property and the bounded uniform partitions of the unity are the basic starting points.

Our aim is to give an example of Q-dense V-separated sequences in the Blaschke-group. As we will see it is easier to give an example of right bounded partition of unity and to give geometrical interpretation of Q-density from right in terms of the hyperbolic metric. Q-density from right means that there is a sequence $(x_i)_{i\in I}$ in $\mathbb{B}$ such that $\bigcup Qx_i = \mathbb{B}$. Separated from right for some compact neighborhood V of the unity means $Vx_j \cap Vx_j = \emptyset,\ j \neq i$. The Q density from the left in general is not the same with the Q-density from right, except when the group is [IN]-group.

In the general theory of atomic decomposition the concept of Q-density from the left is used. This is the reason why we will adapt the discretizing operator to the Q-density from the right in order to obtain atomic decomposition in the weighted Bergman spaces.

Recall that the hyperbolic distance of two points from the unit disc is given by

$$\beta(z,w) = \frac{1}{2}\log\frac{1+g(z,w)}{1-g(z,w)},\ g(z,w) = \left|\frac{z-w}{1-\overline{w}z}\right| = \left|B_{(w,1)}(z)\right|,$$

and the hyperbolic disc or Bergman disc of radius $r > 0$ and center b is

$$D(b,r) = \{z \in \mathbb{D} : \beta(z,b) < r\}.$$

Lemma 1 ([35]). *Let us consider $r > 0$ and $Q = Q_1 \times \mathbb{T}$, where $Q_1 = \{z \in \mathbb{D} : |z| < \tanh r\}$. Then there exists a sequence $x_n = (b_n, -1) \in \mathbb{B}$ which is Q-dense from the right, i.e. $\bigcup Qx_n = \mathbb{B}$ and V-separated from right, i.e. $Vx_n \cap Vx_m = \emptyset$, and there is also a corresponding right bounded uniform partition of the unity corresponding to $\{x_n\}$.*

Due to Lemma 2.28 in [43] p. 63, there exists a sequence $(b_k)_{k\in\mathbb{N}}$ and Borel sets D_k satisfying the following conditions

- $D(b_k, \frac{r}{4}) \subset D_k \subset D(b_k, r)$,
- $D_m \cap D_n = \phi$,
- $\mathbb{D} = \bigcup D_k$.

Then $B_{(b_k,-1)}(\{z \in \mathbb{D} : |z| < \tanh\frac{r}{4}\}) \subset D_k \subset B_{(b_k,-1)}(\{z \in \mathbb{D} : |z| < \tanh r\})$. Let us consider $\psi_k = \chi_{D_k\times\mathbb{T}}$ the characteristic function of the set $D_k \times \mathbb{T}$ then $\Psi = \{\psi_i\}_{i\in I}$ is a bounded uniform partition of unity from right of size Q. Indeed, for all $i \in I$

- $0 \leq \psi_i(x) \leq 1$,
- $\operatorname{supp}\psi_i \subset Qx_i$,
- $\sum_i \psi_i(x) = 1,\ x \in \mathbb{B}$.
- $\sup_{z\in\mathbb{B}} \#\{i \in I : z \in Qx_i\} < \infty$ for any $Q \subset \mathbb{B}$ compact.

□

We shall consider the set of Q-bounded uniform partitions of unity from right (Q-RBUPUs) as a net directed by inclusion of the associated neighborhoods, and write $|\Psi| \to 0$ if these neighborhoods run through a neighborhood base of identity. (having in mind that $2|\Psi|$ describes the maximal diameter of the support of the functions ψ_i).

2.6 Properties of the Voice Transform of the Blaschke Group

In this section we will study the integrability of the voice transform given by (18). Let us denote by $g = 1_{\mathbb{D}}$ the constant 1 function on the unit disc. It turns out that for $\gamma > 0$ every function from the minimal Möbius invariant space $\mathcal{B}_1$ and $g = 1_{\mathbb{D}}$ satisfies the integrability condition. Applying the theory of atomic decomposition we can find new atoms for these spaces.

We observe that the voice transform of the Blaschke group generated by the representation of this group on the weighted Bergman space A^2_γ given by formula (18) can be expressed by the weighted Bergman projection operator in the following way:

$$V_g f(a^{-1}) = \langle f, \pi^\gamma(a^{-1})g\rangle_\gamma = e^{\frac{\gamma+2}{2}\psi}(1-|b|^2)^{\frac{\gamma+2}{2}} P_\gamma(f \cdot \overline{g(B_a)}), \tag{20}$$

$$(a = (b, e^{i\psi}) \in \mathbb{B},\ f, g \in A^2_\gamma).$$

The minimal Möbius invariant space of analytic functions (see [1, 2]), denoted by $\mathcal{B}_1$, contains exactly the analytic functions on the unit disc which admit the representation

$$g(z) = \sum_{j=0}^{\infty} \lambda_j \frac{z - b_j}{1 - \overline{b_j} z}, \quad |b_j| \le 1, \quad \sum_{j=0}^{\infty} |\lambda_j| < \infty.$$

It is easy to prove that for $1 \le p$ and $-1 < \gamma$ the space $\mathcal{B}_1$ is included in A^p_γ.

Theorem 6 ([35]). *If $\gamma > 0$, then for every $g \in \mathcal{B}_1 \cup \{1_{\mathbb{D}}\}$ the integrability condition is satisfied, i.e.:*

$$\int_{\mathbb{B}} |V_g g(a^{-1})|\, dm(a) < \infty.$$

Proof. For $g = 1_{\mathbb{D}} \in A^2_\gamma$, using (20) we get:

$$V_g g(a^{-1}) = e^{\frac{\gamma+2}{2}\psi}(1-|b|^2)^{\frac{\gamma+2}{2}} P_\gamma(g \cdot \overline{g(B_a)})$$

$$= e^{\frac{\gamma+2}{2}\psi}(1-|b|^2)^{\frac{\gamma+2}{2}} \int_{\mathbb{D}} \frac{1}{(1-\overline{z}b)^{\gamma+2}} dA_\gamma(z) = e^{\frac{\gamma+2}{2}\psi}(1-|b|^2)^{\frac{\gamma+2}{2}}.$$

Then

$$\int_{\mathbb{B}} |V_g g(a^{-1})| m(a) = \int_{\mathbb{D}} (1-|b|^2)^{\frac{\gamma+2}{2}} \frac{1}{(1-|b|^2)^2} dA(b)$$

$$= \int_0^1 (1-r)^{\frac{\gamma}{2}-1} dr = \frac{2}{\gamma} < \infty.$$

For $g \in \mathcal{B}_1$ we have the following estimate

$$|V_g g(a^{-1})| = \left| e^{\frac{\gamma+2}{2}\psi} (1-|b|^2)^{\frac{\gamma+2}{2}} P_\gamma(g \cdot \overline{g(B_a)}) \right|$$

$$\leq (1-|b|^2)^{\frac{\gamma+2}{2}} \int_{\mathbb{D}} \left(\sum_j^\infty |\lambda_j| \right)^2 \frac{1}{|1-\bar{z}b|^{\gamma+2}} dA_\gamma(z)$$

$$= (1-|b|^2)^{\frac{\gamma+2}{2}} \left(\sum_j^\infty |\lambda_j| \right)^2 I_{\gamma,0}(b).$$

Due to Theorem A, when $|b| \to 1^-$ we have $I_{\gamma,0}(b) \sim \log \frac{1}{1-|b|^2}$. For $\gamma > 0$

$$\int_{\mathbb{D}} (1-|b|^2)^{\frac{\gamma+2}{2}} \log \frac{1}{1-|b|^2} \frac{1}{(1-|b|^2)^2} dA(b) = -\int_0^1 (1-r)^{\frac{\gamma-2}{2}} \log(1-r) dr$$

$$= -\left[\frac{2}{\gamma}(1-y)^{\frac{\gamma}{2}} \log(1-y) - \left(\frac{2}{\gamma}\right)^2 (1-y)^{\frac{\gamma}{2}} \right]_0^1 = \frac{4}{\gamma^2}.$$

From this it follows that

$$\int_{\mathbb{B}} |V_g g(a^{-1})| \, dm(a) < +\infty.$$

Consequently we proved that for $\gamma > 0$ the space $\mathcal{B}_1 \cup \{1_{\mathbb{D}}\}$ is a subset of $\mathcal{A}_1$.

□

From now on we choose the parameter function g always from the space $\mathcal{B}_1 \cup \{1_{\mathbb{D}}\}$, we also restrict the domain of the definition of the voice transform for $a = (b, 1) \in \mathbb{B}$. Using Theorem B it can be shown that the voice transform $V_g f$ can be defined not only for f belonging to A_γ^2 but also under some assumptions on the parameters $V_g f$ has sense for $f \in A_\delta^p$.

Theorem 7 ([35]). *Let us fix the function g from $\mathcal{B}_1 \cup \{1_{\mathbb{D}}\}$. If $-1 < \gamma, \delta < +\infty$, $1 \leq p$, $(\delta + 1) < (\gamma + 1)p$, then for every $f \in A_\delta^p$ the voice transform is well defined. If $a = (b, 1) \in \mathbb{B}$, then*

$$V_g f(a^{-1}) = V_g f(-b, 1) = (1-|b|^2)^{\frac{\gamma+2}{2}} F_1(b),$$

where $F_1(b) \in A_\delta^p$, and

$$\lim_{|b| \to 1^-} (1-|b|^2)^{\frac{\delta+2}{p} - \frac{\gamma+2}{2}} |V_g f(b)| = 0.$$

For $\gamma = \delta$ and $p = 2$ it follows that, if $f \in A^2_\gamma$, then

$$\lim_{|b|\to 1^-} |V_g f(b)| = 0.$$

The next theorem gives information about the set

$$\mathcal{H}_1^1 = \{f \in A^2_\alpha : V_g f \in L^1(\mathbb{B})\}.$$

Theorem 8 ([35]). *Let $g \in \mathcal{B}_1 \cup \{1\}$, $\gamma > 0$, $p \geq 1$ and $p > \max\left\{\frac{\delta+1}{\gamma+1}, \frac{4+2\delta}{\gamma}\right\}$, then for every $f \in A^p_\delta$ the voice transform $V_g f$ is integrable, i.e., $A^p_\delta \subset \mathcal{H}_1^1$.*

As an immediate consequence of this theorem we get that for $\gamma = \delta > 0$, $p > 2 + \frac{4}{\gamma}$ we have that $A^p_\gamma \subset \mathcal{H}_1^1$.

2.7 Application of the Feichtinger–Gröchenig Theory

We are now ready to use coorbit theory in order to obtain atomic decompositions in weighted Bergman spaces. As a special case we reobtain well-known atomic decompositions in the weighted Bergman spaces, but in addition some new atomic decompositions can be presented. As we have mentioned earlier in the Blaschke group it is easier to give Q-RBUPU, it is more convenient to compute the voice transform given by (18) in $a^{-1} \in \mathbb{B}$. The reproducing formula (19), taking into account that the Blaschke group is unimodular, can be written as follows

$$V_g f(y^{-1}) = \int_{\mathbb{B}} V_g f(x^{-1}) V_g g(x \circ y^{-1})\, dm(x), \quad f, g \in A^2_\gamma,\ g \neq 0,\ \|Ag\| = 1. \tag{21}$$

Let us denote by $F(y^{-1}) = V_g f(y^{-1})$, $G(y^{-1}) = V_g g(y^{-1})$, then the reproducing formula (21) is a convolution operator T, $TF = F \star G$. To discretize this for $F, G \in L^1(\mathbb{B})$ by means of Q-RBUPU we will use the modified version of the operator given by (8), namely let us denote by

$$T_\Psi F(y^{-1}) = \sum_i \langle F, \psi_i \rangle L_{x_i^{-1}} G(y^{-1}), \quad F, G \in L^1(\mathbb{B}), \tag{22}$$

which is composed of a coefficients mapping $F \to (\lambda_i)_{i\in I}$ with $\lambda_i = \langle F, \psi_i \rangle = \int_{\mathbb{B}} F(y^{-1}) \psi_i(y)\, dm(y)$ and a convolution operator

$$(\lambda_i)_{i\in I} \to \sum_i \lambda_i L_{x_i^{-1}} G = (\sum_i \lambda_i \delta_{x_i^{-1}}) \star G.$$

Our aim is to approximate the convolution operator $TF = F \star G$ by the modified operator (22). Analogous to Lemma 4.3 from [19] one can prove that:

(i) For $F \in L^1(\mathbb{B})$ the sequence of coefficients $(\lambda_i)_{i\in I}$ given by $\lambda_i = \langle F, \psi_i\rangle = \int_{\mathbb{B}} F(y^{-1})\psi_i(y)\,dm(y)$ belongs to ℓ^1, and the norms of the linear operators $F \to (\lambda_i)_{i\in I}$ are uniformly bounded.
(ii) Given $G \in L^1(\mathbb{B})$, $(\lambda_i)_{i\in I} \in \ell^1$ and any family $X = (x_i)_{i\in I}$ in $\mathcal{G}$ on has

$$F(y^{-1}) = \sum_i \lambda_i L_{x_i^{-1}} G(y^{-1}) \in L^1(\mathbb{B}),$$

the sum being absolutely convergent in $L^1(\mathbb{B})$, and there is a universal constant C_1 such that $\|F\|_1 \leq C_1\|(\lambda_i)_{i\in I}\|_1$.

There is valid also the analogue of Lemma 4.5 from [19] the only differences in the proof arise because of Q-RBUPU.

Lemma 2 ([35]). *The net set $\{T_\Psi\}$ of Q-RBUPU, directed according to inclusions of the neighborhoods Q to $\{e = (0,1)\}$, is norm convergent as operators on $L^1(\mathbb{B})$:*

$$\lim_{|\Psi|\to 0} |||T_\Psi - T|||_1 = 0.$$

This lemma implies that the modified discretization operator is also invertible if Q is sufficiently small. From the invertibility and Theorem 8 we get the following atomic decomposition for A_δ^p:

Theorem 9 ([35]). *Let us suppose that $\gamma > 0,\ p \geq 1$ and $p > \max\left\{\frac{\delta+1}{\gamma+1}, \frac{4+2\delta}{\gamma}\right\}$ and $g \in \mathcal{B}_1 \cup \{1_{\mathbb{D}}\}$, $\|Ag\| = 1$. Then for any g there exists a neighborhood Q of the identity and a constant $C_1 > 0$ both depending only on g such that for every Q-dense family $(x_i)_{i\in I}$ from right of the Blaschke group every $f \in A_\delta^p$ can be written as*

$$f(z) = \sum_i \lambda_i \pi^\gamma(x_i^{-1})g(z) \quad \textit{with} \quad \sum_i |\lambda_i| \leq C_1\|f\|_{\mathcal{H}_1^1},$$

the series is absolutely convergent in $\mathcal{H}_1^1$. The coefficients depend linearly on f, namely $\lambda_i = \int_{\mathbb{D}} T_\Psi^{-1}(V_g f(y^{-1}))\psi_i(y)dA(y)$.

The above theorem gives an atomic decomposition for every $f \in A_\delta^p$ with atoms $\pi^\gamma(x_i^{-1})g$, $g \in \mathcal{B}_1 \cup \{1_{\mathbb{D}}\}$. But how is this atomic decomposition result related to the well known atomic decompositions obtained by complex techniques?

The Q-density from right of the set $\{x_i = (b_i, -1)\}_{i\in I}$ in the language of the complex analysis is equivalent to the ϵ-net property of $\{b_i\}_{i\in I}$, with $\epsilon = \tanh r$ (see [30] p. 172). From Lemma 8. ([17] p. 188) it follows that the lower density of the set $\{b_i\}$ satisfies the following estimate:

$$D^-(\{b_i\}) \geq \frac{(1-\tanh r)^2}{2\tanh^2 r}.$$

Using Theorem 5.23 from [30] p. 161, we have that a separated sequence $\{b_i\}$ is a sampling sequence for A^p_δ if and only if

$$D^-(\{b_i\}) > \frac{\delta+1}{p}.$$

Let us choose r so small that

$$\frac{(1-\tanh r)^2}{2\tanh^2 r} > \frac{\delta+1}{p},$$

then $\{b_i\}$ is a sampling sequence for A^p_δ.

For the special case $g = 1_{\mathbb{D}}$ we obtain the following atomic decomposition: if $\gamma > 0,\ p \geq 1$ and $p > \max\left\{\frac{\delta+1}{\gamma+1}, \frac{4+2\delta}{\gamma}\right\}$, then for every $f \in A^p_\delta \subset \mathcal{H}^1_1$:

$$f = \sum \lambda_i(f)\pi^\gamma(x_i^{-1})1_{\mathbb{D}} = \sum \lambda_i(f)\frac{(1-|b_i|^2)^{\frac{\gamma+2}{2}}}{(1-\overline{b_i}z)^{\gamma+2}},$$

holds, which is very similar to the atomic decompositions obtained by complex analysis techniques (see [43], p. 69). In the classical atomic decomposition results atoms of following form arise (see [43], p. 69):

$$\frac{(1-|b_i|^2)^a}{(1-\overline{b_i}z)^b}.$$

Only the existence of the coefficients $(\lambda_i(f))_i$ is proved and their ℓ^p norm is controlled as well as the convergence is in A^p_δ norm. In our case we have ℓ^1 information about the coefficients instead of ℓ^p information and the convergence is in the stronger $\mathcal{H}^1_1$ norm instead of A^p_δ.

Using coorbit theory we also find that more general atoms for the weighted Bergman spaces. In fact, every nonzero function $g \in \mathcal{B}_1 \cup \{1_{\mathbb{D}}\}$ generates an atomic decomposition for $f \in A^p_\delta$ with atoms of the form

$$(\pi^\gamma(x_i^{-1})g)_i,$$

and the coefficients of the atomic decomposition are exactly given in terms of f and the voice transform by the following formula $\lambda_i(f) = \int_{\mathbb{D}} T_\Psi^{-1}(V_g f(y^{-1}))\psi_i(y)dA(y)$.

2.8 Coorbit Theory for Dual Pairs

Christensen and Olafson in the paper [4] gave examples of coorbit spaces for which the space $\mathcal{A}_w$ is the zero space, yet both the construction of $\mathcal{C}o(Y)$ and atomic decomposition results yield nontrivial results. In the paper [5] they presented a generalized coorbit theory which is able to cover the cases when the integrability condition is not satisfied. The idea is to replace the space $\mathcal{H}_w^1$ with a Fréchet space S. For square integrable Lie groups the space of smooth vectors is a natural choice. As an example the smooth vectors of the discrete series representation of a subgroup of $SL_2(\mathbb{R})$ are used to a complete characterization of the Bergman spaces of holomorphic functions on the unit disc. In what follows we give a brief overview of generalized coorbit theory and its consequences for the weighted Bergman spaces over the unit disc.

The original theory required that the initial representation was irreducible, unitary, and integrable. As a consequence not all Bergman spaces could be described as coorbit spaces. Their approach relies on duality arguments, which are often verifiable in cases where integrability fails. Moreover it does not require the representation to be irreducible or even come from a unitary representation on a Hilbert space. Replacing the integrability criteria with duality also has the advantage that the reproducing kernel need not provide a continuous projection from a larger Banach function space.

In the original coorbit theory the integrability of the representation is used to define the intermediate space $\mathcal{H}_w^1$. This is done in order to get a large enough pool of distributions $\mathcal{H}_w^{1\sim}$ to be able to define the coorbit space. In applications the Banach space $\mathcal{H}_w^1$ is often replaced by a Fréchet space invariant under the representation. Let S be a Fréchet space and let S^* be the space of continuous conjugate linear functionals on S equipped with the weak topology. We assume that S is continuously imbedded and weakly dense in S^*. The conjugate dual pairing of elements $v \in S$ and $v \in S^*$ will be denoted by $\langle v, v' \rangle$. Let $\mathcal{G}$ be a locally compact group with a fixed left Haar measure dx, and assume that (π, S) is a representation of $\mathcal{G}$. Also assume that the representation is continuous. As usual define the contragradient representation (π^*, S^*) by $\langle \pi^*(x)v', v \rangle = \langle v', \pi(x^{-1})v \rangle$.

Then π^* is a continuous representation of $\mathcal{G}$ on S^*. For a fixed vector $u \in S$ define the linear map $W_u : S^* \to C(\mathcal{G})$ by $W_u(v')(x) = \langle \pi^*(x)v', v \rangle$. The map W_u is called the voice transform or the wavelet transform. Instead of initial representation was irreducible, unitary, and integrable let us suppose that the vector u is an analyzing or cyclic vector, i.e.:

Definition 3. Let Y be a left invariant Banach Space of Functions on $\mathcal{G}$. A nonzero vector $u \in S$ is *cyclic vector* if the following properties are satisfied:

(R1) the reproducing formula $W_u(v) * W_u(u) = W_u(v)$ is true for all $v \in S$
(R2) the space Y is stable under convolution with $W_u(u)$ and $F \to F * W_u(u)$ is continuous

(R3) if $F = F * W_u(u) \in \boldsymbol{Y}$, then the mapping $S \ni v \to \int_{\mathscr{G}} F(x)\langle \pi^*(x)u, v\rangle dx \in \mathbb{C}$ is in S^*

(R4) the mapping $S^* \ni \phi \to \int_{\mathscr{G}} \langle \phi, \pi(x)u\rangle \langle \pi^*(x)u, u\rangle dx \in \mathbb{C}$ is weakly continuous.

Let us consider u a cyclic vector (satisfying assumptions R1–R4). The coorbit spaces of **Y** with respect u are defined as follows

$$\boldsymbol{Co}_S^u \boldsymbol{Y} = \{\phi \in S^* : W_u(\phi) \in \boldsymbol{Y}\} \tag{23}$$

equipped with the norm $\|\phi\| = \|W_u(\phi)\|_{\boldsymbol{Y}}$.

Crhistensen and Olafson in the paper [5] proved that under these general assumptions the coorbit spaces of $\boldsymbol{Y}$ with respect to u own similar properties as the classical $\boldsymbol{Co}(\boldsymbol{Y})$ coorbit spaces.

Theorem 10 ([5]). *Assume that $\boldsymbol{Y}$ and u satisfy all assumption from the previous definition, then the following properties hold:*

(1) $W_u(v) * W_u(u) = W_u(v)$ for $v \in \boldsymbol{Co}_S^u \boldsymbol{Y}$.
(2) The space $\boldsymbol{Co}_S^u \boldsymbol{Y}$ is a π^*-invariant Banach space.
(3) $W_u : \boldsymbol{Co}_S^u \boldsymbol{Y} \to \boldsymbol{Y}$ intertwines π^* and left translation.
(4) If left translation is continuous on $\boldsymbol{Y}$, then π^* acts continuously on $\boldsymbol{Co}_S^u \boldsymbol{Y}$.
(5) $\boldsymbol{Co}_S^u \boldsymbol{Y} = \{\pi^*(F)u : F \in \boldsymbol{Y}, F = F * W_u(u)\}$.
(6) $W_u : \boldsymbol{Co}_S^u \boldsymbol{Y} \to \boldsymbol{Y} * W_u(u)$ is an isometric isomorphism.

It is also shown that besides a reproducing formula a duality requirement is sufficient for the construction of coorbit spaces.

The coorbit theory of Feichtinger and Gröchenig is a special case of this general coorbit theory for dual pairs. In this classical case the proof is based on properties of Wiener amalgam spaces over locally compact groups. These spaces were used to verify properties (R2) and (R3). Sometimes it is easier to prove these properties by duality without the use of the Wiener amalgam machinery. Theorem 4.2(i) in [20] states that $\boldsymbol{Co}(\boldsymbol{Y})$ is continuously included in $\mathcal{H}_w^{1\sim}$. Theorem 4.5.13(d) in [37] states further that $\mathcal{H}_w^1$ is continuously included in $\boldsymbol{Co}(\boldsymbol{Y})$. In general $S \not\subseteq \boldsymbol{Co}_S^u \boldsymbol{Y}$, since, for example, the coorbit space $\boldsymbol{Co}_{\mathcal{H}}^u L^1(\mathscr{G})$ for an integrable representation does not contain $\mathcal{H}$. It is an open problem if the inclusion $\boldsymbol{Co}_S^u \boldsymbol{Y} \hookrightarrow S^*$ is continuous for general coorbit theory.

The general results regarding coorbit spaces for dual pairs imply that a large class of weighted Bergman spaces on the unit disc can be identified with coorbit spaces of the smooth vectors with respect to the cyclic vector $u = 1_{\mathbb{D}}$ and this vector also generates atomic decomposition results.

Let $a > 0, b \in \mathbb{R}$ and from now on let us consider $\mathscr{G} \subseteq SL_2(\mathbb{R})$ the connected subgroup of upper triangular matrices, i.e.,

$$\mathscr{G} = \left\{ \begin{pmatrix} a & b \\ 0 & a^{-1} \end{pmatrix} : a > 0, b \in \mathbb{R} \right\}$$

with left-invariant measure $\frac{dadb}{a^2}$. Through the Cayley transform this group can be identified with $\mathbb{S}U(1,1)$. Every $\mathbf{g} \in \mathbb{S}U(1,1)$ can be represented as

$$\mathbf{g} = \begin{pmatrix} \alpha & \beta \\ \overline{\beta} & \overline{\alpha} \end{pmatrix} = \frac{1}{2}\begin{pmatrix} a + a^{-1} + ib & b + i(a - a^{-1}) \\ b - i(a - a^{-1}) & a + a^{-1} - ib \end{pmatrix} \quad a > 0,\, b \in \mathbb{R}.$$

The discrete series representation of $\mathbb{S}U(1,1)$ on the weighted Bergman space is given by

$$\Pi^{\gamma+2}\left(\begin{pmatrix} \alpha & \beta \\ \overline{\beta} & \overline{\alpha} \end{pmatrix} f\right)(z) := \frac{1}{(-\overline{\beta}z + \alpha)^{\gamma+2}} f\left(\frac{\overline{\alpha}z - \beta}{-\overline{\beta}z + \alpha}\right) = \Pi^{\gamma+2}_{(a,b)}(f)(z). \tag{24}$$

The representations $\Pi^{\gamma+2}$ are square integrable for all $\gamma > -1$ and integrable for $\gamma > 0$. Let us denote by $W_u^{\gamma+2}$ the voice transform induced by $\Pi^{\gamma+2}$ and with parameter $u = 1_{\mathbb{D}}$.

The coefficients corresponding to $u = 1_{\mathbb{D}}$ are

$$W_u^{\gamma+2}(u)(a,b) = \langle u, \Pi^{\gamma+2}_{(a,b)} u \rangle_\gamma = 2^{\gamma+2}(a + a^{-1} - ib)^{-\gamma-2}.$$

Consider the submultiplicative weight $w_r = 2^r[(a + a^{-1})^2 + b^2]^{r/2}$, for $r \geq 0$ let $L_r^p(\mathcal{G})$ denote space

$$L_r^p(\mathcal{G}) = \left\{ F : \|F\|_{L_r^p(\mathcal{G})} = \left(\int |F(a,b) w_r(a,b)|^p \frac{dadb}{a^2} \right)^p < \infty \right\}.$$

The smooth vectors of the representation $\Pi^{\gamma+2}$ will be denoted by $\mathcal{H}^\infty_{\gamma+2}$. This is a Fréchet space and has the following characterization: they are the power series $\sum_{k=0}^\infty a_k z^k$ for which there for any m exists a constant C such that

$$|a_k|^2 \leq C \binom{\gamma + k + 1}{k} (1 + k)^{-m}.$$

The conjugate dual $\mathcal{H}^{-\infty}_{\gamma+2}$ of this space consists of formal power series $\sum_{k=0}^\infty b_k z^k$ for which there is an m and a constant C such that

$$|b_k|^2 \leq C \binom{\gamma + k + 1}{k} (1 + k)^{m}.$$

Theorem 4.3 and Lemma 4.4 of [5] imply that for $-\gamma < r + 2/p < 2 + \gamma$ the spaces $\boldsymbol{Co}^u_{\mathcal{H}^\infty_{\gamma+2}} L^p_r$ are nonzero $\Pi^{\gamma+2}$ invariant Banach spaces and if $f \in A^p_{(\gamma+2-r)p/2}(\mathbb{D})$, then $f \in \mathcal{H}^{-\infty}_{\gamma+2}$.

Moreover, the following result holds true:

Theorem 11 ([5]). *For $1 < (\gamma + 2 - r)p/2 < (\gamma + 1)p + 1$ the space $A^p_{(\gamma+2-r)p/2}(\mathbb{D})$ corresponds to the coorbit space $\boldsymbol{Co}^u_{\mathcal{H}^\infty_{\gamma+2}} L^p_r$, with equivalence of norms.*

Define the sequence space

$$\ell^p_r = \left\{ (\lambda_i) : \|(\lambda_i)\|_{\ell^p_r} = \left(\sum |\lambda_i w_r(x_i)|^p \right)^{1/p} < \infty \right\}.$$

Theorem 12 ([5]). *Let $V \subset U$ be compact neighborhoods of the identity of $\mathbb{S}U(1,1)$. Assume that the points $\{x_i\}$ are V-separated and U-dense. Then there exists U and a corresponding partition of unity $\{\psi_i\}$ for which $supp(\psi_i) \subseteq x_i U$ such that the following are true*

(a) The map $\ell^p_r \ni (\lambda_i) \to \sum_i \lambda_i L_{x_i} W_u^{\gamma+2} \in L^p_r(\mathcal{G}) * W_u^{\gamma+2}$ is continuous.
(b) The map $L^p_r(\mathcal{G}) * W_u^{\gamma+2} \ni F \to (F(x_i))_{i \in I} \in \ell^p_r$ is continuous.
(c) The map $L^p_r(\mathcal{G}) * W_u^{\gamma+2} \ni F \to \left(\int_{\mathcal{G}} F(x)\psi_i(x)dx \right)_{i \in I} \in \ell^p_r$ is continuous.

As in [25] sums are understood as limits of the net of partial sums over finite subsets with convergence in $L^p_r(\mathcal{G})$.

The following theorem gives reconstruction formulas that provide atomic decompositions for the generalized coorbit spaces.

Theorem 13 ([5]). *We can choose a compact neighborhood Q of unity in $\mathcal{G}$, Q-dense points $\{x_i\}$ in $\mathcal{G}$, and a corresponding partition of unity $\{\psi_i\}$ for which $supp(\psi_i) \subseteq x_i Q$ such that the operators $T_\psi, S_\psi, U_\psi : L^p_r(\mathcal{G}) * W_u^{\gamma+2} \to L^p_r(\mathcal{G}) * W_u^{\gamma+2}$ defined below are invertible with continuous inverses*

(a) $S_\psi F = \sum_i F(x_i)\psi_i * W_u^{\gamma+2}$,
(b) $U_\psi F = \sum_i c_i F(x_i) L_{x_i} W_u^{\gamma+2}$, with $c_i = \int \psi_i$
(c) $T_\psi F = \sum_i (\int F(x)\psi_i(x)dx) L_{x_i} W_u^{\gamma+2}$.

In the proof of the above result the use of integrability is avoided. This allows to cover cases of weighted Bergman spaces A^p_γ, $-1 < \gamma \leq 0$ which could not be treated with the initial coorbit theory. It should be mentioned that Feichtinger and Gröchenig anticipated the possibility of a coorbit construction for certain nonintegrable representations, but the emphasis in [19] was more on having atomic decompositions for extensive families of Banach spaces.

3 Coorbit Theory: Applications and Outlook

3.1 Coorbit Spaces and Toeplitz Operators

Having a transform domain (such as the range for the STFT or of the continuous wavelet transform) with a regularizing reproducing kernel, or a normed space of analytic functions, it is clear that the multiplication with a general function (even if it is smooth) will in almost all cases inevitably lead to functions outside of the domain of the RKH (reproducing kernel Hilbert space) under discussion. Hence it is natural to investigate operators arising by first applying a pointwise multiplication (with a function or distribution) and then the orthogonal projection. Since one gets exactly the classical Toeplitz operators if one starts with the analytic Hardy space such operators are often called *Toeplitz operators*, while they are called wavelet resp. STFT-multipliers (in analogy with Fourier multipliers) for the prototypical examples of coorbit theory. Especially in the STFT-context it is also possible to multiply with discrete (unbounded or bounded) measure, obtaining then Gabor multipliers. Alternatively they can be described as linear combinations of projection operators onto the Gabor atoms, arising from a given atom g by applying suitable TF-shifts. Still another viewpoint on the same subject is through the so-called Wick-calculus, which is well described, e.g., in [8] or [3].

While both the theory of STFT-multipliers and Gabor multipliers and the theory of Toeplitz operators on Bergman spaces are well developed (almost independently), the setting described above suggests that much can be said about questions of the following type:

- What is the quality of the Toeplitz operator (e.g., Hilbert-Schmidt), given the quality of the multiplier (upper symbol) and the quality of the reproducing kernel (the Gabor atom, in the case of Gabor multipliers);
- Are the resulting discrete operators in a one-to-one relationship with their multiplier sequences, or more precisely: under which conditions does the family of corresponding projection operators for a Riesz basic sequence.
- What are the mapping properties of such Toeplitz operators, especially if the multipliers are weight functions showing a definitive global behavior. Recent results in this direction are given in very recent papers such as [27, 28].

On the other hand we need modulation spaces to describe the corresponding Toeplitz operators (see [42]).

3.2 Coorbit Spaces in a Wider Context

In the last years many interesting results connected to coorbit theory were published by several authors. For instance, we mention the following papers: [4–7, 9, 10, 12, 13, 22, 26, 32, 37–40]. Without any claim for completeness of the list let us comment on some of these results potentially interesting for the readers.

In the paper [6] Christensen, Mayeli, and Olafsson identified homogeneous Besov spaces on stratified Lie groups as coorbit spaces and use this to derive atomic decomposition for Besov spaces. Later they gave a coorbit description and atomic decompositions for such Besov spaces (see [7]).

In the papers [9, 10] Dahlke, Steidl, and Teschke obtained results connected to coorbit spaces and Banach frames on homogeneous spaces with applications to the sphere. In their approach the parameter space of the transform is not anymore a group but a homogeneous space.

In the paper [12] Dahlke, Fornasier, Rauhut, Steidl, and Teschke treat other generalizations and specific applications of the coorbit space theory based on group representations modulo quotients. They show that the general theory applied to the affine Weyl-Heisenberg group gives rise to families of smoothness spaces that can be identified with alpha-modulation spaces. A more direct approach towards atomic decompositions for alpha-modulation spaces is taken by Fornasier in [24].

Dahlke, Teschke, and Stingl in [11] are concerned with the analysis and decomposition of medical multichannel data. They present a signal processing technique that reliably detects and separates signal components such as mMCG, fMCG, or MMG by involving the spatiotemporal morphology of the data provided by the multisensor geometry of the so-called multichannel superconducting quantum interference device (SQUID) system. The mathematical building blocks are coorbit theory, multi–modulation frames, and the concept of joint sparsity measures. Combining the ingredients, they end up with an iterative procedure (with component-dependent projection operations) that delivers the individual signal components.

Dahlke, Kutyniok, Steidl, Teschke introduce the shearlet group in [13] and establish the square-integrability of its representation in $L^2(\mathbb{R}^2)$. They studied the relationships of the shearlet transform with the coorbit space theory. It is shown that all the conditions to construct the coorbit spaces can be satisfied, thus establishing the new family of shearlet coorbit spaces. To derive the associated Banach frames for these spaces, suitable discrete subsets, called *U-dense sets*, had to be constructed and an additional integrability condition had to be satisfied. They proved that the Schwartz space is contained in the shearlet coorbit spaces. The power of best n-term approximation schemes based on the shearlet frames is shown to depend on the smoothness of the signal as measured in a second shearlet coorbit space. The performance of the new frame algorithms is investigated by applying them to some test images.

In the paper [14] Dahlke, Steid, Teschke show that compactly supported functions with sufficient smoothness and some vanishing moments can serve as analyzing vectors for shearlet coorbit space. This approach has been used to prove embedding theorems for subspaces of shearlet coorbit spaces resembling shearlets on the cone into Besov spaces. The embedding relations of traces of these subspaces with respect to the real axes have been established. An interesting conjecture has been suggested that the traces of shearlet coorbit spaces on R^3 with respect to two-dimensional hyperplanes are again shearlet coorbit spaces. They also studied the multivariate shearlet transform, shearlet coorbit spaces, and their structural properties (see [15]).

Dahlke, Häuser, Teschke in the paper [16] studied the coorbit space theory for the Toeplitz shearlet transform.

M. Fornasier and H. Rauhut in the paper [23] studied continuous frames, function spaces, and discretization problems related to the coorbit theory.

Gröchenig and Piotrowski studied the notion of molecules in coorbit spaces. The main result states that if an operator, originally defined on an appropriate space of test functions, maps atoms to molecules, then it can be extended to a bounded operator on coorbit spaces. For time-frequency molecules they recover some boundedness results on modulation spaces, for time-scale molecules they obtain the boundedness on homogeneous Besov spaces (see [26]).

In the book [31] Kutyniok (ed.) et al. devoted a chapter to the generalization of the continuous shearlet transform to higher dimensions as well as to the construction of associated smoothness spaces and to the analysis of their structural properties, respectively. To construct canonical scales of smoothness spaces, so-called shearlet coorbit spaces, and associated atomic decompositions and Banach frames they prove that the coorbit space theory is applicable for the proposed shearlet setting. For the two-dimensional case they show that for large classes of weights, variants of Sobolev embeddings exist. Furthermore, they prove that for natural subclasses of shearlet coorbit spaces which in a certain sense correspond to "cone-adapted shearlets" there exist embeddings into homogeneous Besov spaces. Moreover, the traces of the same subclasses onto the coordinate axis can again be identified with homogeneous Besov spaces. These results are based on the characterization of Besov spaces by atomic decompositions and rely on the fact that shearlets with compact support can serve as analyzing vectors for shearlet coorbit spaces. Finally, they demonstrate that the proposed multivariate shearlet transform can be used to characterize certain singularities.

Mantoiu in the paper [32] published results concerning the quantization rules, Hilbert algebras, and coorbit spaces for families of bounded operators.

In paper [37] Rauhut is concerned with the construction of atomic decompositions and Banach frames for subspaces of certain Banach spaces consisting of elements which are invariant under some symmetry group. The construction is established via a generalization of the well-established coorbit theory. Examples include radial wavelet-like atomic decompositions and frames for radial Besov–Triebel–Lizorkin spaces and radial Gabor frames and atomic decompositions for radial modulation spaces.

In paper [38] he generalized the classical coorbit space theory to quasi-Banach spaces. As a main result it provides atomic decompositions for coorbit spaces defined with respect to quasi-Banach spaces. These atomic decompositions are used to prove fast convergence rates of best n-term approximation schemes. The abstract theory is applied to time-frequency analysis of modulation spaces for the full range of parameters $M^{p,qm}, 0 < p, q \leq \infty$.

In paper [39] Rauhut and Ullrich presented a generalized coorbit space theory for the inhomogeneous function spaces of Besov–Lizorkin–Triebel type. This is of interest because the classical coorbit theory does not cover such spaces.

It was recognized by Fornasier and Rauhut (see [23]) that one may replace coherent states related to the group representation by more general abstract continuous frames. Instead of assumptions on the generating atom (which in turn imply decay properties of G) in this more general situation the properties of the reproducing kernel have to be taken into account. They significantly extend this abstract coorbit space theory to treat a wider variety of coorbit spaces. A unified approach towards atomic decompositions and Banach frames with new results for general coorbit spaces is presented, viewing inhomogeneous Besov–Lizorkin–Triebel spaces of various types of interest as coorbit spaces. The paper contains a number of old and new wavelet characterizations based on explicit smoothness, decay, and vanishing moment assumptions of the respective wavelet.

In a recent paper (see [40]) Romero gives a characterization of coorbit spaces using phase-space covers. He showed that coorbit spaces can be characterized in terms of arbitrary phase-space covers, which are families of phase-space multipliers associated with partitions of unity. This generalizes previously known results for time-frequency analysis to include time-scale decompositions.

References

1. J. Arazy, Some aspects of the minimal, Möbius-invariant space of analytic functions on the unit disc., Interpolation spaces and allied topics in analysis, Proc. Conf., Lund/Swed. 1983, Lect. Notes Math. **1070**, 24–44 (1984)
2. J. Arazy, S. Fisher, J. Peetre, Möbius invariant function spaces., J. Reine Angew. Math. **363**, 110–145 (1985)
3. P. Boggiatto, E. Cordero, K. Gröchenig, Generalized anti-Wick operators with symbols in distributional Sobolev spaces. Integr. Equat. Oper. Theory **48**(4), 427–442 (2004)
4. J.G. Christensen, G. Olafson, Examples of coorbit spaces for dual pairs. Acta Appl. Math. **107**(1–2), 25–48 (2009)
5. J.G. Christensen, G. Olafsson, Coorbit spaces for dual pairs. Appl. Comput. Harmon. Anal. **31**(2), 303–324 (2011)
6. J.G. Christensen, A. Mayeli, G. Olafsson, Coorbit description and atomic decomposition of Besov spaces. Arxiv preprint arXiv:1110.6676 (2011)
7. J.G. Christensen, A. Mayeli, G. Olafsson, Coorbit description and atomic decomposition of Besov spaces. Numer. Funct. Anal. Opt. **33**(7–9), 847–871 (2012)
8. E. Cordero, L. Rodino, Wick calculus: a time-frequency approach. Osaka J. Math. **42**(1), 43–63 (2005)
9. S. Dahlke, G. Steidl, G. Teschke, Coorbit spaces and Banach frames on homogeneous spaces with applications to the sphere. Adv. Comput. Math. **21**(1–2), 147–180 (2004)
10. S. Dahlke, G. Steidl, G. Teschke, Weighted coorbit spaces and Banach frames on homogeneous spaces. J. Fourier Anal. Appl. **10**(5), 507–539 (2004)
11. S. Dahlke, G. Teschke, K. Stingl, Coorbit theory, multi-α-modulation frames, and the concept of joint sparsity for medical multichannel data analysis, 2008, http://user.hs-nb.de/~teschke/ps/55.pdf
12. S. Dahlke, M. Fornasier, H. Rauhut, G. Steidl, G. Teschke, Generalized coorbit theory, Banach frames, and the relation to $alpha$-modulation spaces. Proc. Lond. Math. Soc. **96**(2), 464–506 (2008)
13. S. Dahlke, G. Kutyniok, G. Steidl, G. Teschke, Shearlet coorbit spaces and associated Banach frames, Appl. Comput. Harmon. Anal. **27**(2), 195–214 (2009)
14. S. Dahlke, G. Steidl, G. Teschke, Shearlet coorbit spaces: Compactly supported analyzing shearlets, traces and embeddings. J. Fourier Anal. Appl. **17**(6), 1232–1255 (2011)
15. S. Dahlke, G. Steidl, G. Teschke, Multivariate shearlet transform, shearlet coorbit spaces and their structural properties, http://user.hs-nb.de/~teschke/ps/73.pdf
16. S. Dahlke, S. Häuser, G. Teschke, Coorbit space theory for the Toeplitz shearlet transform, Int. J. Wavelets Multiresolut. Inf. Process. **10**(04), p.1250037, 13 p. (2012)

17. P. Duren, A. Schuster, *Bergman Spaces*, vol. 100 of Mathematical Surveys and Monographs. (Amer. Math. Soc. (AMS), Providence, RI, 2004)
18. H.G. Feichtinger, On a new Segal algebra. Monatsh. Math. **92**, 269–289 (1981)
19. H.G. Feichtinger, K. Gröchenig, A unified approach to atomic decompositions via integrable group representations. Lect. Note Math. **1302**, 52–73 (1988)
20. H.G. Feichtinger, K. Gröchenig, Banach spaces related to integrable group representations and their atomic decompositions, I. J. Funct. Anal. **86**(2), 307–340 (1989)
21. H.G. Feichtinger, K. Gröchenig, Banach spaces related to integrable group representations and their atomic decompositions, II. Monatsh. Math. **108**(2–3), 129–148 (1989)
22. M. Fornasier, Nota on the coorbit theory of $alpha$- modulation spaces, 2005
23. M. Fornasier, H. Rauhut, Continuous frames, function spaces, and the discretization problem. J. Fourier Anal. Appl. **11**(3), 245–287 (2005)
24. M. Fornasier, Banach frames for α-modulation spaces. Appl. Comput. Harmon. Anal. **22**(2), 157–175 (2007)
25. K. Gröchenig, Describing functions: atomic decompositions versus frames. Monatsh. Math. **112**(3), 1–41 (1991)
26. K. Gröchenig, M. Piotrowski, Molecules in coorbit spaces and boundedness of operators. Studia Math. **192**(1), 61–77 (2009)
27. K. Gröchenig, J. Toft, Isomorphism properties of Toeplitz operators in time-frequency analysis. J. d'Analyse **114**(1), 255–283 (2011)
28. K. Gröchenig, J. Toft, The range of localization operators and lifting theorems for modulation and Bargmann-Fock spaces. Trans. Am. Math. Soc. 23 (2012)
29. C. Heil, D.F. Walnut, Continuous and discrete wavelet transforms. SIAM Rev. **31**, 628–666 (1989)
30. H. Hedenmalm, B. Korenblum, K. Zhu, *Theory of Bergman Spaces* (Springer, New York, 2000)
31. G. Kutyniok, et al. (eds.) *Shearlets. Multiscale Analysis for Multivariate Data* (Birkhäuser, Boston, MA). Appl. Numer. Harmonic Anal. 105–144 (2012)
32. M. Mantoiu, Quantization rules, Hilbert algebras and coorbit spaces for families of bounded operators I. *The Abstract Theory*. Arxiv preprint arXiv:1203.6347 (2012)
33. M. Pap, F. Schipp, The voice transform on the Blaschke group III. Publ. Math. Debrecen **75**(1–2), 263–283 (2009)
34. M. Pap, The voice transform generated by a representation of the Blaschke group on the weighted Berman spaces. *Ann. Univ. Sci. Budapest. Sect. Comp.* **33**, 321–342 (2010)
35. M. Pap, Properties of the voice transform of the Blaschke group and connections with atomic decomposition results in the weighted Bergman spaces. J. Math. Anal. Appl. **389**(1), 340–350 (2012)
36. A. Perälä, J. Taskinen, J. Virtanen, New results and open problems on Toeplitz operators in Bergman spaces. New York J. Math. **17a**, 147–164 (2011)
37. H. Rauhut, Banach frames in coorbit spaces consisting of elements which are invariant under symmetry groups. Appl. Comput. Harmon. Anal. **18**(1), 94–122 (2005)
38. H. Rauhut, Coorbit space theory for quasi-Banach spaces. Studia Math. **180**(3), 237–253 (2007)
39. H. Rauhut, T. Ullrich, Generalized coorbit space theory and inhomogeneous function spaces of Besov-Lizorkin-Triebel type. J. Funct. Anal. **11**, 3299–3362 (2011)
40. J.L. Romero, Characterization of coorbit spaces with phase-space covers. J. Funct. Anal. **262**(1), 59–93 (2012)
41. F. Schipp, W. Wade, *Transforms on Normed Fields*, Leaflets in Mathematics (Janus Pannonius University, Pecs, 1995)
42. J. Toft, P. Boggiatto, Schatten classes for Toeplitz operators with Hilbert space windows on modulation spaces. Adv. Math. **217**(1), 305–333 (2008)
43. K. Zhu, *Spaces of Holomorphic Functions in the Unit Ball*, vol. 226 of Graduate Texts in Mathematics (Springer, New York, NY, 2005)

Quadrature Domains and Their Two-Phase Counterparts

Stephen J. Gardiner and Tomas Sjödin

Abstract This survey describes recent advances on quadrature domains that were made in the context of the ESF Network on Harmonic and Complex Analysis and its Applications (2007–2012). These results concern quadrature domains, and their two-phase counterparts, for harmonic, subharmonic and analytic functions.

Keywords Analytic function • Harmonic function • Partial balayage • Quadrature domain • Subharmonic function

Mathematics Subject Classification (2010). Primary 31B05, secondary 30E20.

1 Introduction

A bounded open set $\Omega \subset \mathbb{R}^N$ $(N \geq 2)$ is called a quadrature domain for harmonic functions with respect to a (signed Radon) measure μ if $\operatorname{supp}\mu \subset \Omega$ and

$$\int_\Omega h d\lambda = \int h d\mu \quad \text{for all integrable harmonic functions } h \text{ on } \Omega, \tag{1}$$

S.J. Gardiner (✉)
School of Mathematical Sciences, University College Dublin, Belfield, Dublin 4, Ireland
e-mail: stephen.gardiner@ucd.ie

T. Sjödin
Department of Mathematics, Linköping University, 581 83 Linköping, Sweden
e-mail: tosjo@mai.liu.se

A. Vasil'ev (ed.), *Harmonic and Complex Analysis and its Applications*, Trends in Mathematics, DOI 10.1007/978-3-319-01806-5_5,

where λ denotes Lebesgue measure. The above quadrature identity can be considered, after division by $\lambda(\Omega)$, as an analogue of the familiar ball mean value property of harmonic functions. Alternatively, if we take $N = 3$ and consider harmonic functions of the form $h(x) = |x - y|^{-1}$, where $y \notin \Omega$, it can be viewed as a statement about the gravitational (or electrostatic) equivalence of two mass (or charge) distributions. Quadrature domains have many interesting connections with other avenues of mathematical research, for example Hele–Shaw flow. An introductory survey may be found in Gustafsson and Shapiro [11].

The purpose of this article is to present an overview of some recent advances on quadrature domains that were obtained as part of the programme of the ESF Network on Harmonic and Complex Analysis and its Applications. We will consider quadrature domains, and their two-phase counterparts, for harmonic functions, subharmonic functions, and for analytic functions in the complex plane. There are two main sections, which we now briefly outline.

Let ω be an open set such that $\text{supp}\mu \subset \omega \subset\subset \Omega$. It is not difficult to see that, if Ω is a quadrature domain for harmonic functions with respect to μ, then it is also a quadrature domain with respect to the balayage (or sweeping out) of μ onto $\partial\omega$. However, a different type of balayage of μ, onto Lebesgue measure rather than onto the boundary of a region, turns out to be an even more useful tool in the construction of quadrature domains. Section 2 begins by briefly recalling the basic properties of this technique of partial balayage. It then proceeds to describe the resolution of a question about measure substitution (that is, whether the signed measure μ in (1) can always be replaced by a positive measure), progress on the classical exterior inverse problem of potential theory, and the recently introduced notion of a harmonic ball.

Section 3 concerns two-phase quadrature domains, which arose out of recent work on the two-phase membrane problem. We will offer an introduction to such quadrature domains, followed by a description of how they can be constructed using an analogue of partial balayage, and also some uniqueness results. This section concludes with a discussion of the corresponding notion for analytic functions in the complex plane and the associated analogue of the Schwarz function.

2 (One-Phase) Quadrature Domains

2.1 *Basic Definitions*

The notion of a quadrature domain for harmonic functions was recalled in the introduction. If $N = 2$, and we replace harmonic functions by analytic functions (again required to be integrable on Ω), then Ω is called a quadrature domain for analytic functions with respect to μ. In this case it is natural to allow μ to be a complex measure. Alternatively, if

$$\int_{\Omega} s d\lambda \geq \int s d\mu \quad \text{for all integrable subharmonic functions } s \text{ on } \Omega,$$

then Ω is called a quadrature domain for subharmonic functions.

Quadrature domains may equivalently be defined using Newtonian potentials, as follows. We denote by $U\mu$ the Newtonian (or logarithmic, if $N = 2$) potential of μ, normalized so that $-\Delta U\mu = \mu$ in the sense of distributions. Now let Ω be a bounded open subset of $\mathbb{R}^N$, let μ be a measure with compact support in Ω, and define

$$u = U\mu - U(\lambda|_{\Omega}). \tag{2}$$

We write $\Omega^c = \mathbb{R}^N \backslash \Omega$. Then

- Ω is a quadrature domain for analytic functions (where $N = 2$) if and only if $|\nabla u| = 0$ in Ω^c;
- Ω is a quadrature domain for harmonic functions if and only if $u = |\nabla u| = 0$ in Ω^c;
- Ω is a quadrature domain for subharmonic functions if and only if $u \geq 0$ in $\mathbb{R}^N$, $u = 0$ in Ω^c.

When $N = 2$ it is often more natural to use complex notation ($\mathbb{C}$ instead of $\mathbb{R}^2$). In this case there is a further equivalent formulation of the concept of a quadrature domain for analytic functions. Namely, $\Omega \subset \mathbb{C}$ is a quadrature domain for analytic functions with respect to some measure if and only if it has a one-sided Schwarz function S; that is, there is a function $S \in C(\overline{\Omega \backslash K})$, analytic on $\Omega \backslash K$, where K is some compact subset of Ω, such that $S(z) = \bar{z}$ on $\partial\Omega$. If Ω is a quadrature domain for analytic functions with respect to a measure μ, then we can take $S = \bar{z} - 4\partial u$, where u is as defined above.

2.2 Existence Theory, Partial Balayage

It is easy to find examples of measures μ with respect to which there can be no quadrature domains: we can simply take $\mu = \frac{1}{2}\lambda|_D$, where D is an open bounded set, for instance. Roughly speaking, non-existence occurs when the measure μ is insufficiently concentrated. The method of partial balayage allows us to make such assertions more precise. In this section μ will denote a positive Radon measure with compact support in $\mathbb{R}^N$.

We now recall, without proofs, some basic facts about the notion of partial balayage, which was originally developed by Gustafsson and Sakai [10]. A recent exposition of it may be found in [7]. There is also a close relationship between partial balayage and weighted equilibrium measures (see [14] or [13]).

For an open set $D \subset \mathbb{R}^N$ and a positive measure μ with compact support in D we define

$$V_D\mu(x) = \sup\left\{ v(x) : v \text{ is subharmonic on } D \text{ and } v \leq U\mu + \frac{|\cdot|^2}{2N} \text{ on } \mathbb{R}^N \right\} - \frac{|x|^2}{2N}$$

and then put $B_D\mu = -\Delta V_D\mu$. It turns out that there is a measure ν such that

$$B_D\mu = \lambda|_{\omega(D,\mu)} + \mu|_{\omega(D,\mu)^c} + \nu = \lambda|_{\Omega(D,\mu)} + \mu|_{\Omega(D,\mu)^c} + \nu,$$

where

$$\omega(D,\mu) = \{V_D\mu < U\mu\}$$

and

$$\Omega(D,\mu) = \bigcup\{U : U \subset D \text{ open and } B_D\mu = \lambda \text{ in } U\},$$

and these are both bounded open subsets of D. (Clearly $V_D\mu = U\mu$ on D^c.) Further,

$$B_D\mu \leq \lambda \text{ on } D \quad \text{and} \quad \nu \geq 0,$$

and ν is supported by $\partial D \cap \partial\omega(D,\mu)$. We remark that $\omega(D,\mu) \subset \Omega(D,\mu)$ and that this inclusion may be strict, even when $\operatorname{supp}\mu \subset \Omega(D,\mu)$. These sets clearly increase as D increases and as μ increases. It will be convenient to define

$$W_D\mu = U\mu - V_D\mu,$$

whence $W_D\mu$ is lower semicontinuous, $-\Delta W_D\mu \geq \mu - \lambda$ on D and $W_D\mu \geq 0$ on $\mathbb{R}^N$. Finally, if $D = \mathbb{R}^N$, we will abbreviate the above notation to $V\mu$, $B\mu$, $\omega(\mu)$, $\Omega(\mu)$, and $W\mu$, respectively. In this case $\nu = 0$.

If there is a quadrature domain Ω for subharmonic functions with respect to μ, it is easy to see that, up to a Lebesgue null set, $\Omega = \Omega(\mu)$. Thus partial balayage yields this essentially unique set if it exists. If μ is singular with respect to Lebesgue measure, then the remainder term $\mu|_{\Omega(\mu)^c}$ is zero, since $\mu|_{\Omega(\mu)^c} \leq \lambda$. Unfortunately, it is harder to say when μ will have compact support in $\Omega(\mu)$. One case that is easy to handle is as follows.

Theorem 2.1. *Suppose that $U\mu = \infty$ on* suppμ. *Then μ has compact support in $\Omega(\mu)$, which equals $\omega(\mu)$, and this is the only quadrature domain for subharmonic functions with respect to μ.*

Thus we get absolute uniqueness here, not merely up to Lebesgue null sets.

2.3 Integrability of Positive Harmonic Functions and the Measure Substitution Problem

In this section we will discuss the problems solved in [5], concerning quadrature domains for harmonic functions. Before that we will make a related, but more straightforward, observation about quadrature domains for subharmonic functions. Throughout this section μ denotes a signed measure.

We first recall some basic facts about the Martin boundary that will be useful (see Chapter 8 of [1] for a detailed account). Let Ω be Greenian and $G_\Omega(x, y)$ denote its Green function. Further, let (ω_n) be an exhaustion of Ω by (for convenience) smoothly bounded open sets; thus, $\omega_1 \subset \overline{\omega}_1 \subset \omega_2 \subset \ldots \subset \Omega$ and $\cup_{n=1}^{\infty} \omega_n = \Omega$. We fix a reference point $x_0 \in \omega_1$.

Given a positive harmonic function h on Ω it is easy to see that the reduced function u_n of h over $\overline{\omega}_n$, which equals h in $\overline{\omega}_n$ and solves the Dirichlet problem in $\Omega \backslash \overline{\omega}_n$ with boundary values h on $\partial\omega_n$ and 0 on $\partial\Omega$, is a Green potential, namely

$$u_n(x) = \int G_\Omega(x, y) d\mu_n(y),$$

where the measure μ_n is supported by $\partial\omega_n$. If we now introduce the "Martin kernel"

$$M(x, y) = \frac{G_\Omega(x, y)}{G_\Omega(x_0, y)} \quad (y \neq x_0),$$

and the measure η_n defined by

$$d\eta_n = G_\Omega(x_0, \cdot) d\mu_n,$$

then

$$u_n(x) = \int M(x, y) d\eta_n(y)$$

and

$$\eta_n(\Omega) = u_n(x_0) = h(x_0).$$

A difficulty we face here is that $M(x, \cdot)$ does not in general have a continuous extension to the Euclidean boundary. The Martin compactification $\hat{\Omega} = \Omega \cup \partial^M \Omega$ is constructed to be the smallest compactification such that the functions $M(x, \cdot)$ have continuous extensions up to $\partial^M \Omega$. Such a compactification always exists. (If Ω is a Lipschitz domain, then $\hat{\Omega}$ is homeomorphic to $\overline{\Omega}$.)

The crucial point here is that, since the sequence (η_n) is bounded in norm it has a weak*-cluster point η, which is a measure supported by $\partial^M \Omega$, and it follows from the definition of weak*-convergence that

$$h(x) = \int M(x,y)d\eta(y) \quad (x \in \Omega).$$

This is an analogue of the Poisson integral formula for positive harmonic functions on a ball. In general, the measure η need not be unique, but this is not an issue for our purposes.

Now we can establish the following theorem.

Theorem 2.2. *Let Ω be a quadrature domain for subharmonic functions with respect to μ. Then every positive harmonic function is integrable on Ω.*

Proof. The function u defined by (2) vanishes on Ω^c, whence it follows easily that $u = G_\Omega\mu - G_\Omega(\lambda|_\Omega)$ in Ω. Since Ω is a quadrature domain for subharmonic functions, $u \geq 0$ in Ω, so

$$\int_\Omega G_\Omega(x,y)d\lambda(x) \leq \int G_\Omega(x,y)d\mu(x) \quad (y \in \Omega)$$

and hence

$$\int_\Omega M(x,y)d\lambda(x) \leq \int M(x,y)d\mu(x) \quad (y \in \Omega\backslash\{x_0\}).$$

This inequality remains valid for $y \in \partial^M\Omega$, by Fatou's lemma, continuity of $M(x,\cdot)$ at $\partial^M\Omega$ and the fact that μ has compact support in Ω. If h is a positive harmonic function in Ω, then $h(x) = \int M(x,y)d\eta(y)$ for some measure η on $\partial^M\Omega$, and so

$$\begin{aligned}\int_\Omega h(x)d\lambda(x) &= \int\int_\Omega M(x,y)d\lambda(x)d\eta(y)\\ &\leq \int\int M(x,y)d\mu(x)d\eta(y) = \int h d\mu.\end{aligned}$$

Since μ has compact support in Ω, this last integral is obviously finite, and the theorem is proved. □

In [5] we extended this result to the wider class of quadrature domains for harmonic functions, as we will now explain. The two-dimensional case had earlier been verified using complex variable techniques by Gustafsson, Sakai and Shapiro [9].

Theorem 2.3. *Let Ω be a quadrature domain for harmonic functions with respect to μ. Then every positive harmonic function is integrable over Ω.*

Sketch proof. As in the previous proof, if $y \in \Omega\backslash\{x_0\}$ is such that $u(y) \geq 0$, then

$$\int_\Omega M(x,y)d\lambda(x) \leq \int M(x,y)d\mu(x). \tag{3}$$

However, in this case the set $\{u < 0\}$ may be non-empty, so we need information about its structure. Let $B_r(y)$ denote the open ball of centre y and radius r. Using a blow-up argument, and deep uniform $C^{1,1}$ bounds from Caffarelli, Karp and Shahgholian [3] that apply to functions like u, we were able to prove that there is a constant $C < 1$ such that for each $y \in \{u < 0\}$ the set $B_{Cr}(y) \cap \{u \geq 0\}$ is non-empty, where r is the distance from y to $\partial\Omega$. It follows from the Harnack inequality that, if we choose a point $\tilde{y} \in B_{Cr}(y) \cap \{u \geq 0\}$, then

$$M(x, y) \leq C_1 M(x, \tilde{y}) \quad (x \in \Omega \backslash B_r(y)),$$

where C_1 only depends on N and C. Also,

$$C_2 := \sup\{M(x, y) : x \in \operatorname{supp}(\mu), y \in \partial^M \Omega\} < \infty.$$

For a given point $z \in \partial^M \Omega$ we can choose a sequence (z_n) in Ω which converges to z, and then choose $(\tilde{z}_n)$ as above. By Fatou's lemma and (3)

$$\begin{aligned}\int_\Omega M(x,z)d\lambda(x) &= \int_\Omega \lim_{n\to\infty} M(x, z_n)d\lambda(x)\\ &\leq C_1 \int_\Omega \liminf_{n\to\infty} M(x, \tilde{z}_n)d\lambda(x)\\ &\leq C_1 \liminf_{n\to\infty} \int_\Omega M(x, \tilde{z}_n)d\lambda(x)\\ &\leq C_1 \liminf_{n\to\infty} \int M(x, \tilde{z}_n)d\mu(x) \leq C_1 C_2 \mu(\Omega).\end{aligned}$$

If $h = \int M(\cdot, y)d\eta(y)$ is an arbitrary positive harmonic function on Ω, then

$$\begin{aligned}\int_\Omega h d\lambda &= \int \int_\Omega M(x, y)d\lambda(x)d\eta(y)\\ &\leq C_1 C_2 \mu(\Omega)\eta(\partial^M \Omega) = C_1 C_2 \mu(\Omega) h(x_0).\end{aligned}$$

Hence every positive harmonic function is integrable over Ω. □

We can now give a positive answer to the measure substitution question for quadrature domains for harmonic functions, as posed in [9].

Corollary 2.4. *Let Ω be a quadrature domain for harmonic functions with respect to μ. Then there is a positive measure ν such that Ω is a quadrature domain for harmonic functions also with respect to ν.*

Proof. To prove the existence of ν as stated, let ω_n be an exhaustion of Ω as before. We can arrange that $\operatorname{supp}\mu \subset \omega_1$. Now suppose that, for each n, there is a positive harmonic function h_n on ω_n such that $\int h_n d\mu < 0$. We may, by normalizing, assume that $h_n(x_0) = 1$. By Harnack's convergence theorem and a diagonal sequence argument we deduce that a subsequence of (h_n) converges locally uniformly to some positive harmonic function h_0 on Ω with $h_0(x_0) = 1$. Using Theorem 2.3 we then arrive at the contradictiory conclusion that

$$\int_\Omega h_0 d\lambda = \int h_0 d\mu = \lim_{n\to\infty} \int h_n d\mu \leq 0.$$

Thus there must be some n such that $\int h d\mu \geq 0$ for every positive harmonic function h on ω_n. We recall that the balayage of μ onto $\partial\omega_n$ is the unique measure $\mu^{\omega_n^c}$ such that, for every continuous function f on $\partial\omega_n$, we have

$$\int H_f^{\omega_n} d\mu = \int f d\mu^{\omega_n^c},$$

where $H_f^{\omega_n}$ denotes the solution to Dirichlet's problem in ω_n with boundary values f. Since $f \geq 0$ implies that $H_f^{\omega_n} \geq 0$ by the maximum principle, we see that the measure $\nu = \mu^{\omega_n^c}$ is positive. Finally,

$$\int h d\nu = \int H_h^{\omega_n} d\mu = \int h d\mu = \int_\Omega h d\lambda$$

for every integrable harmonic function h on Ω, so Ω is a quadrature domain for harmonic functions with respect to ν. □

Remark 2.5. It is natural to ask if an analogue of Corollary 2.4 holds also for the subharmonic case. In the case where $u > 0$ on Ω this can be answered affirmatively, as follows. Suppose that

$$\Omega = \{x \in \mathbb{R}^N : u(x) = (U\mu - U(\lambda|_\Omega))(x) > 0\}$$

where suppμ is a compact subset of Ω. We can then choose $\varepsilon > 0$ such that suppμ is contained in

$$\omega = \{x \in \Omega : u(x) > \varepsilon\}.$$

If we define

$$u_\varepsilon(x) = \begin{cases} \varepsilon & (x \in \omega) \\ u(x) & (x \in \Omega \setminus \omega) \end{cases},$$

then it is easy to see that the measure

$$\nu = (-\Delta u_\varepsilon)|_{\partial\omega} + \lambda|_\omega,$$

which is positive and compactly supported in Ω, satisfies

$$u_\varepsilon(x) = U\nu - U(\lambda|_\Omega).$$

Hence Ω is a quadrature domain for subharmonic functions with respect to ν.

2.4 The Exterior Inverse Problem of Potential Theory

A bounded domain Ω in $\mathbb{R}^N$ is called *solid* if $\overline{\Omega}^c$ is connected and $\Omega = \overline{\Omega}^\circ$. The *exterior inverse problem of potential theory*, which dates back to work of Novikov in the 1930s, is as follows:

If Ω_1 and Ω_2 are solid domains in $\mathbb{R}^N$ such that $\lambda|_{\Omega_1}$ and $\lambda|_{\Omega_2}$ produce the same potential in the complement of $\Omega_1 \cup \Omega_2$, must Ω_1 and Ω_2 coincide?

Novikov himself proved that the answer is yes if both domains are assumed to be convex or, more generally, starshaped with respect to a common point. Although it is now suspected that the answer to the general question may be negative, it has long been conjectured that convexity of *one* of the domains should be enough for a positive answer. That this is, in fact, the case was recently proved in [6]. The proof, which relies on partial balayage and the "moving plane" method, will be outlined in this section. We begin by formulating the main result.

Theorem 2.6. *If Ω_1 and Ω_2 are solid domains in $\mathbb{R}^N$, where Ω_2 is convex, and if*

$$U(\lambda|_{\Omega_1}) = U(\lambda|_{\Omega_2}) \ \text{in} \ (\Omega_1 \cup \Omega_2)^c, \tag{4}$$

then $\Omega_1 = \Omega_2$.

In order to describe the moving plane method, we need some notation. Points in $\mathbb{R}^N$ will be denoted by $(x', x_N) \in \mathbb{R}^{N-1} \times \mathbb{R}$, and we will write

$$W_+ = \{(x', x_N) : x_N > 0\},\ W_- = \{(x', x_N) : x_N < 0\},\ H = \{(x', x_N) : x_N = 0\}.$$

Lemma 2.7. *Let μ be a measure with compact support contained in $W_- \cup H$ and let $A = \{x' : (x', 0) \in \Omega(\mu) \cap H\}$. Then there is a continuous function $g : A \to (0, \infty)$, continuously vanishing on ∂A, such that*

$$\Omega(\mu) \cap W_+ = \{(x', x_N) : x' \in A \text{ and } 0 < x_N < g(x')\}.$$

Proofs of this lemma may be found in several papers, for instance [7].

We now proceed to the proof of Theorem 2.6. Let Ω_1 and Ω_2 be solid domains such that (4) holds, where Ω_2 is convex, and let

$$\eta = -\Delta \min\{U(\lambda|_{\Omega_1}), U(\lambda|_{\Omega_2})\}$$

and $\Omega_0 = \Omega(\eta)$. The first step is to prove a straightforward lemma about Ω_0.

Lemma 2.8. *With the above notation, $B\eta = \lambda|_{\Omega_0}$ and $\Omega_1 \cap \Omega_2 \subset \Omega_0$.*

Proof. By standard Sobolev space theory the set of points

$$\{x : U(\lambda|_{\Omega_1})(x) = U(\lambda|_{\Omega_2})(x)\} \cap [(\Omega_1 \backslash \Omega_2) \cup (\Omega_2 \backslash \Omega_1)]$$

has Lebesgue measure zero. Also, since

$$U\eta = U(\lambda|_{\Omega_1 \cap \Omega_2}) + \min\left\{U(\lambda|_{\Omega_1 \backslash \Omega_2}), U(\lambda|_{\Omega_2 \backslash \Omega_1})\right\},$$

we see that $\eta \geq \lambda|_{\Omega_1 \cap \Omega_2}$. Hence η is of the form $\lambda|_D + \eta_1$, where D is an open set containing $\Omega_1 \cap \Omega_2$, and η_1 is positive and singular with respect to Lebesgue measure. Since $B\eta = \lambda|_{\Omega_0} + \eta|_{\Omega_0^c}$ we see that $D \subset \Omega_0$. It follows that $\eta|_{\Omega_0^c} = \eta_1|_{\Omega_0^c} = 0$, since $B\eta \leq \lambda$. □

It is easy to see that $\lambda(\Omega_1) = \lambda(\Omega_2) = \lambda(\Omega_0)$ and, by construction, $U(\lambda|_{\Omega_0}) \leq U(\lambda|_{\Omega_i})$ $(i = 1, 2)$. Further, $\partial\Omega_i \subset \overline{\Omega_0}$ $(i = 1, 2)$, because if there were an open ball B in $\mathbb{R}^N \backslash \overline{\Omega_0}$ with centre in $\partial\Omega_i$, then we would arrive at the impossible situation that $U(\lambda|_{\Omega_i}) - U(\lambda|_{\Omega_0})$ is non-negative and superharmonic on B and vanishes on $B \backslash \Omega_i$ but not on all of B.

Let $\mu = \lambda|_{\Omega_2} + \lambda|_E$, where $E = \Omega_2 \backslash \Omega_0$. Also, let $D = \Omega_0 \cup E = \Omega_0 \cup \Omega_2$. The next step in the proof is to show that $B\mu = \lambda|_D$ and hence $D \subset \Omega(\mu)$. From this we will be able to obtain a contradiction if $\Omega_1 \neq \Omega_2$ by an application of the moving plane method. What follows is a sketch of the remaining proof; full details may be found in [6].

Let

$$A_1 = \Omega_1 \backslash \overline{\Omega_0} = \Omega_1 \backslash \overline{D} \quad \text{and} \quad A_2 = \overline{\Omega_0 \cup \Omega_1 \cup \Omega_2}^{\,c} = \overline{D \cup \Omega_1}^{\,c}.$$

On A_2 we have, by assumption, $U\eta = U(\lambda|_{\Omega_0}) = U(\lambda|_{\Omega_1}) = U(\lambda|_{\Omega_2})$, and so $U\mu = U(\eta + \lambda|_E) = U(\lambda|_D)$. However, on A_1 it is also clear that $U\eta = U(\lambda|_{\Omega_2})$, since $-\Delta U(\lambda|_{\Omega_1}) = \lambda$ and $-\Delta U(\lambda|_{\Omega_0}) = 0$ there. Thus $U\mu = U(\lambda|_D)$ there also. Hence $U\mu = U(\lambda|_D)$ on $(\partial\Omega_1 \cup \overline{D})^c$. It follows by continuity that $U\mu = U(\lambda|_D)$ on $(\overline{D}^{\circ})^c$, and we also have $U\mu \geq U(\eta + \lambda|_E) \geq U(\lambda|_{\Omega_0} + \lambda|_E) = U(\lambda|_D)$ everywhere. Hence $\Omega(\mu) \subset \overline{D}^{\circ}$. But $\lambda(\Omega(\mu)) = \lambda(D)$, so $D \subset \Omega(\mu)$, and these sets differ by at most a Lebesgue null set. Thus $B\mu = \lambda|_D$. Since $\lambda(\partial\Omega_2) = 0$ and the set $\Omega(\mu) \backslash (E \cup \partial\Omega_2)$ is open, the latter is a subset of Ω_0, by construction. It differs from $D \backslash E = \Omega_0$ by at most a Lebesgue null set. Therefore $D \subset \Omega(\mu) \subset D \cup \partial\Omega_2$.

By Lemma 2.7 and the convexity of Ω_2 it is not possible for $\Omega(\mu)$ to have any holes outside the set $\overline{\Omega_2}$. Hence $\Omega_1 \subset \Omega(\mu)$, since $\partial\Omega_1 \subset \overline{\Omega_0}$, as we noted earlier, and so $\Omega_1 \subset \Omega_0$. Using the fact that Ω_1 and Ω_0 have equal Lebesgue measure and are both saturated with respect to Lebesgue measure, it follows that they are identical.

However, this implies that $U(\lambda|_{\Omega_1}) \leq U(\lambda|_{\Omega_2})$ everywhere. Since $U(\lambda|_{\Omega_2}) - U(\lambda|_{\Omega_1})$ is superharmonic on $(\overline{\Omega}_1)^c$, which is connected, and attains the value 0 it follows that $U(\lambda|_{\Omega_1}) = U(\lambda|_{\Omega_2})$ on this set. Thus $\Omega_2 \subset \Omega_1$, and since these sets are both solid and have equal Lebesgue measure they are identical.

2.5 *Harmonic Balls*

The article [17] introduced the concept of harmonic balls and investigated some of their properties, but left open many questions. Of particular interest is the question of uniqueness, which has only been settled for special types of domains R. Here we will merely give the definition and the most basic results.

Let $R \subset \mathbb{R}^N$ be a Greenian domain. Given a subdomain $D \subset R$ we define

$$\tilde{H}_R(D) = \{G_R\mu : \mu \text{ is a signed Radon measure, supp}\mu \subset R\backslash D\},$$

$$\tilde{S}_R(D) = \{G_R\mu : \mu \text{ is a signed Radon measure, supp}\mu \subset R,\ \mu|_D \leq 0\}.$$

Definition 2.9 (Harmonic and Subharmonic balls). Let $x^0 \in R$ and $\alpha > 0$. A domain $D(x^0, \alpha) \subset R$ is called a harmonic ball relative to R if

$$\int_{D(x^0,\alpha)} h(x)\, d\lambda = \alpha h(x^0), \qquad (h \in \tilde{H}_R(D(x^0,\alpha))).$$

A domain $D(x^0, \alpha) \subset R$ is called a subharmonic ball relative to R if

$$\int_{D(x^0,\alpha)} s(x)\, d\lambda \geq \alpha s(x^0), \qquad (s \in \tilde{S}_R(D(x^0,\alpha))).$$

We will call x^0 the centre of the ball, and α its size.

If $\overline{D(x^0,\alpha)} \subset R$ is a harmonic ball, then it coincides with the standard ball of centre x^0 and Lebesgue measure α. This follows easily from the standard mean value property for harmonic functions over balls, and the fact that balls are the only domains with this property (see [12]). We note that if R were not Greenian and we were to use logarithmic potentials in place of Green potentials in our definition, then the only domains reasonably corresponding to the notion of a harmonic ball would be of the form $B_r(x) \cap R$. Thus there is no significant loss of generality in our assumption that R is Greenian.

Subharmonic balls were also studied by Sakai [16], who called them restricted quadrature domains.

Theorem 2.10. *Let $x^0 \in R$ and $\alpha > 0$, let $D \subset R$ be open, and define*

$$u_R = \alpha G_R(x^0, \cdot) - G_R(\lambda|_D).$$

Then

(a) *D is a harmonic ball of centre x^0 and size α if and only if $u_R = 0$ in $R \backslash D$;*
(b) *D is a subharmonic ball of centre x^0 and size α if and only if $u_R \geq 0$ in R and $u_R = 0$ in $R \backslash D$.*

Proof. The "only if" statements are clear since $G_R(\cdot, x) \in \tilde{H}_R(D)$ if $x \in R \backslash D$ and $-G_R(\cdot, x) \in \tilde{S}_R(D)$ for all $x \in R$.

Conversely, we note that, if μ has compact support in $R \backslash D$ and $u_R = 0$ in $R \backslash D$ then, by Fubini's theorem,

$$\alpha G_R \mu(x^0) - \int_D G_R \mu \, d\lambda = \int (\alpha G_R \delta_{x^0} - G_R(\lambda|_D)) d\mu = 0,$$

because the integrand on the right-hand side is identically zero on the support of μ. (Here δ_{x^0} denotes the unit measure concentrated at x^0.)

If, instead, we assume that $\mu \leq 0$ in D, and that $u_R \geq 0$ with equality in $R \backslash D$, then a similar argument yields

$$\alpha G_R \mu(x^0) - \int_D G_R \mu \, d\lambda = \int (\alpha G_R \delta_{x^0} - G_R(\lambda|_D)) d\mu \leq 0.$$

□

In the case where $R = D$, the class $\tilde{H}_R(D)$ only contains the zero function, so R itself will always be a harmonic ball according to our definition. We call this the trivial harmonic ball. If R is small compared to α, then R will be the only candidate for a harmonic ball, so we do not wish to exclude this trivial case. However, this convention requires us to take care when formulating uniqueness results. It is unclear whether, even if we overlook this trivial case, the definition is enough to guarantee some sort of uniqueness in general. It does so in many cases, for instance if R is a half-space.

Regarding existence and uniqueness in the subharmonic case we have the following result, which follows almost immediately from the facts about partial balayage recorded earlier.

Theorem 2.11. *For every $x^0 \in R$ and every $\alpha > 0$ there is a unique, up to a Lebesgue null set, subharmonic ball $D(x^0, \alpha)$.*

Example 2.12. If we let R be the upper halfplane and look at the (unique) subharmonic ball $D(x^0, \alpha)$ with $x^0 = (0, 1)$ and $\alpha = 4\pi$, say, then the ball $B_2(x^0)$, which has area 4π, is not contained in R. In this case we have the strict inclusion $D(x^0, \alpha) \subset B_2(x^0)$, and $\partial D(x^0, \alpha) \cap \partial R \neq \emptyset$. It is not hard to see that this intersection is an interval, and at the two endpoints it is further known that $\partial D(x^0, \alpha) \setminus \partial R$ hits ∂R tangentially in a C^1-way (see [18]).

The most important question about harmonic balls concerns uniqueness. Are there domains R for which there exist two nontrivial harmonic balls with the same centre and size? As mentioned above we know that if R is a half-space, then there is only one non-trivial harmonic ball.

A harmonic ball $D(x^0, \alpha)$ is said to be positive if the balayage of the signed measure $\alpha\delta_{x^0} - \lambda|_{D(x^0,\alpha)}$ onto ∂R is positive. It is shown in [17] that, if R is starshaped with respect to x^0, then there is at most one non-trivial positive harmonic ball $D(x^0, \alpha)$ and this ball is also starshaped with respect to x^0.

3 Two-Phase Quadrature Domains

3.1 Basic Definitions and Integral Representations

Recent work [19, 20] on two-phase free boundary problems has led Emamizadeh, Prajapat and Shahgholian [4] to propose the study of two-phase quadrature domains, which we now will define. Let Ω^+, Ω^- be disjoint bounded open sets in $\mathbb{R}^N$, and μ^+, μ^- be measures with compact supports in Ω^+, Ω^-, respectively. We noted earlier that a (one-phase) quadrature domain Ω for harmonic functions with respect to a measure μ is characterized by the property that the function $U\mu - U(\lambda|_\Omega)$ vanishes, along with its gradient, in Ω^c. We now define the function

$$u = U(\mu^+ - \mu^-) - U(\lambda|_{\Omega^+} - \lambda|_{\Omega^-}), \tag{5}$$

and say that the pair (Ω^+, Ω^-) is a two-phase quadrature domain for harmonic functions with respect to the pair (μ^+, μ^-) if $u = 0$ in $(\Omega^+ \cup \Omega^-)^c$. An important point to note here is that we do not require ∇u to vanish outside $\Omega^+ \cup \Omega^-$. If we had made this additional assumption, then it would have been easy to see that Ω^+ is a one-phase quadrature domain with respect to μ^+, and Ω^- is a one-phase quadrature domain with respect to μ^-. Such pairs will, of course, always give us two-phase quadrature domains provided Ω^+ and Ω^- are disjoint, but the above definition allows for non-trivial additional cases. Here are a couple of examples in the plane where the gradient of u does not vanish on $\partial\Omega^+ \cap \partial\Omega^-$.

Example 3.1. Let $a \geq 0$, let $\Omega^+ = \{|x| < 1\}$ and $\Omega^- = \{1 < |x| < 2\})$. Then (Ω^+, Ω^-) is a two-phase quadrature domain with respect to $(\pi a\delta_0, \mu_a)$, where μ_a has total mass $(2 + a)\pi$ uniformly distributed on the circle $\{\log|x| = (8\log 2 - 3)/(4 + 2a)\}$. This follows readily from the fact that the mean value of $\log|y - \cdot|$ over $\{|x| = r\}$ is given by $\max\{\log|y|, \log r\}$.

Example 3.2. Let $p = (0, 1) \in \mathbb{R}^2$, and $\mu^+ = 4\delta_p, \mu^- = 4\delta_{-p}$. Partial balayage of μ^+ onto λ in $\{y > 0\}$ yields a measure ν on the upper halfplane $\{y \geq 0\}$ such that $\nu|_{\{y>0\}} = \lambda|_{\Omega^+}$ for some bounded domain $\Omega^+ \subset \{y > 0\}$, $\nu|_{\{y=0\}} \neq 0$, and the function $v = U\mu^+ - U\nu$ vanishes outside Ω^+. We now define the reflected domain $\Omega^- = \{(x, y) : (x, -y) \in \Omega^+\}$ and the function

$$u(x,y) = \begin{cases} v(x,y) & \text{if } y \geq 0 \\ -v(x,-y) & \text{if } y < 0 \end{cases}.$$

Then

$$\begin{aligned} u &= U\mu^+ - U\left(\lambda|_{\Omega^+}\right) - U\left(\nu|_{\{y=0\}}\right) \\ &\quad - \left\{U\mu^- - U\left(\lambda|_{\Omega^-}\right) - U\left(\nu|_{\{y=0\}}\right)\right\} \\ &= U(\mu^+ - \mu^-) - U(\lambda|_{\Omega^+} - \lambda|_{\Omega^-}) \end{aligned}$$

and $u = 0$ on $(\Omega^+ \cup \Omega^-)^c$, so (Ω^+, Ω^-) is a two-phase quadrature domain with respect to (μ^+, μ^-).

A pair (Ω^+, Ω^-) will be called a two-phase quadrature domain for subharmonic functions if it is a two-phase quadrature domain for harmonic functions and, in addition,

$$u \geq 0 \text{ in } \Omega^+ \text{ and } u \leq 0 \text{ in } \Omega^-. \tag{6}$$

Later we will also discuss two-phase quadrature domains for analytic functions in the plane.

Below we will use the fact that, if ν is a finite positive measure on a bounded open set Ω, and we define the "Dirichlet modification" of $U\nu$ with respect to Ω by

$$v = \begin{cases} H^{\Omega}_{U\nu} & \text{in } \Omega \\ U\nu & \text{in } \Omega^c \end{cases},$$

then $v = U\nu^{\Omega^c}$ outside a polar set, where ν^{Ω^c} is, as before, the balayage of ν onto $\partial\Omega$.

When it comes to discussing quadrature identities associated with two-phase quadrature domains, we have to identify a suitable test class of functions. One possibility is to introduce a suitable potential space (analogous to the spaces $\tilde{H}_R(D)$ and $\tilde{S}_R(D)$ in Sect. 2.5), as is done in [17], which will give complete equivalence between integral representations and the definition based on potentials. However, the quadrature identity then becomes almost trivial (unlike the one-phase case). Below we will see that it is essentially enough to work with harmonic functions on $\Omega^+ \cup \Omega^-$ which are continuous up to the boundary.

Theorem 3.3. *Let Ω^+, Ω^- be disjoint bounded open sets and μ^+, μ^- be measures with compact supports in Ω^+, Ω^-, respectively.*

(a) If (Ω^+, Ω^-) is a two-phase quadrature domain for harmonic functions with respect to (μ^+, μ^-), then

$$\int h d(\mu^+ - \mu^-) = \int_{\Omega^+} h d\lambda - \int_{\Omega^-} h d\lambda \tag{7}$$

for every $h \in C(\overline{\Omega^+ \cup \Omega^-})$ that is harmonic on $\Omega^+ \cup \Omega^-$.

(b) If (7) holds, then there are polar sets Z_1, Z_2 such that $(\Omega^+ \cup Z_1, \Omega^- \cup Z_2)$ is a two-phase quadrature domain for harmonic functions with respect to (μ^+, μ^-).

Proof. (a) Let $h \in C(\overline{\Omega^+ \cup \Omega^-})$, where h is harmonic on $\Omega^+ \cup \Omega^-$. By hypothesis

$$U\left(\lambda|_{\Omega^+} - \lambda|_{\Omega^-}\right) = U(\mu^+ - \mu^-) \quad \text{on } (\Omega^+ \cup \Omega^-)^c. \tag{8}$$

We can form the Dirichlet modification of each side with respect to $\Omega = \Omega^+ \cup \Omega^-$, and then take distributional Laplacians to see that

$$(\lambda|_{\Omega^+})^{(\Omega^+)^c} - (\lambda|_{\Omega^-})^{(\Omega^-)^c} = (\mu^+)^{(\Omega^+)^c} - (\mu^-)^{(\Omega^-)^c}. \tag{9}$$

We now integrate $h|_{\partial\Omega}$ against these measures. Since $H_h^{\Omega} = h$, we deduce (7).

(b) [Sketch] Suppose that (7) holds, and let Ω_0 be a Greenian domain containing $\overline{\Omega^+ \cup \Omega^-}$. It follows easily from (7), applied to the functions

$$h_y = U\delta_y - G_{\Omega_0}(y, \cdot) \quad (y \in \Omega_0)$$

(suitably defined at y), that the function u defined by (5) can be expressed as

$$u = G_{\Omega_0}(\mu^+ - \lambda|_{\Omega^+}) - G_{\Omega_0}(\mu^- - \lambda|_{\Omega^-}) \quad \text{in } \Omega_0. \tag{10}$$

Let

$$E = \left\{x : (\Omega^+ \cup \Omega^-)^c \text{ is non-thin at } x\right\}$$

and $y \in E \cap \Omega_0$. It can be shown that $G_{\Omega_0}(y, \cdot)$ may be approximated from below by potentials v_n which are continuous on Ω_0 and harmonic on $\Omega^+ \cup \Omega^-$, whence $u(y) = 0$ by (7) and (10). From this it follows by continuity that $u = 0$ on $\overline{E}$. Since $\overline{E}^c$ is open, contains $\Omega^+ \cup \Omega^-$, and differs from it by at most a polar set, we thus have a two-phase quadrature domain of the stated form.

□

It is not difficult to see why there must be an exceptional set in part (b) of the above result. For, if (7) holds and we remove a point y from the open set Ω^+, then (7) remains true for the modified open set by a standard removable singularity theorem for harmonic functions. However, the equality in (8) would break down at y in general.

The corresponding result for two-phase quadrature domains for subharmonic functions is given below.

Theorem 3.4. *Let Ω^+, Ω^- be disjoint bounded open sets and μ^+, μ^- be measures with compact supports in Ω^+, Ω^- respectively.*

(a) If (Ω^+, Ω^-) is a two-phase quadrature domain for subharmonic functions with respect to (μ^+, μ^-), then

$$\int s d(\mu^+ - \mu^-) \le \int_{\Omega^+} s d\lambda - \int_{\Omega^-} s d\lambda \text{ for every } s \in C(\overline{\Omega^+ \cup \Omega^-}) \quad (11)$$

that is subharmonic on Ω^+ and superharmonic on Ω^-.

(b) If (11) holds, then there are polar sets Z_1, Z_2 such that $(\Omega^+ \cup Z_1, \Omega^- \cup Z_2)$ is a two-phase quadrature domain for subharmonic functions with respect to (μ^+, μ^-).

Proof. (a) Let $s \in C(\overline{\Omega^+ \cup \Omega^-})$, where s is subharmonic on Ω^+ and superharmonic on Ω^-, and let $\nu = \Delta s$ on $\Omega^+ \cup \Omega^-$. Further, let h_+, h_- be the solutions to the Dirichlet problem on Ω^+, Ω^-, respectively, with boundary data s. Using (9) and the fact that (8), (6) imply that

$$G_{\Omega^+}(\mu^+ - \lambda|_{\Omega^+}) \ge 0 \text{ on } \Omega^+, \text{ and } G_{\Omega^-}(\mu^- - \lambda|_{\Omega^-}) \ge 0 \text{ on } \Omega^-,$$

we deduce that

$$\begin{aligned}
\int s d(\mu^+ - \mu^-) &= \int (h_+ - G_{\Omega^+}(\nu|_{\Omega^+}))\, d\mu^+ - \int (h_- - G_{\Omega^-}(\nu|_{\Omega^-}))\, d\mu^- \\
&= \int s d(\mu^+)^{(\Omega^+)^c} - \int s d(\mu^-)^{(\Omega^-)^c} \\
&\quad - \int_{\Omega^+} G_{\Omega^+}\mu^+ d\nu + \int_{\Omega^-} G_{\Omega^-}\mu^- d\nu \\
&\le \int s d(\lambda|_{\Omega^+})^{(\Omega^+)^c} - \int s d(\lambda|_{\Omega^-})^{(\Omega^-)^c} \\
&\quad - \int_{\Omega^+} G_{\Omega^+}(\lambda|_{\Omega^+}) d\nu + \int_{\Omega^-} G_{\Omega^-}(\lambda|_{\Omega^-}) d\nu \\
&= \int_{\Omega^+} (h_+ - G_{\Omega^+}(\nu|_{\Omega^+}))\, d\lambda - \int_{\Omega^-} (h_- - G_{\Omega^-}(\nu|_{\Omega^-}))\, d\lambda \\
&= \int_{\Omega^+} s d\lambda - \int_{\Omega^-} s d\lambda.
\end{aligned}$$

(b) We know from the corresponding case of Theorem 3.3 that there are disjoint open sets D^+, D^- containing Ω^+, Ω^-, respectively, such that $D^+ \backslash \Omega^+, D^- \backslash \Omega^-$ are polar and the function u vanishes on $(D^+ \cup D^-)^c$. Now let $x \in \Omega^+$ and choose $n_0 \in \mathbb{N}$ such that $U\delta_x \leq n_0$ outside Ω^+. Then the function $s = -\min\{U\delta_x, n\}$ is subharmonic on Ω^+ and harmonic on Ω^- whenever $n \geq n_0$. We can thus apply (11) and let $n \to \infty$ to see that $u \geq 0$ on Ω^+, and hence on D^+. Similarly, $u \leq 0$ on D^-, so the result follows □

3.2 *Integrability of Positive Harmonic Functions on Two-Phase Quadrature Domains and the Measure Substitution Problem*

Two natural questions to ask about two-phase quadrature domains concern the integrability of positive harmonic functions over such domains, and the possibility of substituting positive measures for signed measures as we have verified in the one-phase case.

The situation regarding two-phase quadrature domains for subharmonic functions is straightforward. Let (Ω^+, Ω^-) be a two-phase quadrature domain for subharmonic functions with respect to signed measures (μ^+, μ^-), and let u be as in (5). Then we know that in Ω^+ we have $u = G_{\Omega^+}\mu^+ - G_{\Omega^+}(\lambda|_{\Omega^+}) \geq 0$, and by the same argument as in the one-phase situation it can be proved that any positive harmonic function is integrable over Ω^+. We can deal with Ω^- similarly, and hence obtain the integrability of positive harmonic functions over such domains.

When it comes to the measure substitution problem, if we assume that $\Omega^+ = \{x \in \mathbb{R}^n : u(x) > 0\}$ and $\Omega^- = \{x \in \mathbb{R}^N : u(x) < 0\}$, then we can find positive measures (η^+, η^-) such that (Ω^+, Ω^-) is a two-phase quadrature domain for subharmonic functions with respect to (η^+, η^-) by defining

$$u_\varepsilon(x) = \begin{cases} u(x) & (|u(x)| \leq \varepsilon) \\ \varepsilon \operatorname{sign} u(x) & (|u(x)| > \varepsilon) \end{cases},$$

where ε is chosen small enough, and then proceeding as in the one-phase situation.

The case of two-phase quadrature domains for harmonic functions is more problematic. The first difficulty we encounter is the lack of known regularity results for the boundaries of such domains, so the one-phase approach is not available to prove that every positive harmonic function is integrable over such a domain. Secondly, it is unclear if the sweeping of, say, μ^+ onto $\partial\Omega^+$ is a positive measure, which is necessary for the existence of a new measure η^+ with compact support in Ω^+ that gives the same integral when applied to positive harmonic functions continuous up to the boundary of Ω^+. It might perhaps still be possible to replace the pair (μ^+, μ^-) with a pair of positive measures (η^+, η^-) since all relevant identities involve both measures μ^+, μ^-, but the problem then becomes substantially different from the one-phase case.

What can be said is that, if the positive harmonic functions on Ω^+ that belong to $C(\overline{\Omega^+})$ are dense in the space of all positive harmonic functions on Ω^+ (which, for instance, is the case if Ω^+ is a Lipschitz domain), and if the sweeping $(\mu^+)^{(\Omega^+)^c}$ is positive, then there is a positive measure η^+ such that

$$\int_\Omega h d\mu^+ = \int_\Omega h d\eta^+$$

for all positive harmonic functions h in Ω^+ (see Theorem 5.1 in [21]). Hence, in particular, if both the boundaries of Ω^+ and Ω^- are Lipschitz and both sweepings $(\mu^+)^{(\Omega^+)^c}$ and $(\mu^-)^{(\Omega^-)^c}$ are positive, then there is a pair of positive measures (η^+, η^-) such that (Ω^+, Ω^-) is a two-phase quadrature domain for harmonic functions with respect to this pair.

Using the methods from [5] one could relax this to the assumption that the boundaries are regular enough locally around any point of $\partial\Omega^+ \cap \partial\Omega^-$ (which is known to be true [19, 20] in the special case when $\Omega^+ = \{x \in \mathbb{R}^n : u(x) > 0\}$ and $\Omega^- = \{x \in \mathbb{R}^N : u(x) < 0\}$). However, it is unclear if the boundaries are sufficiently regular in the more general harmonic case.

3.3 Construction

In [8] a "two-phase partial balayage" technique was introduced to construct two-phase quadrature domains from suitable pairs of positive measures (μ^+, μ^-) with compact support in $\mathbb{R}^N$. We will briefly indicate this construction below, and refer to [8] for further details. We need some terminology and conventions.

The fine topology on $\mathbb{R}^N$ is the coarsest topology for which all subharmonic functions are continuous. (An account of it may be found in Chapter 7 of [1].) It is connected with the notion of thinness by the fact that Ω is a fine neighbourhood of $x \in \Omega$ if and only if the set $\mathbb{R}^N \backslash \Omega$ is thin at x. A function s is called δ-subharmonic (on $\mathbb{R}^N$) if it can be expressed as $s = s_1 - s_2$, where s_1, s_2 are subharmonic functions. If s is δ-subharmonic, then Δs is (locally) a signed measure μ_s (and, of course, $\mu_s \geq 0$ if and only if s is subharmonic). A δ-subharmonic function $s = s_1 - s_2$ will be undefined on the polar set Z where $s_1 = -\infty = s_2$. However, it can be shown that s has a fine limit at $|\mu_s|$-almost every point as well as being finely continuous everywhere outside Z. We assign s this limiting value wherever it exists. With this convention we can reformulate a result of Brezis and Ponce [2] as follows. (A short proof of it may be found in [8].)

Theorem 3.5 (Kato's Inequality). *If s is a δ-subharmonic function, then $\Delta s^+ \geq (\Delta s)|_{\{s \geq 0\}}$.*

Our main focus in this section is the following question: given measures μ^+, μ^- with disjoint compact support, how can we construct a two-phase quadrature domain for (sub)harmonic functions with respect to (μ^+, μ^-)? (We might reasonably hope to be able to do this at least when μ^+, μ^- are sufficiently concentrated.) Let $\mu = \mu^+ - \mu^-$. Motivated by the case of one-phase partial balayage, it is natural to seek a function u such that

$$-\Delta u = \mu^+ - \lambda|_{\Omega^+} - (\mu^- - \lambda|_{\Omega^-}),$$

where

$$\Omega^+ = \{u > 0\} \text{ and } \Omega^- = \{u < 0\};$$

that is, we seek a function u with compact support such that $-\Delta u = \eta(u, \mu)$, where we make the temporary definition

$$\eta(u, \mu) = \mu^+ - \lambda|_{\{u>0\}} - \left(\mu^- - \lambda|_{\{u<0\}}\right).$$

How might we construct such a function? We proceed by considering associated upper and lower classes of functions. In the familiar case of the Dirichlet problem on a (bounded) open set Ω for $f \in C(\partial\Omega)$, the Perron method involves consideration of superharmonic functions v on Ω such that $\liminf_{x\to y} v(x) \geq f(y)$ for all $y \in \partial\Omega$, and subharmonic functions w on Ω such that $\limsup_{x\to y} w(x) \leq f(y)$ for all $y \in \partial\Omega$; and the maximum principle guarantees that any such v, w satisfy $v \geq w$ on Ω. In our context, the role of the maximum principle is played by Kato's inequality, as we will see in the proof of the following key lemma.

Lemma 3.6. *Let v, w be δ-subharmonic functions with compact supports. If*

$$-\Delta v \geq \eta(v, \mu) \text{ and } -\Delta w \leq \eta(w, \mu),$$

then $v \geq w$.

Proof. Let $u = w - v$. Then

$$\begin{aligned}\Delta u &\geq \eta(v, \mu) - \eta(w, \mu)\\ &= \mu^+ - \lambda|_{\{v>0\}} - \left(\mu^- - \lambda|_{\{v<0\}}\right)\\ &\quad - \left\{\mu^+ - \lambda|_{\{w>0\}} - \left(\mu^- - \lambda|_{\{w<0\}}\right)\right\}\\ &= \lambda|_{\{w>0\}} - \lambda|_{\{v>0\}} + \lambda|_{\{v<0\}} - \lambda|_{\{w<0\}},\end{aligned}$$

so $(\Delta u)|_{\{u\geq 0\}} \geq 0$. It follows from Kato's inequality that $\Delta u^+ \geq 0$, whence u^+ is subharmonic. Since u has compact support, we conclude that $u^+ \equiv 0$, and hence $w \leq v$. □

It turns out that, for our proposed construction, we need to work with a slightly more complicated formula for $\eta(u,\mu)$, namely

$$\eta(u,\mu) = (\mu^+ - \lambda)^+ - (\mu^+ - \lambda)^-|_{\{u>0\}} \\ - \left\{(\mu^- - \lambda)^+ - (\mu^- - \lambda)^-|_{\{u<0\}}\right\}.$$

The proof of the above lemma is easily adapted for this definition.

It is also not hard to show that if u is a δ-subharmonic function with compact support and $-\Delta u \geq \eta(u,\mu)$, then $u \geq -W\mu^-$. We now consider the collection

$$\tau_\mu = \{u \text{ is } \delta\text{-subharmonic: } -\Delta u \geq \eta(u,\mu) \text{ and } u \geq -W\mu^-\}.$$

Although it is not immediately obvious, it can be shown that τ_μ has a least element, which we denote by $\overline{W}\mu$. It has compact support. Further, if w is a δ-subharmonic function with compact support and $-\Delta w = \eta(w,\mu)$, then $w = \overline{W}\mu$. Writing

$$\Omega^+ = \{\overline{W}\mu > 0\} \text{ and } \Omega^- = \{\overline{W}\mu < 0\},$$

we now list a sample of the results that can be deduced.

Corollary 3.7. *If*

$$\operatorname{supp}(\mu^+) \subset \Omega^+ \quad \textit{and} \quad \operatorname{supp}(\mu^-) \subset \Omega^-,$$

then (Ω^+, Ω^-) *is a two-phase quadrature domain for subharmonic functions with respect to* (μ^+, μ^-).

Corollary 3.8. *Suppose that* μ^+, μ^- *have disjoint compact supports, and that* $U\mu^+ = \infty$ *on* $\operatorname{supp}(\mu^+)$ *and* $U\mu^- = \infty$ *on* $\operatorname{supp}(\mu^-)$. *Then* (Ω^+, Ω^-) *is a two-phase quadrature domain for subharmonic functions with respect to* (μ^+, μ^-).

Corollary 3.9. *Let* μ^+, μ^- *be positive measures with disjoint compact supports such that*

$$\overline{\omega(\mu^-)} \cap \operatorname{supp}(\mu^+) = \emptyset, \quad \overline{\omega(\mu^+)} \cap \operatorname{supp}(\mu^-) = \emptyset,$$

and

$$\limsup_{r\to 0+} \frac{\mu^+(B_r(x))}{\lambda(B_r(x))} > 2^N \quad (x \in \operatorname{supp}(\mu^+)),$$
$$\limsup_{r\to 0+} \frac{\mu^-(B_r(y))}{\lambda(B_r(y))} > 2^N \quad (y \in \operatorname{supp}(\mu^-)).$$

Then (Ω^+, Ω^-) *is a two-phase quadrature domain for subharmonic functions with respect to* (μ^+, μ^-).

Finally, we have some uniqueness results.

Theorem 3.10. *If* (Ω^+, Ω^-) *is a two-phase quadrature domain for subharmonic functions with respect to* (μ^+, μ^-), *then*

$$\Omega^+ = \{\overline{W}\mu > 0\} \cup \operatorname{supp}(\mu^+) \cup Z_1 \ \textit{ and } \ \Omega^- = \{\overline{W}\mu < 0\} \cup \operatorname{supp}(\mu^-) \cup Z_2,$$

where $\lambda(Z_1) = 0 = \lambda(Z_2)$. *In particular, such two-phase quadrature domains are unique up to* λ*-null sets.*

Proof. Let $U = \{u > 0\}$, where $u = U(\mu^+ - \lambda|_{\Omega^+}) - U(\mu^- - \lambda|_{\Omega^-})$. Clearly $U \subset \Omega^+$. Then

$$(-\Delta u)|_{\Omega^+ \setminus U} \leq (-\Delta \min\{u, 0\})\,|_{\Omega^+} = 0$$

by Kato's inequality, since $u \geq 0$ on Ω^+. Hence $(-\Delta u)|_{\Omega^+ \setminus U} \ll \lambda$, since $(-\Delta u)|_{\Omega^+} = \mu^+ - \lambda|_{\Omega^+}$. Similarly, $(-\Delta u)|_{\Omega^- \setminus V} \ll \lambda$, where $V = \{u < 0\}$. On the other hand, $(-\Delta u)^{\{u=0\}} = 0$, so $(-\Delta u)|_{\tilde{U}^c \cup \tilde{V}^c} \perp \lambda$ since swept measures are known to be singular with respect to λ. (Here $\tilde{U}$ denotes $\{x : U^c \text{ is thin at } x\}$.) Hence $(-\Delta u)|_{\Omega^+ \setminus U} = 0 = (-\Delta u)|_{\Omega^- \setminus V}$. We conclude that

$$\begin{aligned} -\Delta u &= \mu^+ - \lambda|_{\Omega^+} - (\mu^- - \lambda|_{\Omega^-}) \\ &= (\mu^+ - \lambda)^+ - (\mu^+ - \lambda)^-|_{\{u>0\}} - \left((\mu^- - \lambda)^+ - (\mu^- - \lambda)^-|_{\{u<0\}}\right) \\ &= \eta(u, \mu), \end{aligned}$$

so $u = \overline{W}\mu$. Since

$$0 = -\Delta u = -\lambda \ \text{ on the set } \ Z_1 = \Omega^+ \setminus \left(\{\overline{W}\mu > 0\} \cup \operatorname{supp}(\mu^+)\right),$$

we see that $\lambda(Z_1) = 0$, as required. A similar argument applies to Ω^-. □

Theorem 3.11. *If* (Ω^+, Ω^-) *is a two-phase quadrature domain for subharmonic functions with respect to* (μ^+, μ^-) *satisfying the strict inequalities*

$$u > 0 \text{ in } \Omega^+ \text{ and } u < 0 \text{ in } \Omega^-,$$

then it is unique and is given by the above construction; that is,

$$\Omega^+ = \{\overline{W}\mu > 0\}, \quad \Omega^- = \{\overline{W}\mu < 0\}.$$

Proof. In this case we have $U = \Omega^+$, so $\Omega^+ = \{\overline{W}\mu > 0\}$, and similarly $\Omega^- = \{\overline{W}\mu < 0\}$. □

3.4 Two-Phase Quadrature Domains for Analytic Functions and the Two-Phase Schwarz Function

We conclude this article by briefly discussing the notion of two-phase quadrature domains for analytic functions in the complex plane $\mathbb{C}$. As in the cases of harmonic and subharmonic functions treated in the previous section the question arises as to what test class of functions is most appropriate. We saw above for those cases that, provided the boundaries are nice enough, we can use functions continuous up to the boundary. We will adopt a similar approach here.

Definition 3.12. A pair (Ω^+, Ω^-) of plane open sets is said to be a two-phase quadrature domain for analytic functions with respect to a (complex Radon) measure μ if $\mathrm{supp}\mu \subset \Omega$, where $\Omega = \Omega^+ \cup \Omega^-$, and

$$\int_{\Omega^+} f d\lambda - \int_{\Omega^-} f d\lambda = \int f d\mu$$

for every $f \in C(\overline{\Omega})$ that is analytic on Ω.

There are limitations with using functions continuous up to the boundary here. If we had used these as our test class in the definition of a one-phase quadrature domain Ω for analytic functions with respect to some measure μ, then the removal of a rectifiable curve γ with compact support in $\Omega\backslash\mathrm{supp}\mu$ would still leave us with a quadrature domain with respect to μ, because a function which is continuous in $\overline{\Omega\backslash\gamma}$ and analytic in $\Omega\backslash\gamma$ is automatically analytic in Ω. For two-phase quadrature domains the situation is, in a sense, even worse, as we will see below. It remains a challenge to find some more appropriate test class, particularly if we continue to require that two-phase quadrature domains for harmonic functions should also be two-phase quadrature domains for analytic functions.

We now introduce a two-phase analogue of the Schwarz function, which was defined in Sect. 2.1.

Definition 3.13. Let Ω^+, Ω^- be disjoint bounded open sets, let $\Omega = \Omega^+ \cup \Omega^-$ and $\Gamma = \partial\Omega^+ \cap \partial\Omega^-$. If there are compact subsets $C^\pm \subset \Omega^\pm$ and functions $S^\pm \in C(\overline{\Omega^\pm})$, analytic in $\Omega^\pm \setminus C^\pm$, such that

$$S^\pm(z) = \pm\overline{z} \quad (z \in \partial\Omega^\pm \setminus \Gamma),$$
$$S^+(z) - S^-(z) = 2\overline{z} \quad (z \in \Gamma),$$

then we say that (S^+, S^-) is a two-phase Schwarz function for (Ω^+, Ω^-).

All information of interest concerning the two-phase Schwarz function is encapsulated in the function

$$S(z) = \begin{cases} S^+(z) \ (z \in \overline{\Omega^+}\backslash\Gamma) \\ S^-(z) \ (z \in \overline{\Omega^-}\backslash\Gamma), \end{cases}$$

and we note that

$$\lim_{w\to z, w\in\Omega^+} S(w) - \lim_{w\to z, w\in\Omega^-} S(w) = 2\bar{z} \qquad (z \in \partial\Omega^+ \cap \partial\Omega^-).$$

We will also refer to this function S as the two-phase Schwarz function. (The only difference here is that we do not have any values on Γ, where the function S typically would have discontinuities).

We also note that, if there are functions $S^\pm \in C(\overline{\Omega^\pm \backslash C^\pm})$ analytic on $\Omega^\pm \backslash C^\pm$ and satisfying the boundary conditions in the above definition, then (by enlarging $C^\pm$ slightly if necessary) we may extend $S^\pm$ to become continuous, and even smooth, in all of $\Omega^\pm$. In particular, we see that having a two-phase Schwarz function is a local property of the boundaries $\partial\Omega^\pm$. Further, since the distribution $(\bar{\partial}S)|_\Omega$ has compact support in Ω, a standard mollification argument yields a measure μ which has compact support in Ω and gives the same result when applied to functions analytic in Ω.

The following theorem relates two-phase quadrature domains for analytic functions to two-phase Schwarz functions.

Theorem 3.14. *If (Ω^+, Ω^-) is a pair of disjoint bounded open sets such that $\partial\Omega^\pm$ are piecewise C^1, then it has a two-phase Schwarz function if and only if it is a two-phase quadrature domain for analytic functions with respect to some measure.*

Proof. We first assume that (Ω^+, Ω^-) has a two-phase Schwarz function S. Let f be analytic in $\Omega = \Omega^+ \cup \Omega^-$ and continuous on $\overline{\Omega}$. By assumption $\bar{\partial}S = \mu$ has compact support in Ω. By Stokes' theorem and the smoothness assumption on the boundaries,

$$\begin{aligned} 2i\int_{\Omega^+} f d\lambda - 2i\int_{\Omega^-} f d\lambda &= \int_{\partial\Omega^+} f(z)\bar{z}dz + \int_{\partial\Omega^-} -f(z)\bar{z}dz \\ &= \int_{\partial\Omega^+} f(z)S(z)dz + \int_{\partial\Omega^-} f(z)S(z)dz \\ &= 2i\int f d\mu. \end{aligned}$$

(Here we have used the fact that the boundaries $\partial\Omega^+$ and $\partial\Omega^-$ have opposite orientations on $\partial\Omega^+ \cap \partial\Omega^-$, and so the quantity $S^+ - S^-$ is irrelevant on this set.)

Conversely, we assume that (Ω^+, Ω^-) is a two-phase quadrature domain with respect to μ, and define

$$u = U\mu - U(\lambda|_{\Omega^+}) + U(\lambda|_{\Omega^-}).$$

By assumption we have $\partial u = 0$ in $\overline{D}^c$. By continuity this equality extends to $\overline{\overline{D}^c}$ and, by our hypothesis on $\partial\Omega^\pm$, this set contains $\partial\Omega^\pm \backslash \Gamma$. If we define

$$S^{\pm} = \pm\bar{z} - 4\partial u \text{ in } \overline{\Omega}^{\pm},$$

then it is easy to see that the requirements in the definition of a two-phase Schwarz function are met. □

Unfortunately, being a two-phase quadrature domain for analytic functions does not imply any regularity of the boundaries of the domains $\partial\Omega^{\pm}$, as the following example will show. This reflects the remark in the above proof that, under mild hypotheses, a function analytic in $\Omega^{+} \cup \Omega^{-}$ and continuous on the closure of this set will automatically be analytic also on the interior of the set $\overline{\Omega^{+} \cup \Omega^{-}}$.

Example 3.15. Let Ω^{+} be a simply connected plane domain with (reasonably) smooth boundary. Let $\Omega^{-} = \omega(2\lambda|_{\Omega^{+}})\backslash\Omega^{+}$. Then it is easy to see that

$$\int_{\Omega^{+}} h d\lambda - \int_{\Omega^{-}} h d\lambda = 0$$

for all integrable harmonic functions h on $D = (\overline{\Omega^{+} \cup \Omega^{-}})^{\circ} = \omega(2\lambda|_{\Omega^{+}})$. Note that $\partial\Omega^{+} \subset D$ in this case. Further, as above, all functions f which are continuous in $\overline{\Omega^{+} \cup \Omega^{-}}$ and analytic in $\Omega^{+} \cup \Omega^{-}$ are automatically analytic in D. Therefore

$$\int_{\Omega^{+}} f d\lambda - \int_{\Omega^{-}} f d\lambda = 0 \tag{12}$$

for all functions which are continuous on $\overline{D}$ and analytic in $\Omega^{+} \cup \Omega^{-}$. Since the assumption that $\partial\Omega^{+}$ is smooth can be substantially relaxed, we cannot deduce any general results regarding the regularity of the boundary merely from the quadrature identity (12).

This should be contrasted with the results of [19, 20], which show that for a two-phase quadrature domain for subharmonic functions the boundary consists locally of one or two graphs that are C^1 (but not, in general, $C^{1,\alpha}$). The regularity in the case of two-phase quadrature domains for harmonic functions is still open. Also, we note that on the set $\partial\Omega^{\pm}\backslash\Gamma$ we are locally in the one-phase situation, which is treated in [15], and so this part of the boundary is locally real analytic, apart from possibly a finite number of singularities.

References

1. D.H. Armitage, S.J. Gardiner, *Classical Potential Theory* (Springer, London, 2001)
2. H. Brezis, A.C. Ponce, Kato's inequality when Δu is a measure. C. R. Acad. Sci. Paris Ser. I **338**, 599–604 (2004)
3. L.A. Caffarelli, L. Karp, H. Shahgholian, Regularity of a free boundary with application to the Pompeiu problem. Ann. Math. (2) **151**, 269–292 (2000)

4. B. Emamizadeh, J.V. Prajapat, H. Shahgholian, A two phase free boundary problem related to quadrature domains. Potential Anal. **34**, 119–138 (2011)
5. S.J. Gardiner, T. Sjödin, Quadrature domains for harmonic functions. Bull. Lond. Math. Soc. **39**, 586–590 (2007)
6. S.J. Gardiner, T. Sjödin, Convexity and the exterior inverse problem of potential theory. Proc. Am. Math. Soc. **136**, 1699–1703 (2008)
7. S.J. Gardiner, T. Sjödin, Partial balayage and the exterior inverse problem of potential theory. In: *Potential theory and stochastics in Albac*, ed. by D. Bakry, et al. (Bucharest, Theta, 2009), pp. 111–123
8. S.J. Gardiner, T. Sjödin, Two-phase quadrature domains. J. Anal. Math. **116**, 335–354 (2012)
9. B. Gustafsson, M. Sakai, H.S. Shapiro, On domains in which harmonic functions satisfy generalized mean value properties. Potential Anal. **7**, 467–484 (1997)
10. B. Gustafsson, M. Sakai, Properties of some balayage operators, with applications to quadrature domains and moving boundary problems. Nonlinear Anal. **22**, 1221–1245 (1994)
11. B. Gustafsson, H.S. Shapiro, What is a quadrature domain? *Quadrature domains and their applications*, 1–25. Oper. Theory Adv. Appl., vol. 156 (Birkhäuser, Basel, 2005)
12. Ü. Kuran, On the mean-value property of harmonic functions. Bull. Lond. Math. Soc. **4**, 311–312 (1972)
13. J. Roos, Weighted Potential Theory and Partial Balayage. MSc thesis in Mathematics, KTH, 2011
14. E.B. Saff, V. Totik, *Logarithmic Potentials with External Fields* (Springer, Berlin, 1997)
15. M. Sakai, Regularity of a boundary having a Schwarz function. Acta Math. **166**, 263–297 (1991)
16. M. Sakai, Restriction, localization and microlocalization, *Quadrature domains and their applications*, 195–205. Oper. Theory Adv. Appl., vol. 156 (Birkhäuser, Basel, 2005)
17. H. Shahgholian, T. Sjödin, Harmonic balls and the two-phase Schwarz function. Complex Var. Elliptic Eqn. **58**, 837–852 (2013)
18. H. Shahgholian, N. Uraltseva, Regularity properties of a free boundary near contact points with the fixed boundary. Duke Math. J. **116**, 1–34 (2003)
19. H. Shahgholian, N. Uraltseva, G.S. Weiss, The two-phase membrane problem—regularity of the free boundaries in higher dimensions. Int. Math. Res. Not. IMRN **8**(Art. ID rnm026), 16 pp (2007)
20. H. Shahgholian, G.S. Weiss, The two-phase membrane problem—an intersection-comparison approach to the regularity at branch points. Adv. Math. **205**, 487–503 (2006)
21. T. Sjödin, Quadrature identities and deformation of quadrature domains. *Quadrature domains and Their Applications*, 239–255. Oper. Theory Adv. Appl., vol. 156 (Birkhäuser, Basel, 2005)

Exponential Transforms, Resultants and Moments

Björn Gustafsson

Abstract We give an overview of some recent developments concerning harmonic and other moments of plane domains, their relationship to the Cauchy and exponential transforms, and to the meromorphic resultant and elimination function. The paper also connects to certain topics in mathematical physics, for example domain deformations generated by harmonic gradients (Laplacian growth) and related integrable structures.

Keywords Cauchy transform • Elimination function • Exponential transform • Laplacian growth • Moment • Polubarinova–Galin equation • Quadrature domain • Resultant • String equation

Subject Classification: Primary: 30-02; Secondary: 13P15, 30E05, 31A05, 76D27.

1 Introduction

In the present article we will focus on developments concerning harmonic and other moments of a domain in the complex plane, the exponential transform and, in the case of a quadrature domain, the relation between these objects and the resultant and elimination function. Much of the material is based on joint work with Vladimir Tkachev, Mihai Putinar, Ahmed Sebbar and is in addition inspired by ideas of Mark Mineev-Weinstein, Paul Wiegmann, Anton Zabrodin, and others.

The organization of the paper is as follows. Section 2 gives the basic definitions of Cauchy and exponential transforms, including extended versions in four complex

B. Gustafsson (✉)
Department of Mathematics, KTH, 100 44, Stockholm, Sweden
e-mail: gbjorn@kth.se

A. Vasil'ev (ed.), *Harmonic and Complex Analysis and its Applications*, Trends in Mathematics, DOI 10.1007/978-3-319-01806-5_6,

variables. In Sect. 3 it is shown how these generate harmonic and exponential moments. In the case of quadrature domains (equivalently, algebraic domains or finitely determined domains), there are strong algebraic relationships between the various transforms and moments, and this is exposed in Sect. 4. Most of the material in Sects. 2–4 is by now relatively classical, being based on developments in the period 1970–2000 within complex analysis and operator theory. For example, the exponential transform came out as a by-product from the theory of hyponormal operators.

Central for the paper is Sect. 5 on the meromorphic resultant, which was introduced in [19]. One main message is that, in the case of quadrature domains, the elimination function, which is defined by means of the meromorphic resultant (and hence is a purely algebraic object), turns out to be the same as the exponential transform (which is an analytic object). In Sect. 6, some potential theoretic interpretations of the resultant are given.

In the last three sections, Sects. 7–9, we connect the previously discussed material to some quite exciting and relatively recent developments in mathematical physics. It much concerns deformations (or variations) of domains when the harmonic moments are used as coordinates (in the simply connected case). Such deformations fit into the framework of Laplacian growth processes, or moving boundaries for Hele–Shaw flows. More exactly, they can naturally be thought of as evolutions driven by harmonic gradients. We mostly restrict to finite dimensional subclasses of domains, namely polynomial (of a fixed degree) conformal images of the unit disk, this is in order to keep the presentation transparent and rigorous. A nice feature then is that the Jacobian determinant for the transition between harmonic moment coordinates and the coordinates provided by the coefficients of the polynomial mapping functions can be made fully explicit, (49), in terms of a resultant involving the mapping function. This result is due to O. Kuznetsova and V. Tkachev [25, 46].

The further results concern the string equation (52) (which is equivalent to the Polubarinova–Galin equation for a Hele–shaw blob), integrability properties of the harmonic moments (16), and a corresponding prepotential (18), which is the logarithm of a τ-function and which in some sense explains the mentioned integrability properties. Finally, we give Hamiltonian formulations of the evolution equation for the conformal map under variation of the moments (68). All these matters are due to M. Mineev-Weinstein, P. Wiegmann, A. Zabrodin and others, see, for example, [24, 27, 29, 53], but we try to explain everything in our own words and using the particular settings of the present paper.

1.1 List of Notations

Below is a list of general notations that will be used.

- $\mathbb{D} = \{\zeta \in \mathbb{C} : |\zeta| < 1\}, \mathbb{D}(a, r) = \{\zeta \in \mathbb{C} : |\zeta - a| < r\}$.
- $\mathbb{P} = \mathbb{C} \cup \{\infty\}$, the Riemann sphere.

- $dm = dm(z) = dx \wedge dy = \frac{1}{2i} d\bar{z} \wedge dz$. $(z = x + iy)$.
- $f^*(\zeta) = \overline{f(1/\bar{\zeta})}$.

2 The Cauchy and Exponential Transforms

The *logarithmic potential* of a measure μ with compact support in $\mathbb{C}$ is

$$U_\mu(z) = \int \log \frac{1}{|z - \zeta|} d\mu(\zeta), \tag{1}$$

satisfying $-\Delta U_\mu = 2\pi\mu$ in the sense of distributions. The gradient of U_μ can, modulo a constant factor and a complex conjugation, be identified with the *Cauchy transform*

$$C_\mu(z) = -\frac{2}{\pi} \frac{\partial U_\mu(z)}{\partial z} = -\frac{1}{\pi} \int \frac{d\mu(\zeta)}{\zeta - z},$$

which may be naturally viewed as a differential: $C_\mu(z)dz$. From these potentials, or fields, μ can be recovered by

$$\mu = \frac{\partial C_\mu}{\partial \bar{z}} = -\frac{1}{2\pi} \Delta U_\mu. \tag{2}$$

We shall mainly deal with measures of the form $\mu = \chi_\Omega m$, where m denotes Lebesgue measure in $\mathbb{C}$ and $\Omega \subset \mathbb{C}$ is a bounded open set. Then we write U_Ω, C_Ω in place of U_μ and C_μ. On writing C_Ω as

$$C_\Omega(z) = \frac{1}{2\pi i} \int_\Omega \frac{d\zeta \wedge d\bar{\zeta}}{\zeta - z} = \frac{1}{2\pi i} \int_\Omega \frac{d\zeta}{\zeta - z} \wedge d\bar{\zeta}.$$

one is naturally led to consider the more symmetric "double Cauchy transform,"

$$C_\Omega(z, w) = \frac{1}{2\pi i} \int_\Omega \frac{d\zeta}{\zeta - z} \wedge \frac{d\bar{\zeta}}{\bar{\zeta} - \bar{w}}, \tag{3}$$

which is much richer than the original transform. It may be viewed as a double differential: $C_\mu(z, w)\, dz d\bar{w}$. After exponentiation it gives the by now quite well-studied [5, 13, 34, 36, 37] *exponential transform* of Ω:

$$E_\Omega(z, w) = \exp C_\Omega(z, w).$$

The original Cauchy transform reappears when specializing one variable at infinity:

$$C_\Omega(z) = \operatorname*{Res}_{w=\infty} C_\Omega(z,w)\,d\bar{w} = -\lim_{w\to\infty} \bar{w}C_\Omega(z,w).$$

As a substitute for (2) we have, for the double transform,

$$\frac{\partial^2 C_\Omega(z,w)}{\partial\bar{z}\partial w} = -\pi\delta(z-w)\chi_\Omega(z)\chi_\Omega(w).$$

Even the double Cauchy transform is not fully complete. It contains the Cauchy kernel $\frac{d\zeta}{\zeta-z}$, which is a meromorphic differential on the Riemann sphere with a pole at $\zeta = z$, but it has also a pole at $\zeta = \infty$. It is natural to make the latter pole visible and movable. This has the additional advantage that one can avoid the two Cauchy kernels which appear in the definitions of the double Cauchy transform, and the exponential transform, to have coinciding poles (namely at infinity). Thus one arrives naturally at the extended (or four variable) Cauchy and exponential transforms:

$$C_\Omega(z,w;a,b) = \frac{1}{2\pi \mathrm{i}}\int_\Omega \Big(\frac{d\zeta}{\zeta-z} - \frac{d\zeta}{\zeta-a}\Big)\wedge\Big(\frac{d\bar\zeta}{\bar\zeta-\bar w} - \frac{d\bar\zeta}{\bar\zeta-\bar b}\Big), \tag{4}$$

$$E_\Omega(z,w;a,b) = \exp C_\Omega(z,w;a,b) = \frac{E_\Omega(z,w)E_\Omega(a,b)}{E_\Omega(z,b)E_\Omega(a,w)}. \tag{5}$$

Clearly $C_\Omega(z,w) = C_\Omega(z,w;\infty,\infty)$, $E_\Omega(z,w) = E_\Omega(z,w;\infty,\infty)$. If the points z, w, a, b are taken to be all distinct, then both transforms are well defined and finite for any open set Ω in the Riemann sphere $\mathbb{P}$. For example, $E_{\mathbb{P}}(z,w;a,b)$ turns out to be the modulus squared of the cross-ratio:

$$E_{\mathbb{P}}(z,w;a,b) = |(z:a:w:b)|^2, \tag{6}$$

where, according to the classical definition [2],

$$(z:a:w:b) = \frac{(z-w)(a-b)}{(z-b)(a-w)} \tag{7}$$

with the variables in this order. To prove (6), one may use the formula for the two-variable exponential transform for the disk $\Omega = \mathbb{D}(0,R)$ when both variables are inside, namely (see [13])

$$E_{\mathbb{D}(0,R)}(z,w) = \frac{|z-w|^2}{R^2 - z\bar{w}} \quad (z,w\in\mathbb{D}(0,R)),$$

insert this into the last member of (5) and let $R\to\infty$. A different proof of (6) will be indicated in Sect. 6.

We record also the formula for the exponential transform for an arbitrary disk $\mathbb{D}(a, R)$ when both variables are outside ($z, w \in \mathbb{C} \setminus \mathbb{D}(a, R)$):

$$E_{\mathbb{D}(a,R)}(z, w) = 1 - \frac{R^2}{(z-a)(\bar{w}-\bar{a})}. \tag{8}$$

See again [13].

From (6) it is immediate that for any domain $\Omega \subset \mathbb{P}$,

$$E_{\Omega}(z, w; a, b) E_{\mathbb{P}\setminus\Omega}(z, w; a, b) = |(z : a : w : b)|^2 \tag{9}$$

(clearly E_D makes sense even if the set D is not open). The two-variable exponential transform $E_{\mathbb{P}\setminus\Omega}(z, w)$ is identically zero if Ω is bounded, but there is still a counterpart of (9) in this case. It is

$$E_{\Omega}(z, w) \cdot \frac{1}{H_{\Omega}(z, w)} = |z - w|^2, \tag{10}$$

where $H_{\Omega}(z, w)$ is the *interior exponential transform*, which is a renormalized version of one over the exponential transform of the exterior domain, more precisely

$$H_{\Omega}(z, w) = \lim_{R\to\infty} \frac{1}{R^2 E_{\mathbb{D}(0,R)\setminus\Omega}}.$$

We consider this function only for $z, w \in \Omega$, and then it is analytic in z and antianalytic in w.

When the integral in the definition of the exponential transform is transformed into a boundary integral it turns out that the formulas for $E_{\Omega}(z, w)$ (with the variables outside) and $H_{\Omega}(z, w)$ become exactly the same:

$$E_{\Omega}(z, w) = \exp\left[\frac{1}{\mathrm{i}\pi} \int_{\partial\Omega} \log|\zeta - z| \, d\overline{\log(\zeta - w)}\right] \quad (z, w \in \mathbb{C} \setminus \Omega),$$

$$H_{\Omega}(z, w) = \exp\left[\frac{1}{\mathrm{i}\pi} \int_{\partial\Omega} \log|\zeta - z| \, d\overline{\log(\zeta - w)}\right] \quad (z, w \in \Omega).$$

Here one can easily identify the modulus and the angular part, for example for $E_{\Omega}(z, w)$:

$$|E_{\Omega}(z, w)| = \exp\left[-\frac{1}{\pi} \int_{\partial\Omega} \log|\zeta - z| \, d\arg(\zeta - w)\right],$$

$$\arg E_{\Omega}(z, w) = -\frac{1}{\pi} \int_{\partial\Omega} \log|\zeta - z| \, d\log|\zeta - w|$$

($z, w \in \mathbb{C} \setminus \Omega$). These formulas open up for geometrical interpretations.

The exponential transform originally arose as a side product in the theory of hyponormal operators on a Hilbert space, see [5, 28, 30], and the inner product in the Hilbert space then automatically forces the transform to have certain positive definiteness properties, namely the two-variable transforms $1/E_\Omega$ and H_Ω are positive definite (as kernels) in all $\mathbb{C}$, and $1 - E_\Omega$ is positive semidefinite outside Ω. See [15] for direct proofs. For the four-variable exponential transform we similarly have, for example, that $1/E$ is positive definite in the sense that

$$\sum_{k,j} \frac{\lambda_k \overline{\lambda_j}}{E_\Omega(z_k, z_j; a_k, a_j)} \geq 0$$

for any finite tuples $(z_j, a_j) \in \mathbb{C}^2$, $\lambda_j \in \mathbb{C}$, and with strict inequality unless all λ_j are zero.

Aside from operator theory, the ideas of the exponential transform appear implicitly as a tool in the theory of boundaries of analytic varieties, see [3, 21] (Chap. 19).

3 Harmonic, Complex, and Exponential Moments

The Cauchy and exponential transforms are generating functions for series of moments of a bounded domain $\Omega \subset \mathbb{C}$. Specifically we have the *complex moments* $M_{kj} = M_{kj}(\Omega)$,

$$M_{kj} = \frac{1}{\pi} \int_\Omega z^k \bar{z}^j \, dm(z) = \frac{1}{2\pi \mathrm{i}} \int_\Omega z^k \bar{z}^j \, d\bar{z} dz \quad (k, j \geq 0),$$

which in view of the Weierstrass approximation theorem completely determine Ω up to nullsets, and the more restricted *harmonic moments* $M_k = M_k(\Omega)$:

$$M_k = M_{k0} = \frac{1}{\pi} \int_\Omega z^k \, dm(z) = \frac{1}{2\pi \mathrm{i}} \int_\Omega z^k \, d\bar{z} dz \quad (k \geq 0). \tag{11}$$

The latter do not completely determine the domain, not even in the simply connected case (see [41]). However, for simply connected domains with analytic boundaries they are at least sensible for local variations of the domain, i.e., we have a local one-to-one correspondence

$$\Omega \leftrightarrow (M_0, M_1, M_2, \dots). \tag{12}$$

A precise statement in this respect is the following: there exists a compact subset $K \subset \Omega$ such that any Jordan domain $D \supset K$ with $M_k(D) = M_k(\Omega)$ for all $k \geq 0$ necessarily agrees with Ω. See [12], Corollary 3.10, for a proof.

It is actually not a priori obvious that the moments M_k ($k \geq 0$) are independent of each other (for example, the complex moments M_{kj} are definitely not independent, given that they are moments of domains), but as will be discussed in Sects. 7 and 8, it is indeed possible to define deformations of Ω which change one of the M_k without changing the other ones. Further discussions can be found, for example, in [6, 40].

The generating properties for the harmonic and complex moments are

$$C_\Omega(z) = \sum_{k=0}^{\infty} \frac{M_k}{z^{k+1}}, \tag{13}$$

$$C_\Omega(z,w) = \sum_{k,j\geq 0} \frac{M_{kj}}{z^{k+1}\bar{w}^{j+1}}.$$

Finally we introduce the *exponential moments* $B_{kj} = B_{kj}(\Omega)$ having $1 - E(z,w)$ as generating function, hence being defined by

$$1 - \exp\left[-\sum_{k,j\geq 0} \frac{M_{kj}}{z^{k+1}\bar{w}^{j+1}}\right] = \sum_{k,j\geq 0} \frac{B_{kj}}{z^{k+1}\bar{w}^{j+1}}. \tag{14}$$

The positive semidefiniteness of $1 - E(z,w)$ implies that also the infinite matrix (B_{kj}) is positive semidefinite. The extended (four-variable) exponential transform similarly generates some four index moments, see [28, 32, 33, 36, 37].

In the remaining part of this section we shall assume that Ω is a simply connected domain with $0 \in \Omega$ and such that $\partial\Omega$ is an analytic curve (without singularities). By writing (11) as a boundary integral,

$$M_k = \frac{1}{2\pi \mathrm{i}} \int_{\partial\Omega} z^k\, \bar{z} dz, \tag{15}$$

the harmonic moments become meaningful also for $k < 0$, and these can be considered as functions of the moments for $k \geq 0$, as far as local variations are concerned. Thus we may write

$$M_{-k} = M_{-k}(\Omega) = M_{-k}(M_0, M_1, M_2, \dots)$$

for $k > 0$.

In a series of papers by I. Krichever, A. Marshakov, M. Mineev-Weinstein, P. Wiegmann and A. Zabrodin, for example [24, 27, 29, 53], the so extended moment sequence has been shown to enjoy remarkable integrability properties, for example (in present notation)

$$\frac{1}{k}\frac{\partial M_{-k}}{\partial M_j} = \frac{1}{j}\frac{\partial M_{-j}}{\partial M_k} \quad (k, j \geq 1). \tag{16}$$

This can be explained in terms of the presence of a prepotential $\mathcal{F}(\Omega)$ such that

$$\frac{1}{k} M_{-k}(\Omega) = \frac{\partial \mathcal{F}(\Omega)}{\partial M_k}. \tag{17}$$

In fact, for any sufficiently large $R > 0$ (it is enough that $\overline{\Omega} \subset \mathbb{D}(0, R)$), the energy functional

$$\mathcal{F}(\Omega) = \mathcal{F}_R(\Omega) = \frac{1}{\pi^2} \int_{\mathbb{D}(0,R)\setminus\Omega} \int_{\mathbb{D}(0,R)\setminus\Omega} \log \frac{1}{|z - \zeta|} dm(z) dm(\zeta) \tag{18}$$

serves as a prepotential. The exponential of $\mathcal{F}(\Omega)$ can be identified with a certain τ-function which appears as a partition function in mathematical models in statistical mechanics, see [23]. An indication of the proof of (17), and hence (16), will be given in Sect. 8.

Remark 3.1. The dependence of M_{-k} and $\mathcal{F}(\Omega)$ on the M_j is actually not analytic, so the above partial derivatives are really Wirtinger derivatives:

$$\frac{\partial}{\partial M_j} = \frac{1}{2}\left(\frac{\partial}{\partial \operatorname{Re} M_j} - \mathrm{i} \frac{\partial}{\partial \operatorname{Im} M_j}\right),$$

$$\frac{\partial}{\partial \bar{M}_j} = \frac{1}{2}\left(\frac{\partial}{\partial \operatorname{Re} M_j} + \mathrm{i} \frac{\partial}{\partial \operatorname{Im} M_j}\right).$$

On extending the power series (13) for the Cauchy transform to a full Laurent series by means of the negative moments one gets, at least formally, the *Schwarz function* of $\partial\Omega$:

$$S(z) = \sum_{k=-\infty}^{\infty} \frac{M_k}{z^{k+1}}. \tag{19}$$

Even though the full series here need not converge anywhere it can be given the following precise meaning: the negative part of the series defines a germ of an analytic function at the origin and the positive part a germ of an analytic function at infinity. When $\partial\Omega$ is analytic the domains of analyticity of these two analytic functions overlap, and the overlap region contains $\partial\Omega$. Thus $S(z)$, as written above, makes sense and is analytic in a neighborhood of $\partial\Omega$. Moreover, it satisfies

$$S(z) = \bar{z} \quad \text{for } z \in \partial\Omega. \tag{20}$$

This can be seen by realizing that the positive part of the power series (19) defines the Cauchy transform $C_\Omega(z)$, which can be written, for large z,

$$\sum_{k=0}^{\infty} \frac{M_k}{z^{k+1}} = -\frac{1}{2\pi \mathrm{i}} \int_{\partial\Omega} \frac{\bar{\zeta}\, d\zeta}{\zeta - z} = C_\Omega(z).$$

Similarly, the negative part of the series agrees with the power series expansion of a boundary integral which can be interpreted as the Cauchy transform of the exterior domain:

$$\sum_{k=-\infty}^{-1} \frac{M_k}{z^{k+1}} = -\frac{1}{2\pi \mathrm{i}} \int_{\partial(\mathbb{P}\setminus\Omega)} \frac{\bar{\zeta}\, d\zeta}{\zeta - z} = C_{\mathbb{P}\setminus\Omega}(z)$$

for z small. Now (20) follows from a well-known jump formula for Cauchy integrals.

4 Finitely Determined Domains

In order to exhibit transparent links between some of the previously defined objects we shall now work in a basically algebraic framework. This will mean that we shall work with classes of domains involving only finitely many parameters, specifically *finitely determined domains* in the terminology of [34], in other words *quadrature domains* [1, 17, 42, 44] or *algebraic domains* [48]. In this section our domains need not be simply connected, they will from outset be just bounded domains in the complex plane (bounded open set would work equally well).

There are many definitions or equivalent characterizations of quadrature domains, the most straightforward in the context of the present paper perhaps being the following: a bounded domain Ω in the complex plane is a (classical) quadrature domain (or algebraic domain), if the exterior Cauchy transform is a rational function, i.e., if there exists a rational function $R(z)$ such that

$$C_\Omega(z) = R(z) \quad \text{for all } z \in \mathbb{C} \setminus \Omega. \tag{21}$$

Below is a list of equivalent requirements on a bounded domain $\Omega \subset \mathbb{C}$. Strictly speaking, the last three items, (c)–(e), are insensitive to changes of Ω by nullsets, but one may achieve equivalence in the pointwise sense by requiring that the domain Ω considered is complete with respect to area measure m, i.e., that all points $a \in \mathbb{C}$ such that $m(\mathbb{D}(a, \varepsilon) \setminus \Omega) = 0$ for some $\varepsilon > 0$ have been adjoined to Ω.

(a) There exist finitely many points $a_k \in \Omega$ and coefficients $c_{kj} \in \mathbb{C}$ such that

$$\int_\Omega h\, dm = \sum_{k=1}^{m} \sum_{j=0}^{n_k - 1} c_{kj} h^{(j)}(a_k) \tag{22}$$

for every function h which is integrable and analytic in Ω. This is the original definition of a quadrature domain, used in [1], for example.

(b) There exists a meromorphic function $S(z)$ in Ω, extending continuously to $\partial\Omega$ with

$$S(z) = \bar{z} \quad \text{for } z \in \partial\Omega. \tag{23}$$

This $S(z)$ will be the Schwarz function of $\partial\Omega$ [7,44].

(c) The exponential transform $E_\Omega(z, w)$ is, for z, w large, a rational function of the form

$$E_\Omega(z,w) = \frac{Q(z,\bar{w})}{P(z)\overline{P(w)}}, \tag{24}$$

where P and Q are polynomials in one and two variables, respectively.

(d) Ω is determined by a finite sequence of complex moments $(M_{kj})_{0\le k,j\le N}$ (see Example 4.1 below for a clarification of the meaning of this).

(e) For some positive integer N there holds

$$\det(B_{kj})_{0\le k,j\le N} = 0.$$

Basic references for (c)–(e) are [13,34–36].

When the above conditions hold then the minimum possible number N in (d) and (e), the degree of P in (c) and the number of poles (counting multiplicities) of $S(z)$ in Ω, all coincide with the *order* of the quadrature domain, i.e., the number $N = \sum_{k=1}^{m} n_k$ in (22). For Q, see more precisely below.

If Ω is simply connected, the above conditions (a)–(e) are also equivalent to that any conformal map $f : \mathbb{D} \to \Omega$ is a rational function. If Ω is multiply connected, then it is necessarily finitely connected (in fact, the boundary will, up to finitely many points, be exactly the algebraic curve $Q(z,\bar{z}) = 0$ with Q as in (c) above), and there is a beautiful extension, due to D. Yakubovich, of the aforementioned statement: $\partial\Omega$ is traced out (on $\partial\mathbb{D}$) by the eigenvalues of a rational normal matrix function $F : \mathbb{D} \to \mathbb{C}^{N\times N}$ which is holomorphic in $\mathbb{D}$; see [54] for details.

The positive definiteness properties of the exponential transform (see end of Sect. 2) imply that when Ω is a quadrature domain of order N then $Q(z,w)$ admits the following representation [14]:

$$Q(z,\bar{w}) = P(z)\overline{P(w)} - \sum_{k=0}^{N-1} P_k(z)\overline{P_k(w)}.$$

Here each $P_k(z)$ is a polynomial of degree k exactly. Recall that $P(z) = P_N(z)$ has degree N, and one usually normalizes it to be monic (thereby also making $Q(z,\bar{w})$ uniquely determined). Thus the rational form (24) of the exponential transform can be expressed as

$$1 - E_\Omega(z,w) = \sum_{k=0}^{N-1} \frac{P_k(z)\overline{P_k(w)}}{P(z)\overline{P(w)}}.$$

With the same polynomials, the rational form (21) of the Cauchy transform is precisely

$$C_\Omega(z) = \frac{P_{N-1}(z)}{P(z)}.$$

Note, however, that the double Cauchy transform $C_\Omega(z,w)$ will not be rational, only its exponential will be.

Example 4.1. The following example should clarify the meaning of a domain Ω being finitely determined (determined by finitely many moments M_{kj}), as in the equivalent condition (d) above. Suppose that, by some measurements, we happen to know the moments M_{00}, M_{10}, M_{01}, M_{11}, but we do not know from which domain they come. Generally speaking, there will be infinitely many domains whose moment sequence starts out with just these four moments, unless of course the data are contradictory so that no domain has them. However, for exceptional choices of the data it happens that these four moments (or, in general, finitely many moments) determine the domain uniquely. This happens exactly for quadrature domains.

Assume, for example, that

$$M_{00} = M_{10} = M_{01} = 4, \quad M_{11} = 12, \tag{25}$$

and let us try to compute the exponential moments B_{kj} as far as possible. We have

$$1 - \exp\left[-\sum_{k,j\geq 0} \frac{M_{kj}}{z^{k+1}\bar{w}^{j+1}}\right] = 1 - \exp\left[-\frac{4}{z\bar{w}} - \frac{4}{z^2\bar{w}} - \frac{4}{z\bar{w}^2} - \frac{12}{z^2\bar{w}^2} - \dots\right]$$

$$= \frac{4}{z\bar{w}} + \frac{4}{z^2\bar{w}} + \frac{4}{z\bar{w}^2} + \frac{12}{z^2\bar{w}^2} + \dots - \frac{1}{2!}\left(-\frac{4}{z\bar{w}} - \dots\right)^2 - \frac{1}{3!}\left(-\frac{4}{z\bar{w}} - \dots\right)^3 - \dots$$

$$= \frac{4}{z\bar{w}} + \frac{4}{z^2\bar{w}} + \frac{4}{z\bar{w}^2} + \frac{4}{z^2\bar{w}^2} + \text{higher order terms.}$$

Thus we see that

$$B_{00} = B_{10} = B_{01} = B_{11} = 4.$$

Clearly the determinant $\det(B_{kj})_{0\leq k,j\leq N}$ vanishes for $N = 1$. The theory (equivalent condition (e) above) then tells that the moments come from a quadrature domain of order one, i.e., a disk. To determine which disk one uses (14) together with (24). Arguing in general one first realizes that

$$P(z)\overline{P(w)}\sum_{k,j\geq 0}\frac{B_{kj}}{z^{k+1}\bar{w}^{j+1}} = P(z)\overline{P(w)} - Q(z,\bar{w}), \tag{26}$$

in particular the left member is a polynomial. Setting $P(z) = \sum_{j=0}^{N}\alpha_j z^j$ and chosing $\alpha_N = 1$ (normalization), the vanishing of the coefficient of $1/z\bar{w}$ in the left member gives

$$\sum_{k,j=0}^{N}\alpha_j\bar{\alpha}_k B_{kj} = 0.$$

This, by the way, explains the condition $\det(B_{kj})_{0\leq k,j\leq N} = 0$. It also gives the equation

$$\sum_{j=0}^{N} B_{kj}\alpha_j = 0 \quad (k = 0,\ldots,N)$$

for the coefficients of $P(z)$. In our particular example this equation gives $\alpha_0 = -\alpha_1 = -1$, hence

$$P(z) = z - 1.$$

Next, $Q(z,\bar{w})$ is easily obtained from (26):

$$Q(z,\bar{w}) = P(z)\overline{P(w)} - P(z)\overline{P(w)}\sum_{k,j\geq 0}\frac{B_{kj}}{z^{k+1}\bar{w}^{j+1}}$$

$$= (z\bar{w} - z - \bar{w} + 1) - (z\bar{w} - z - \bar{w} + 1)\left(\frac{4}{z\bar{w}} + \frac{4}{z^2\bar{w}} + \frac{4}{z\bar{w}^2} + \frac{4}{z^2\bar{w}^2} + \ldots\right)$$

$$= z\bar{w} - z - \bar{w} + 1 - 4 + \text{terms containing negative powers of } z \text{ or } \bar{w}$$

$$= z\bar{w} - z - \bar{w} - 3 = (z-1)(\bar{w}-1) - 4.$$

Thus we have identified Ω as the disk

$$\Omega = \{z \in \mathbb{C} : |z-1|^2 < 4\}.$$

No other domain (or open set) has the first four moments given by (25).

5 The Meromorphic Resultant and the Elimination Function

In this section we review the definitions of the meromorphic resultant and the elimination function, as introduced in [19], referring to that paper for any details. If f is a meromorphic function on any compact Riemann surface M, we denote by (f) its divisor of zeros and poles, symbolically $(f) = f^{-1}(0) - f^{-1}(\infty)$ (written in additive form). If D is any divisor and g is any function, we denote by $\langle D, g\rangle$ the additive action of D on g, and by $g(D)$ the multiplicative action of g on D. See the example below for clarification.

Now the *meromorphic resultant* between f and g is, by definition,

$$\mathcal{R}(f, g) = g((f)) = e^{\langle (f), \log g\rangle}$$

whenever this makes sense (the resultant is undefined if expressions like $0 \cdot \infty$, $\frac{0}{0}$, $\frac{\infty}{\infty}$ appear when the expression in the middle member is evaluated). In the last expression, $\log g$ refers to arbitrarily chosen local branches of the logarithm. When defined, $\mathcal{R}(f, g)$ is either a complex number or ∞. As a consequence of the Weil reciprocity theorem [51] the resultant is symmetric:

$$\mathcal{R}(f, g) = \mathcal{R}(g, f).$$

Example 5.1. To illustrate and explain the above definitions we spell out what they look like for the sample divisor

$$D = 1 \cdot (a) + 1 \cdot (b) - 2 \cdot (c).$$

Here $a, b, c \in M$ (any compact Riemann surface) and g is a function on M (typically a meromorphic function, but this is not absolutely necessary as for the definitions). It is possible to think of D as a 0-chain and of g as a 0-form. Then the above definitions amount to

$$\langle D, g\rangle = 1 \cdot g(a) + 1 \cdot g(b) - 2 \cdot g(c) = \int_D g, \tag{27}$$

$$g(D) = e^{\langle D, \log g\rangle} = \frac{g(a)g(b)}{g(c)^2}. \tag{28}$$

We can associate with D the 2-form current

$$\delta_D dx \wedge dy = \delta_a dx \wedge dy + \delta_b dx \wedge dy - 2\delta_c dx \wedge dy, \tag{29}$$

where $\delta_a dx \wedge dy$ denotes the Dirac measure (point mass) at a considered as a 2-form (so that it acts on functions by integration over M). Then (27) can be augmented to

$$\langle D, g\rangle = \int_D g = \int_M g\delta_D dx \wedge dy.$$

If f is a meromorphic function with divisor $(f) = D$, then the above $g(D)$ (in (28)) is also the resultant $\mathcal{R}(f, g) = g((f))$. For example, if $M = \mathbb{P}$ and

$$f(z) = \frac{(z-a)(z-b)}{(z-c)^2},$$

then

$$\mathcal{R}(f, g) = g((f)) = \frac{g(a)g(b)}{g(c)^2}.$$

It is also possible to let functions of two (or more) variables act on divisors. For example, if $g(z, w)$ is such a function and D_1, D_2 are two divisors, say $D_1 = 1 \cdot (a) + 1 \cdot (b) - 2 \cdot (c)$, $D_2 = 3 \cdot (p) - 3 \cdot (q)$, then (by definition)

$$g(D_1, D_2) = \exp\langle D_1 \otimes D_2, \log g\rangle = \frac{g(a,p)^3 g(b,p)^3 g(c,q)^6}{g(a,q)^3 g(b,q)^3 g(c,p)^6}. \tag{30}$$

Next the *elimination function* between two meromorphic functions, f and g, on M is defined by

$$\mathcal{E}_{f,g}(z, w) = \mathcal{R}(f - z, g - w),$$

where $z, w \in \mathbb{C}$ are parameters. It is always a rational function in z and w, more precisely of the form

$$\mathcal{E}_{f,g}(z, w) = \frac{Q(z, w)}{P(z)R(w)}, \tag{31}$$

where Q, P, and R are polynomials, and it embodies the necessary (since M is compact) polynomial relationship between f and g:

$$\mathcal{E}_{f,g}(f(\zeta), g(\zeta)) = 0 \quad (\zeta \in M).$$

There is also an extended elimination function, defined by

$$\mathcal{E}_{f,g}(z,w;a,b) = \mathcal{R}(\frac{f-z}{f-a}, \frac{g-w}{g-b}).$$

To relate the elimination function to the exponential transform one need integral formulas for the elimination function. For this purpose we shall make some computations within the framework of currents (distributional differential forms) on M. As a first issue, if f is a meromorphic function in M we need to make its logarithm a single-valued function almost everywhere on M. This is done as usual by making suitable "cuts" on M, and by choosing a branch of the logarithm, call it $\operatorname{Log} f$, on the remaining set $M \setminus$ (cuts). Its differential in the sense of currents will then be of the form

$$d\operatorname{Log} f = \frac{df}{f} + \text{distributional contributions on the cuts.} \tag{32}$$

If there is a cut along the x-axis, for example, the distributional contribution will be dy times a measure on the cut, that measure being arc length measure times the size of the jump (an integer multiple of $2\pi\mathrm{i}$) over the cut, between the two branches of the logarithm.

The next issue is the identity

$$\frac{1}{2\pi\mathrm{i}} d\left(\frac{df}{f}\right) = \delta_{(f)} dx \wedge dy.$$

See [19] for the simple proof.

Now, proceeding formally we transform the resultant to an integral in the following series of steps:

$$\mathcal{R}(f,g) = g((f)) = e^{\langle (f), \operatorname{Log} g\rangle} = \exp \int_{(f)} \operatorname{Log} g$$

$$= \exp\left[\int_M \operatorname{Log} g\, \delta_{(f)} dx \wedge dy\right] = \exp\left[\frac{1}{2\pi\mathrm{i}} \int_M \operatorname{Log} g\, d\left(\frac{df}{f}\right)\right]$$

$$= \exp\left[\frac{1}{2\pi\mathrm{i}} \int_M \frac{df}{f} \wedge d\operatorname{Log} g\right].$$

It should be remarked that, in the final integral, the only contributions come from the jumps of $\operatorname{Log} g$ (the last term in (32) when stated for $\operatorname{Log} g$) because outside this set of discontinuities the integrand contains $d\zeta \wedge d\zeta = 0$ as a factor. Similarly one gets, for the elimination function,

$$\mathcal{E}_{f,g}(z,w) = \exp\left[\frac{1}{2\pi \mathrm{i}}\int_M \frac{df}{f-z} \wedge d\,\mathrm{Log}\,(g-w)\right]. \tag{33}$$

Now we shall apply (33) in the case that M is the Schottky double of a plane domain Ω, and connect it to the exponential transform of Ω. So let Ω be a finitely connected plane domain with analytic boundary or, more generally, a bordered Riemann surface and let

$$M = \hat{\Omega} = \Omega \cup \partial\Omega \cup \tilde{\Omega}$$

be the *Schottky double* of Ω, i.e., the compact Riemann surface obtained by completing Ω with a backside with the opposite conformal structure, the two surfaces glued together along $\partial\Omega$ (see [8], for example). On $\hat{\Omega}$ there is a natural anticonformal involution $J : \hat{\Omega} \to \hat{\Omega}$ exchanging corresponding points on Ω and $\tilde{\Omega}$ and having $\partial\Omega$ as fixed points.

Let f and g be two meromorphic functions on $\hat{\Omega}$. Then

$$f^* = \overline{(f \circ J)}, \quad g^* = \overline{(g \circ J)}$$

are also meromorphic on $\hat{\Omega}$.

Theorem 5.1. *With Ω, $\hat{\Omega}$, f, g as above, assume in addition that f has no poles in $\Omega \cup \partial\Omega$ and that g has no poles in $\tilde{\Omega} \cup \partial\Omega$. Then, for large z, w,*

$$\mathcal{E}_{f,g}(z,\bar{w}) = \exp\left[\frac{1}{2\pi \mathrm{i}}\int_\Omega \frac{df}{f-z} \wedge \frac{d\overline{g^*}}{\overline{g^*}-\overline{w}}\right].$$

In particular,

$$\mathcal{E}_{f,f^*}(z,\bar{w}) = \exp\left[\frac{1}{2\pi \mathrm{i}}\int_\Omega \frac{df}{f-z} \wedge \frac{d\overline{f}}{\overline{f}-\overline{w}}\right].$$

Proof. For the divisors of $f-z$ and $g-w$ we have, if z, w are large enough, $\mathrm{supp}(f-z) \subset \tilde{\Omega}$, $\mathrm{supp}(g-w) \subset \Omega$. Moreover, $\log(g-w)$ has a single-valued branch, $\mathrm{Log}\,(g-w)$, in $\tilde{\Omega}$. Using that $g = \overline{g^*}$ on $\partial\Omega$ we therefore get, starting with (33),

$$\begin{aligned}\mathcal{E}_{f,g}(z,\bar{w}) &= \exp\left[\frac{1}{2\pi \mathrm{i}}\int_{\hat{\Omega}} \frac{df}{f-z} \wedge d\,\mathrm{Log}\,(g-\bar{w})\right] \\ &= \exp\left[\frac{1}{2\pi \mathrm{i}}\int_\Omega \frac{df}{f-z} \wedge d\,\mathrm{Log}\,(g-\bar{w})\right]\end{aligned}$$

$$= \exp\left[-\frac{1}{2\pi \mathrm{i}}\int_{\partial\Omega}\frac{df}{f-z}\wedge \mathrm{Log}\,(g-\bar{w})\right]$$

$$= \exp\left[-\frac{1}{2\pi \mathrm{i}}\int_{\partial\Omega}\frac{df}{f-z}\wedge \mathrm{Log}\,(\overline{g^*}-\bar{w})\right]$$

$$= \exp\left[\frac{1}{2\pi \mathrm{i}}\int_{\Omega}\frac{df}{f-z}\wedge \frac{d\overline{g^*}}{\overline{g^*}-\bar{w}}\right].$$

as claimed. □

We next specialize to the case that Ω is a quadrature domain. Let $S(\zeta)$ be the Schwarz function of Ω. Then the relation (23) can be interpreted as saying that the pair of functions $S(\zeta)$ and $\bar{\zeta}$ on Ω combines into a meromorphic function on the Schottky double $\hat{\Omega}$ of Ω, namely that function g which equals $S(\zeta)$ on Ω, and equals $\bar{\zeta}$ on $\tilde{\Omega}$.

The function $f = g^* = \overline{g \circ J}$ is then represented by the opposite pair: ζ on Ω, $\overline{S(\zeta)}$ on $\tilde{\Omega}$. Therefore, on Ω we have $f(\zeta) = g^*(\zeta) = \zeta$, and Theorem 5.1 immediately gives

Corollary 5.1. *For any quadrature domain Ω,*

$$E_\Omega(z,w) = \mathcal{E}_{f,f^*}(z,\bar{w}) \qquad (|z|,|w| \gg 1), \tag{34}$$

where f, f^* are the two meromorphic functions on $\hat{\Omega}$ given on Ω by $f(\zeta) = \zeta$, $f^*(\zeta) = S(\zeta)$.

An alternative way of conceiving the corollary is to think of f being defined on an independent surface W, so that $f : W \to \Omega$ is a conformal map. Then Ω is a quadrature domain if and only if f extends to a meromorphic function of the Schottky double $\hat{W}$, and the assertion of the corollary then is that the exponential transform of Ω is given by (34), with the elimination function in the right member now taken in $\hat{W}$.

If Ω is simply connected, we may choose $W = \mathbb{D}$, so that $\hat{W} = \mathbb{P}$ with involution $J : \zeta \mapsto 1/\bar{\zeta}$. Then $f : \mathbb{D} \to \Omega$ is a rational function when (and only when) Ω is a quadrature domain, hence we conclude that $E_\Omega(z,w)$ in this case is the elimination function for two rational functions, $f(\zeta)$ and $f^*(\zeta) = \overline{f(1/\bar{\zeta})}$.

Finally we want to illustrate the effectiveness of the notation (30) by giving a formula for how the exponential transforms under rational conformal maps.

Theorem 5.2. *Let Ω be bounded domain in the complex plane and let F be a rational function which is bounded and one-to-one on Ω. Then for $z, w \in \mathbb{C} \setminus \overline{F(\Omega)}$ we have*

$$E_{F(\Omega)}(z,w) = E_\Omega((F-z),(F-w)). \tag{35}$$

Proof. We have

$$E_{F(\Omega)}(z,w) = \exp\left[\frac{1}{2\pi \mathrm{i}}\int\limits_{F(\Omega)} \frac{d\zeta\wedge d\bar{\zeta}}{(\zeta - z)(\bar{\zeta}-\bar{w})}\right]$$

$$= \exp\left[\frac{1}{2\pi \mathrm{i}}\int\limits_{F(\Omega)} d\log(\zeta - z)\wedge d\overline{\log(\zeta - w)}\right]$$

$$= \exp\left[\frac{1}{2\pi \mathrm{i}}\int\limits_{\Omega} d\log(F(\zeta) - z)\wedge d\overline{\log(F(\zeta) - w)}\right].$$

Let $D_z = (F - z)$ denote the divisor of $F(\zeta) - z$, and consider it as a function $D_z : \mathbb{P}\to\mathbb{Z}$, namely that function which at each point evaluates the order of the divisor (hence is zero at all but finitely many points). Then

$$d\log(F(\zeta) - z) = \sum_{\alpha\in\mathbb{P}} D_z(\alpha)d\log(\zeta-\alpha),$$

$$d\overline{\log(F(\zeta) - w)} = \sum_{\alpha\in\mathbb{P}} D_w(\beta)d\log(\bar{\zeta}-\bar{\beta}).$$

With these observations we can continue the above series of equalities:

$$E_{F(\Omega)}(z,w) = \exp\left[\frac{1}{2\pi \mathrm{i}}\int\limits_{\Omega} d\log(F(\zeta) - z)\wedge d\overline{\log(F(\zeta) - w)}\right]$$

$$= \exp\left[\frac{1}{2\pi \mathrm{i}}\sum_{\alpha\in\mathbb{P}}\sum_{\beta\in\mathbb{P}} D_z(\alpha)D_w(\beta)\int\limits_{\Omega}\frac{d\zeta}{\zeta-\alpha}\wedge\frac{d\bar{\zeta}}{\bar{\zeta}-\bar{\beta}}\right]$$

$$= \prod_{\alpha,\beta\in\mathbb{P}} E_\Omega(\alpha,\beta)^{D_z(\alpha)D_w(\beta)} = E_\Omega(D_z, D_w),$$

which is (35). □

6 Potential Theoretic Remarks

Recall the definition (1) of logarithmic potential of a signed measure μ with compact support in $\mathbb{C}$. If μ has not zero net mass, then the behavior of the potential at infinity is such that it automatically puts an extra point mass at infinity, to the effect that the potential globally (on $\mathbb{P}$) becomes the potential of a measure (again denoted μ) having vanishing net mass. If μ is given by the divisor of a meromorphic (rational) function f, i.e., if

$$d\mu = \delta_{(f)} dx \wedge dy$$

in the notation (29), then

$$U_\mu = -\log|f|.$$

The point we wish to make here is that the resultant $\mathcal{R}(f, g)$ between two rational functions, f and g, has a corresponding interpretation, namely in terms of the mutual energy. In general, the mutual energy between two signed measure μ and ν of zero net mass (on $\mathbb{P}$) is given by

$$I(\mu,\nu) = \iint \log\frac{1}{|z-\zeta|}\, d\mu(z) d\nu(\zeta) = \int U_\mu\, d\nu = \int U_\nu\, d\mu$$
$$= \frac{1}{2\pi}\int dU_\mu \wedge *dU_\nu.$$

Some care is needed here since the above is not always well defined (for example, if μ and ν have common point masses). But this is just the same situation as for the resultant. Now, with f related as above to μ, and similarly g related to ν by $d\nu = \delta_{(g)} dx \wedge dy$, then we have the relationship

$$I(\mu,\nu) = -\log|\mathcal{R}(f, g)|.$$

The proof is just a straightforward computation:

$$|\mathcal{R}(f,g)|^2 = \exp\left[\langle (f), \log g\rangle + \langle (f), \overline{\log g}\rangle\right]$$
$$= \exp\left[2\langle (f), \log|g|\rangle\right] = \exp\left[2\int_{(f)} \log|g|\right]$$
$$= \exp\left[-2\int U^\nu\, d\mu\right] = \exp\left[-2I(\mu,\nu)\right].$$

Similarly, one can relate the discriminant of f to a renormalized self-energy of μ. We refer to [19] for some details concerning this.

All of the above generalize well from $\mathbb{P}$ to an arbitrary compact Riemann surface M. The potential will then be less explicit, but it follows from classical theory (for example Hodge theory) that given any signed measure μ on M with $\int_M d\mu = 0$ there is potential U_μ, uniquely defined up to an additive constant, such that, considering μ as a 2-form current,

$$-d * dU_\mu = 2\pi\mu. \tag{36}$$

Choosing in particular $d\mu = \delta_a dx \wedge dy - \delta_b dx \wedge dy$ for two points $a, b \in M$ gives a fundamental potential that we shall denote $V(z) = V(z, w; a, b)$. Besides depending on a and b it depends on a parameter w, for normalization. The potential is characterized by the singularity structure

$$V(z) = V(z, w; a, b) = -\log|z-a| + \log|z-b| + \text{harmonic}, \tag{37}$$

together with the normalization $V(w, w; a, b) = 0$ of the additive level. It has a few symmetries,

$$V(z, w; a, b) = V(a, b; z, w) = -V(z, w; b, a), \tag{38}$$

and the transitivity property

$$V(z, w; a, b) + V(z, w; b, c) = V(z, w; a, c). \tag{39}$$

If the Riemann surface happens to be symmetric (like the Schottky double), with involution J, then we have in addition

$$V(z, w; a, b) = V(J(z), J(w); J(a), J(b)). \tag{40}$$

As an application, the ordinary Green's function $g(z, \zeta)$ for a plane domain (or bordered Riemann surface) Ω can be obtained from the potential V for $M = \hat{\Omega}$ by

$$g(z, \zeta) = \frac{1}{2} V(z, J(z); \zeta, J(\zeta)) = V(z, w; \zeta, J(\zeta)), \tag{41}$$

where w is an arbitrary point on $\partial\Omega$ and where the second equality follows from (38), (39), (40). Cf. [9], p. 125f. See also [16] for some more details.

The potential V appears implicitly or explicitly in classical texts, e.g., [43, 52]. Here we discuss it briefly because it connects to the exponential transform. To start, it has a certain self-reproducing property, expressed in the identity

$$V(z, w; a, b) = \frac{1}{2\pi} \int_M dV(\cdot, c; z, w) \wedge^* dV(\cdot, c; a, b), \tag{42}$$

valid for arbitrary non-coinciding points $z, w, a, b, c \in M$. Alternatively,

$$V(z,w;a,b) = \frac{\mathrm{i}}{\pi} \int_M \partial V(\cdot,c;z,w) \wedge \bar{\partial} V(\cdot,c;a,b), \tag{43}$$

where $\partial V = \frac{\partial V}{\partial z} dz$, $\overline{\partial} V = \frac{\partial V}{\partial \bar{z}} d\bar{z}$. The proof of (42) is straightforward: the right member becomes, after partial integration, an application of (36) and by using the symmetries (38) and (39),

$$-\frac{1}{2\pi} \int_M V(\cdot,c;z,w) \wedge d * dV(\cdot,c;a,b)$$

$$= V(a,c;z,w) - V(b,c;z,w) = V(z,w;a,b),$$

which is the left member.

In the case of the Riemann sphere, V relates to the cross-ratio (7) by

$$V(z,w;a,b) = -\log|(z:w:a:b)| = -\log\left|\frac{(z-a)(w-b)}{(z-b)(w-a)}\right|. \tag{44}$$

Therefore the extended exponential transform for any domain $\Omega \subset \mathbb{P}$ can be conveniently expressed in terms of V for $\mathbb{P}$, for example

$$E_\Omega(z,w;a,b) = \exp\left[\frac{2}{\mathrm{i}\pi} \int_\Omega \partial V(\cdot,c;z,a) \wedge \overline{\partial} V(\cdot,c;w,b)\right],$$

and for the modulus,

$$|E_\Omega(z,w;a,b)| = \exp\left[-\frac{1}{\pi} \int_\Omega dV(\cdot,c;z,a) \wedge *dV(\cdot,c;w,b)\right].$$

In the case of a compact Riemann surface, $\int_M \partial V(\cdot,c;z,a) \wedge \bar{\partial} V(\cdot,c;w,b)$ is purely imaginary, by (43). In particular we get, for $\Omega = M = \mathbb{P}$, and by using (43) and (44), that

$$E_{\mathbb{P}}(z,w;a,b) = \exp\left[\frac{2}{\mathrm{i}\pi} \int_{\mathbb{P}} \partial V(\cdot,c;z,a) \wedge \overline{\partial} V(\cdot,c;w,b)\right]$$

$$= \exp\left[-2V(z,a;w,b)\right] = |(z:a:w:b)|^2.$$

Thus formula (6), which was asserted and proved in another way in Sect. 2, has been proved again.

7 Moment Coordinates

In this section we restrict again to the case that $\Omega \subset \mathbb{C}$ is a simply connected domain with analytic boundary, and with $0 \in \Omega$. In order to keep the presentation reasonably rigorous we shall, much of the time, actually restrict even more, namely by considering only those domains which are conformal images of the unit disk under polynomial conformal maps,

$$f(\zeta) = \sum_{k=0}^{N} a_k \zeta^{k+1}, \tag{45}$$

with $a_0 > 0$ (normalization) and N fixed. This will be a manifold, call it $\mathcal{M}_{2N+1}$, of real dimension $2N+1$, in fact it is an open subset of $\mathbb{R} \times \mathbb{C}^N \cong \mathbb{R}^{2N+1}$ with $a_0 \in \mathbb{R}$ and $a_j \in \mathbb{C}$ $(1 \leq j \leq N)$ as coordinates.

Conformal images of the unit disk under polynomial conformal maps are exactly the simply connected quadrature domains with the origin as the sole quadrature node, i.e., with $m = 1$, $a_1 = 0$ in the notation of (22), and the order of the quadrature domain then equals the degree of the polynomial ($N+1$ in the above notation). The class of domains we are considering here is slightly smaller than this class of quadrature domains because the assumption of analytic boundary requires our polynomials f to be univalent in a full neighborhood of the closed unit disk. But it has the advantage that $\mathcal{M}_{2N+1}$ will be an ordinary open manifold (without boundary). For algebraic purposes one could equally work well with the just locally univalent polynomials, i.e., those polynomials f in (45) for which $f' \neq 0$ on $\overline{\mathbb{D}}$.

Remark 7.1. The restriction to polynomial f does not change anything in principle, all formulas will look basically the same as in the general case. In the other extreme, without changing much in practice one may pass from the infinite dimensional case to the formal level (with no bothering about convergence) of transfinite functions, see [18].

As real coordinates on $\mathcal{M}_{2N+1}$ we may of course use the real and imaginary parts of the coefficients appearing in (45), namely a_0, $\operatorname{Re} a_j$, $\operatorname{Im} a_j$ $(1 \leq j \leq N)$, but much of what we are going to discuss concern what happens when we switch to M_0, $\operatorname{Re} M_j$, $\operatorname{Im} M_j$ $(1 \leq j \leq N)$ as local coordinates.

The map (or change of coordinates) $(a_0, a_1, a_2, \ldots) \mapsto (M_0, M_1, M_2, \ldots)$ is explicitly given by Richardson's formula [39],

$$M_k = \sum (j_0 + 1) a_{j_0} \cdots a_{j_k} \bar{a}_{j_0 + \ldots + j_k + k}. \tag{46}$$

The summation here goes over all multi-indices $(j_0, \ldots, j_k) \geq (0, \ldots, 0)$ for which $j_0 + \ldots + j_k + k \leq N$. It is convenient to set $a_j = 0$ for $j > N$, and then the upper bound is not needed. (And with $a_j = 0$ for $j < 0$ the lower bound is not needed either.) The formula (46) is proved by pulling (15) back to the unit circle by the map (45) and then using the residue theorem (after having replaced $\bar{f}$ by f^*).

It is clear from (46) that $M_{N+1} = M_{N+2} = \cdots = 0$ when $a_{N+1} = a_{N+2} = \cdots = 0$, and it follows from basic facts about quadrature domains that the converse statement holds as well (when f is univalent in a full neighborhood of the closed unit disk). It is virtually impossible to invert (46) in any explicit way. Even for $N = 1$ one would need to solve a third degree algebraic equation in order to express a_0, a_1 in terms of M_0, M_1. And the map $(a_0, a_1, a_2, \dots) \mapsto (M_0, M_1, M_2, \dots)$ is (generally speaking) not one-to-one, as the example [41] (and several similar examples, cf. [55]) shows. The author does not know, however, of any such example in the present setting of polynomial conformal maps. C. Ullemar [47] gives an example of two polynomials of degree three, one univalent and one only locally univalent, which give rise to the same moments.

Remark 7.2. Even when $M_{N+1} = M_{N+2} = \cdots = 0$, the negative moments M_{-1}, $M_{-2}, \dots$ generally make up a full infinite sequence (of nonzero numbers). This is related to the fact that when Ω is a quadrature domain, then its complement is almost never a quadrature domain (one can easily make sense to the notion of an unbounded quadrature domain, or a quadrature domain in $\mathbb{P}$).

A deformation of a domain Ω as above corresponds to a smooth curve $t \mapsto \Omega(t)$ with $\Omega(0) = \Omega$ (equivalently, $t \mapsto f(\zeta, t)$) in $\mathcal{M}_{2N+1}$, and its velocity at $t = 0$ is a vector in the real tangent space of $\mathcal{M}_{2N+1}$ at the point in question. This tangent space has, in terms of the coordinates introduced above, two natural bases, namely

$$\frac{\partial}{\partial a_0}, \quad \frac{\partial}{\partial \operatorname{Re} a_j}, \quad \frac{\partial}{\partial \operatorname{Im} a_j}$$

$$\frac{\partial}{\partial M_0}, \quad \frac{\partial}{\partial \operatorname{Re} M_j}, \quad \frac{\partial}{\partial \operatorname{Im} M_j}$$

$(1 \le j \le N)$, respectively.

It is natural to consider also the corresponding complexified tangent space, obtained by allowing complex coefficients in front of the above basis vectors. For this tangent space we also have the bases

$$\frac{\partial}{\partial a_0}, \quad \frac{\partial}{\partial a_j}, \quad \frac{\partial}{\partial \bar{a}_j}$$

$$\frac{\partial}{\partial M_0}, \quad \frac{\partial}{\partial M_j}, \quad \frac{\partial}{\partial \overline{M}_j}.$$

Similar considerations apply for the cotangent spaces, i.e., we have the two natural bases, da_0, da_j, $d\bar{a}_j$ and dM_0, dM_j, $d\bar{M}_j$, etc. It should be noted, however, that complex tangent vectors such as $\frac{\partial}{\partial M_j}$ ($j \geq 1$) do not really correspond to velocity vectors for deformations of domains, only vectors in the real tangent space do.

The dependence of f on the coefficients $a_0, \ldots, a_N$ is certainly analytic by (45), but as can be understood from the appearance of conjugations in (46), the dependence on $M_0, \ldots, M_N$ is no longer analytic. Therefore we prefer to write this dependence as

$$f(\zeta) = f(\zeta; \bar{M}_N, \ldots, \bar{M}_1, M_0, M_1, \ldots, M_N), \tag{47}$$

or just $f(\zeta; M)$, with M shorthand for the list of moments:

$$M = (\bar{M}_N, \ldots, \bar{M}_1, M_0, M_1, \ldots, M_N). \tag{48}$$

Despite the implicit nature of the dependence of f on the moments, several interesting general statements can be made, and we want to highlight a couple of them. The first is due to O. Kuznetsova and V. Tkachev [25], V. Tkachev [46], who were able to compute the Jacobian determinant for the transition between the above sets of coordinates, thereby confirming a conjecture of C. Ullemar [47]. The beautiful result can be expressed in terms of the ratio between the volume forms as follows:

$$d\bar{M}_N \wedge \ldots d\bar{M}_1 \wedge dM_0 \wedge dM_1 \wedge \cdots \wedge dM_N = \tag{49}$$

$$=2a_0^{N^2+3N+1}\mathcal{R}(f', f'^*) \cdot d\bar{a}_N \wedge \ldots d\bar{a}_1 \wedge da_0 \wedge da_1 \wedge \cdots \wedge da_N. \tag{50}$$

Here $\mathcal{R}(f', f'^*)$ is the meromorphic resultant between f' and f'^*, as discussed in Sect. 5. When f is univalent (or locally univalent), then $\mathcal{R}(f', f'^*) \neq 0$, hence the coordinate transition is locally one-to-one.

The second general statement is actually a series of results obtained by M. Mineev-Weinstein, P. Wiegmann, A. Zabrodin, with later contributations also from other authors, [24, 27, 29, 53]. We shall below, and in the next two sections, discuss some of these results, namely the string equation (52), the prepotential (18) and integrability of moments (16) (we will provide a proof of (17)) and, finally, Hamiltonian formulations of the evolution equations for the mapping functions (68).

The results are partly formulated in terms of a *Poisson bracket*, which for any two functions $f = f(\zeta; M)$, $g = g(\zeta; M)$, analytic for ζ in a neighborhood of the unit circle and also depending on the moments M, is defined by

$$\{f, g\} = \zeta\frac{\partial f}{\partial \zeta}\frac{\partial g}{\partial M_0} - \zeta\frac{\partial g}{\partial \zeta}\frac{\partial f}{\partial M_0}. \tag{51}$$

As a first choice here we take f to be the conformal map (45) and $g = f^*$, the reflection of f in the unit circle: $f^*(\zeta; M) = \overline{f(1/\bar{\zeta}; M)}$. Since we are assuming that Ω has analytic boundary, f and f^* are both analytic in some neighborhood of $\partial\mathbb{D}$, hence so is $\{f, f^*\}$. Now it turns out that it is much better than so, in fact $\{f, f^*\}$ is analytic in the whole Riemann sphere, hence is constant. This remarkable fact is sometimes referred to as the *string equation*, which more precisely reads

$$\{f, f^*\} = 1. \tag{52}$$

As for the proof (which follows [53], etc.) one first notices that $f^* = S \circ f$ (S is the Schwarz function), since this holds on $\partial\mathbb{D}$. Since also $S = S(z; M)$ depends on the moments this spells out to

$$f^*(\zeta; M) = S(f(\zeta; M); M).$$

Using hence the chain rule when computing $\frac{\partial f^*}{\partial M_0}$ one arrives at

$$\{f, f^*\} = \zeta \frac{\partial f}{\partial \zeta} \cdot \left(\frac{\partial S}{\partial M_0} \circ f\right).$$

Now we notice from (19) that

$$\frac{\partial S}{\partial M_0}(z; M) = \frac{1}{z} + \text{positive powers of } z,$$

since the coefficients $M_1, M_2, \ldots$ of the remaining negative powers of z in (19) are independent of M_0. From this one sees that $\{f, f^*\}$ is holomorphic in $\mathbb{D}$. Similarly, it is holomorphic in $\mathbb{P} \setminus \overline{\mathbb{D}}$, hence it must in fact be constant, and one easily checks that this constant is one, proving (52).

The string equation (52) may appear to be quite remarkable identity for conformal maps, but actually it can be identified with an equation which was discovered much earlier, namely the *Polubarinova–Galin equation* [10, 20, 31, 49, 50] for the evolution of a Hele–Shaw blob with a source at the origin (Laplacian growth), this taken in combination with the Richardson moment conservation law [39] for such a flow. These considerations will be elaborated in the next section.

8 Domain Deformations Driven by Harmonic Gradients

Continuing on the theme of the previous section, we now discuss general domain variations from the point of view of Laplacian growth. We first consider a, in principle arbitrary, deformation $t \mapsto \Omega(t)$ of a simply connected domain $\Omega = \Omega(0)$. Let V_n denote the speed of $\Gamma(t) = \partial\Omega(t)$ measured in the outward normal direction. We need to assume that all data are real analytic. This can, for example, be expressed by saying that the Schwarz function $S(z, t)$ of $\Gamma(t)$ is real analytic in the parameter

t (besides being analytic in z, in a neighborhood of $\Gamma(t)$). Then, by the Cauchy–Kovalevskaya theorem, there exists, for each t, a harmonic function $g(\cdot,t)$ defined in some neighborhood of $\Gamma(t)$ and satisfying

$$\begin{cases} g(\cdot,t) &= 0 \quad \text{on } \Gamma(t), \\ -\frac{\partial g(\cdot,t)}{\partial n} &= V_n \quad \text{on } \Gamma(t). \end{cases} \tag{53}$$

This function g can actually also be defined directly. In fact, it is easy to see that (53) can be expressed as

$$\frac{\partial S(z,t)}{\partial t} = -4\frac{\partial g(z,t)}{\partial z}$$

holding on $\Gamma(t)$, and hence identically (both members being analytic in z). To be precise, the latter equation actually contains (53) only with the first condition weakened to $g(\cdot,t)$ being constant on $\Gamma(t)$. From the above one gets the formula

$$g(z,t) = -\frac{1}{2}\operatorname{Re}\int_{z_0(t)}^{z}\frac{\partial S}{\partial t}(\zeta,t)d\zeta,$$

where $z_0(t)$ is an arbitrary point on $\Gamma(t)$. Certainly, there will be exist neighborhood of $\Gamma(0)$ in which all the functions $g(\cdot,t)$, for $|t|$ sufficiently small, are defined.

Near $\Gamma(0)$ we have the usual cartesian coordinates (x,y) $(z = x + iy)$, but from g we also get another pair, (t,θ), namely

$$\begin{cases} t = t(z) \quad \text{determined by } z \in \Gamma_{t(z)}, \\ \theta = -*g(z,t(z)). \end{cases}$$

Here $*g$ is the conjugate harmonic function of g. We shall also write $d\theta = -*dg$, so this star is actually the Hodge star. Thus $t = t(z)$ is the time when $\Gamma(t)$ arrives at the point z, and θ is a coordinate along each $\Gamma(t)$. In order for $t(z)$ to be well defined, at least locally, we need to assume that $V_n \neq 0$.

Fix a value of t and introduce auxiliary coordinates (n,s) in a neighborhood of $\Gamma(t)$ such that, on $\Gamma(t)$, $\frac{\partial}{\partial s}$ is the tangent vector and $\frac{\partial}{\partial n}$ is the outward normal vector (thus s is arc length along $\Gamma(t)$). Then we have (on $\Gamma(t)$)

$$*dg = -\frac{\partial g}{\partial s}dn + \frac{\partial g}{\partial n}ds = \frac{\partial g}{\partial n}ds,$$

and so

$$d\theta = -*dg(z,t) = V_n ds.$$

On the other hand (still on $\Gamma(t)$),

$$\frac{\partial t}{\partial s} = 0, \quad \frac{\partial t}{\partial n} = \frac{1}{V_n},$$

so that

$$dt = \frac{1}{V_n} dn.$$

It follows that $dt \wedge d\theta = dn \wedge ds$ on $\Gamma(t)$. But since $(\frac{\partial}{\partial n}, \frac{\partial}{\partial s})$ is just a rotation of $(\frac{\partial}{\partial x}, \frac{\partial}{\partial y})$, we also have $dn \wedge ds = dx \wedge dy$. This gives $dt \wedge d\theta = dx \wedge dy$ on $\Gamma(t)$, which does not involve the auxiliary coordinates (n, s). Hence the latter equation actually holds in a full neighborhood of $\Gamma(t)$. Equivalently,

$$\frac{\partial(x, y)}{\partial(t, \theta)} = 1. \tag{54}$$

We summarize.

Proposition 8.1. *Any real analytic deformation $t \mapsto \Gamma(t)$ can be described, locally in regions where $V_n \neq 0$ and in terms of local coordinates (t, θ) introduced as above, by the "string equation"*

$$dt \wedge d\theta = dx \wedge dy. \tag{55}$$

Compare discussions in, for example, [22]. The reason for calling (55) a string equation will become clear below.

Now we change the point of view. In the above considerations, the normal velocity V_n of the boundary was the given data (essentially arbitrary, except for regularity requirements) and the harmonic function g was introduced as a secondary object. *Laplacian growth*, on the other hand, starts with a rule for prescribing a harmonic function $g = g_\Omega$, associated with any given domain Ω, defined in the domain (or at near the boundary of it) and vanishing on the boundary. Then one asks the boundary to move with the normal velocity

$$V_n = -\frac{\partial g}{\partial n}.$$

Thus g now works as the generating function which forces the motion of $\partial\Omega$, and setting again $\theta = - * g$, (55) remains valid in regions where $V_n \neq 0$. Typically, g is specified by suitable source terms in Ω or by boundary conditions on some fixed boundaries.

The classical case is that $0 \in \Omega$ and that $g = g_\Omega$ is the Green's function of Ω with pole at $z = 0$:

$$\begin{cases} g(z, t) = -\log|z| + \text{harmonic} & (z \in \Omega), \\ g(z, t) = 0 & (z \in \partial\Omega). \end{cases} \tag{56}$$

When pulled back to the unit disk by $f : \mathbb{D} \to \Omega$ (normalized conformal map) one then gets $g(f(\zeta), t) = -\log|\zeta|$, and hence $\theta = -*g(f(\zeta), t) = \arg\zeta$. Thus θ will simply be the angular variable on the unit circle. In addition, one sees that $d\theta$ can be interpreted as harmonic measure (on $\partial\mathbb{D}$, and on $\partial\Omega$). This classical version of Laplacian growth is connected to the harmonic moments in that it is characterized by the moments M_k, $k \geq 1$, being constants of motions ("integrals"), while M_0 evolves linearly in time. Precisely, with g defined by (56) we will have (as functions of t)

$$\begin{cases} M_0 = 2t + \text{constant}, \\ M_k = \text{constant} \quad (k \geq 1). \end{cases} \tag{57}$$

These conservation laws were discovered by S. Richardson [39], and they are easy to verify directly. In fact, one shows that

$$\frac{d}{dt}\int_{\Omega(t)} h dm = 2\pi h(0)$$

for any function h which is harmonic in a region containing the closure of $\Omega(t)$, and then (57) is obtained by choosing $h(z) = z^k$ $(k \geq 0)$.

Now we can compare (52) with (54) or (55): since $\{f, f^*\}$ is an analytic function, the restriction of (52) to $\partial\mathbb{D}$ contains all information. And on $\partial\mathbb{D}$ we have, on setting $\zeta = e^{i\theta}$ and writing $f(e^{i\theta}, t) = x + iy$,

$$\{f, f^*\} = 2\,\mathrm{Im}\left[\frac{\partial f}{\partial\theta}\frac{\partial \bar{f}}{\partial M_0}\right] = 2\frac{\partial(x, y)}{\partial(M_0, \theta)}.$$

Since $2\frac{\partial}{\partial M_0} = \frac{\partial}{\partial t}$ by (57) it follows that (52) is the same as (54), or (55). This was the reason for calling (55) a string equation.

Remark 8.1. A common variant of the above is that Ω is a domain in the Riemann sphere, $\infty \in \Omega$, and that g is the Green's function with pole at infinity.

Remark 8.2. Most of the above considerations also apply in higher dimensions, the main difference being that, in n real dimensions, $d\theta = -*dg$ will be an $(n-1)$-form and that (55) will take the form $dt \wedge d\theta = dx_1 \wedge \cdots \wedge dx_n$ (the volume form). Another generalization, elliptic growth, is discussed in [22].

Now we shall connect Laplacian growth evolutions to the derivatives $\frac{\partial}{\partial M_j}$, $\frac{\partial}{\partial \bar{M}_j}$. These evolutions will not be monotonic, so we must allow now V_n to change sign, and hence to be zero at some points of the boundaries. Let L denote a real-valued differential operator with constant coefficients acting on functions in $\mathbb{C}$, for example $L = \frac{\partial^k}{\partial x^k}$. Then, given $\Omega(0)$ with analytic boundary, there exists locally an evolution $t \mapsto \Omega(t)$ such that

$$\frac{d}{dt}\int_{\Omega(t)} h\,dm = 2\pi(Lh)(0) \tag{58}$$

for every function h which is harmonic in a region containing the closure of $\Omega(t)$. This is a version of Laplacian growth, in fact it is the evolution driven by the harmonic function

$$g_L(z) = \{L_a g(z,a)\}_{a=0},$$

where $g(z,a) = -\log|z-a| + \text{harmonic}$ is the ordinary Green's function with pole at a, and the differential operator $L_a = L$ acts on that variable (the location of the pole). Thus, g_L is harmonic in Ω, has a multipole singularity at the origin, and vanishes on $\partial\Omega$.

It is actually not easy to prove that evolutions $\Omega(t)$ driven by a given harmonic gradient really exist, but it can be done under the right assumptions (analyticity of the initial boundary is usually necessary, for example). An example of such a proof is given in [38]. Other approaches are presented in [45] and [26]. In the algebraic framework of the previous section, i.e., with $\Omega \in \mathcal{M}_{2N+1}$, the full evolution will stay in $\mathcal{M}_{2N+1}$ provided the order k of the differential operator L is at most N, and the existence proof is much easier, see [11]. In fact, the existence and uniqueness can even be read off from (49).

For the rest of this section, and also for the next section, even if it does not change anything in practice it is useful to think of being in the finite dimensional setting of $\mathcal{M}_{2N+1}$, hence assuming that $k \le N$.

To prove the assertion that g_L really achieves (58), recall that the evolution $t \mapsto \Omega(t)$ is defined by the outward normal velocity of the boundary being $V_n = -\frac{\partial g_L(z)}{\partial n}$. Since $-\Delta g(\cdot,a) = 2\pi\delta_a$ we have

$$-\Delta g_L = -\{\Delta L_a g(\cdot,a)\}_{a=0} = -\{L_a \Delta g(\cdot,a)\}_{a=0} = 2\pi L\delta_0.$$

With h harmonic this gives

$$\frac{d}{dt}\int_{\Omega(t)} h\,dm = \int_{\partial\Omega(t)} V_n h\,ds = -\int_{\partial\Omega(t)} h\frac{\partial g_L}{\partial n}\,ds$$

$$= -\int_{\Omega(t)} h\Delta g_L\,dm = 2\pi Lh(0),$$

proving (58).

Useful particular choices of L above are, for $k \ge 1$,

$$L_1 = \frac{\partial^k}{\partial x^k}, \quad L_2 = \frac{\partial^k}{\partial x^{k-1}\partial y}.$$

We may allow h in (58) to be complex-valued, and choosing $h(z) = z^j$ and evaluating at $z = 0$ one gets

$$\{L_1(z^k)\}_{z=0} = k!, \quad \{L_2(z^k)\}_{z=0} = \mathrm{i} \cdot k!, \tag{59}$$

and $\{L_i(z^j)\}_{z=0} = 0$ $(i = 1, 2)$ in all remaining cases. Similarly, with $h(z) = \bar{z}^j$ we have

$$\{L_1(\bar{z}^k)\}_{z=0} = k!, \quad \{L_2(\bar{z}^k)\}_{z=0} = -\mathrm{i} \cdot k!, \tag{60}$$

and $\{L_i(\bar{z}^j)\}_{z=0} = 0$ in the remaining cases.

In what follows, we will denote the time coordinates for the two evolutions corresponding to L_1 and L_2 by different letters, namely t_1 and t_2, respectively. It is a consequence of (58) and (59), (60) that the evolution $t_1 \mapsto \Omega(t_1)$ generated by g_{L_1} has the property that

$$\frac{d}{dt_1} M_k = \frac{d}{dt_1} \bar{M}_k = 2k!, \quad \frac{d}{dt_1} M_j = \frac{d}{dt_1} \bar{M}_j = 0 \quad (j \neq k).$$

Similarly, for the evolution generated by g_{L_2}:

$$\frac{d}{dt_2} M_k = -\frac{d}{dt_2} \bar{M}_k = \mathrm{i} \cdot 2k!, \quad \frac{d}{dt_2} M_j = \frac{d}{dt_2} \bar{M}_j = 0 \quad (j \neq k),$$

In other words, the tangent vector giving the velocity for the evolution driven by g_{L_1} is

$$\frac{d}{dt_1} = 2k!(\frac{\partial}{\partial M_k} + \frac{\partial}{\partial \bar{M}_k}) = 2k! \frac{\partial}{\partial \operatorname{Re} M_k},$$

and that for g_{L_2} is

$$\frac{d}{dt_2} = 2\mathrm{i}k!(\frac{\partial}{\partial M_k} - \frac{\partial}{\partial \bar{M}_k}) = 2k! \frac{\partial}{\partial \operatorname{Im} M_k}.$$

Now we wish to extract $\frac{\partial}{\partial M_k}$ and $\frac{\partial}{\partial \bar{M}_k}$ from the above relations. We then see that these partial derivatives will correspond to linear combinations with complex (and non-real) coefficients of the two different evolutions, or tangent vectors, hence they will not in themselves correspond to evolutions of domains. Still they make sense, of course, as vectors in the complexified tangent space of $\mathcal{M}_{2N+1}$. Precisely, we get

$$\frac{\partial}{\partial M_k} = \frac{1}{4k!}(\frac{d}{dt_1} - \mathrm{i}\frac{d}{dt_2}),$$

$$\frac{\partial}{\partial \bar{M}_k} = \frac{1}{4k!}(\frac{d}{dt_1} + \mathrm{i}\frac{d}{dt_2}).$$

This is what we need in order to prove (17). Using (18) we get

$$\frac{\partial \mathcal{F}(\Omega)}{\partial M_k} = \frac{\partial}{\partial M_k}\frac{1}{\pi^2}\int_{\mathbb{D}(0,R)\setminus\Omega}\int_{\mathbb{D}(0,R)\setminus\Omega}\log\frac{1}{|z-\zeta|}dm(z)dm(\zeta)$$

$$= \frac{1}{4\pi^2 k!}(\frac{d}{dt_1} - \mathrm{i}\frac{d}{dt_2})\int_{\mathbb{D}(0,R)\setminus\Omega}\int_{\mathbb{D}(0,R)\setminus\Omega}\log\frac{1}{|z-\zeta|}dm(z)dm(\zeta).$$

When computing $\frac{d}{dt_1}$, for example, Ω is to be replaced with the evolution $\Omega(t_1)$, and we evaluate the derivative at $t_1 = 0$ ($\Omega(0) = \Omega$). There are two occurrences of Ω in the expression for $\mathcal{F}(\Omega)$, and $\mathcal{F}(\Omega)$ symmetric in them, so it is enough to differentiate one of the occurrences and then multiply the result by two. Writing $2\log|z-\zeta| = \log(\zeta - z) + \log(\bar{\zeta} - \bar{z})$, and using (58), (15) this gives

$$\frac{d}{dt_1}\Big|_{t_1=0}\int_{\mathbb{D}(0,R)\setminus\Omega(t_1)}\int_{\mathbb{D}(0,R)\setminus\Omega(t_1)}\log|z-\zeta|\,dm(z)dm(\zeta)$$

$$= \int_{\mathbb{D}(0,R)\setminus\Omega}\frac{d}{dt_1}\Big|_{t_1=0}\left(\int_{\mathbb{D}(0,R)} - \int_{\Omega(t_1)}\left(\log(\zeta - z) + \log(\bar{\zeta} - \bar{z})\right)dm(z)\right)dm(\zeta)$$

$$= 2\pi\int_{\mathbb{D}(0,R)\setminus\Omega}\left(\frac{(k-1)!}{\zeta^k} + \frac{(k-1)!}{\bar{\zeta}^k}\right)dm(\zeta)$$

$$= \mathrm{i}\pi(k-1)!\int_{\partial\Omega}\left(\frac{1}{\zeta^k} + \frac{1}{\bar{\zeta}^k}\right)\bar{\zeta}d\zeta = -2\pi^2 k!\left(\frac{1}{k}M_{-k} + \frac{1}{k}\bar{M}_{-k}\right).$$

Similarly, the corresponding derivative with respect to t_2 gives

$$\frac{d}{dt_2}\Big|_{t_2=0}\int_{\mathbb{D}(0,R)\setminus\Omega(t_1)}\int_{\mathbb{D}(0,R)\setminus\Omega(t_1)}\log|z-\zeta|\,dm(z)dm(\zeta)$$

$$= -2\mathrm{i}\pi^2 k!(\frac{1}{k}M_{-k} - \frac{1}{k}\bar{M}_{-k}).$$

Hence, in the combination $\frac{d}{dt_1} - \mathrm{i}\frac{d}{dt_2}$, $\bar{M}_{-k}$ will cancel out, and we get

$$\frac{\partial \mathcal{F}(\Omega)}{\partial M_k} = \frac{1}{k}M_{-k},$$

as desired.

9 Hamiltonians

As a final topic, we wish to explain, in our notations and settings, the Hamiltonian descriptions of the evolution of the mapping function f presented in [53] and related articles. Recall that we work within $\mathcal{M}_{2N+1}$, and we use M as a shorthand notation for the moments, which serve as coordinates on $\mathcal{M}_{2N+1}$, see (48).

Let $W = W(z; M)$ be a primitive function of the Schwarz function $S = S(z; M)$. By (19), the power series expansion of W will be

$$W(z; M) = -\sum_{k \in \mathbb{Z} \setminus \{0\}} \frac{M_k}{k z^k} + M_0 \log z + C(M), \tag{61}$$

where $C(M)$ is a constant. Decomposing W into real and imaginary parts, $W = w + \mathrm{i} * w$, the real part can be made perfectly well defined. In fact, $S(z) = 2\frac{\partial w}{\partial z}$, and since $S(z) = \bar{z}$ on $\partial\Omega$ we see that the real-valued function

$$u(z) = \frac{1}{4}|z|^2 - \frac{1}{2} w(z)$$

satisfies $\frac{\partial u}{\partial z} = 0$ on $\partial\Omega$, hence u is constant on $\partial\Omega$. We shall fix the free additive constant in u, and hence that in w, by requiring that $u = 0$ on $\partial\Omega$. It follows that u then satisfies

$$\begin{cases} \Delta u = 1 & \text{in a neighborhood of } \partial\Omega, \\ u = \frac{\partial u}{\partial z} = 0 & \text{on } \partial\Omega. \end{cases}$$

Next we note from (61) that

$$\frac{\partial W}{\partial M_0} = \log z + \sum_{k>0} \frac{1}{k} \frac{\partial M_{-k}}{\partial M_0} z^k + \frac{\partial C}{\partial M_0}.$$

Thus, as for the real part, $\frac{\partial w}{\partial M_0} = \log|z| + \text{harmonic in } \Omega$. On $\partial\Omega$ we have $\frac{\partial w}{\partial M_0} = -2\frac{\partial u}{\partial M_0} = 0$, because u vanishes to the second order on $\partial\Omega$. Hence it follows that $\frac{\partial w}{\partial M_0}$ is minus the Green's function $g = g_\Omega$ of Ω with pole at the origin:

$$\frac{\partial w}{\partial M_0} = -g. \tag{62}$$

Let $G = g + \mathrm{i} * g$ be the analytic completion of g. This Green's function $G = G(z; M)$ is (like g) a function of $z \in \Omega$ and the moments. By (62),

$$G(z; M) = -\frac{\partial W(z; M)}{\partial M_0}. \tag{63}$$

Recall that $f = f(\zeta; M)$ denotes the conformal map (45) from $\mathbb{D}$ to Ω. Define the (complex-valued) *Hamiltonian function* $H_0 = H_0(\zeta; M)$ associated with M_0 by

$$H_0(\zeta; M) = G(f(\zeta; M); M) = -\log \zeta, \tag{64}$$

the last equality because $G \circ f$ is the Green's function of $\mathbb{D}$. We see that H_0 actually is independent of the moments. Therefore, as $\frac{\partial H_0}{\partial M_0} = 0$, $\zeta \frac{\partial H_0}{\partial \zeta} = -1$ by (64), we trivially arrive at the evolution equation

$$\frac{\partial \varphi}{\partial M_0} = \{\varphi, H_0\}, \tag{65}$$

valid for any smooth function $\varphi = \varphi(\zeta; M)$.

For the higher order moments, there are corresponding identities, and they are more interesting and more selective: they hold only for $\varphi = f$. We first define, for $k \geq 1$,

$$H_k(\zeta; M) = -\frac{\partial W(z; M)}{\partial M_k}, \quad \text{where } z = f(\zeta; M). \tag{66}$$

The above means that the derivative $\frac{\partial}{\partial M_k}$ only acts on the M_k which appears in $W(z; M)$, not that in $f(\zeta; M)$. Always when we write z in place of $f(\zeta; M)$ it shall have this implication. Now, $G(f(\zeta; M); M) = -\log \zeta$ gives again, by differentiating with respect to ζ and M_k,

$$\frac{\partial G}{\partial z}(z; M)\frac{\partial f}{\partial \zeta}(\zeta; M) = -\frac{1}{\zeta},$$

$$\frac{\partial G}{\partial z}(z; M)\frac{\partial f}{\partial M_k}(\zeta; M) + \frac{\partial G}{\partial M_k}(z; M) = 0.$$

Thus, multiplying the latter equation with $\zeta \frac{\partial f}{\partial \zeta}$ and using the first equation gives

$$\frac{\partial f}{\partial M_k}(\zeta; M) = \zeta \frac{\partial f}{\partial \zeta}(\zeta; M)\frac{\partial G}{\partial M_k}(z; M). \tag{67}$$

Here we wish to remove G from the right member, in favor of H_k. From (66) we get

$$\zeta \frac{\partial H_k}{\partial \zeta}(\zeta; M) = -\frac{\partial^2 W}{\partial z \partial M_k}(z; M) \cdot \zeta \frac{\partial f}{\partial \zeta}(\zeta; M)$$

and, using also (63),

$$\zeta\frac{\partial f}{\partial\zeta}(\zeta;M)\frac{\partial H_k}{\partial M_0}(\zeta;M)$$

$$= -\zeta\frac{\partial f}{\partial\zeta}(\zeta;M)\frac{\partial^2 W}{\partial z\partial M_k}(z;M)\frac{\partial f}{\partial M_0}(\zeta;M) - \zeta\frac{\partial f}{\partial\zeta}(\zeta;M)\frac{\partial^2 W}{\partial M_0\partial M_k}(z;M)$$

$$= \zeta\frac{\partial H_k}{\partial\zeta}(\zeta;M)\frac{\partial f}{\partial M_0}(\zeta;M) + \zeta\frac{\partial f}{\partial\zeta}(\zeta;M)\frac{\partial G}{\partial M_k}(z;M).$$

The last term coincides with the right member in (67). Thus substituting into (67) we get

$$\frac{\partial f}{\partial M_k} = \{f, H_k\}, \tag{68}$$

which is the evolution equation for f we wanted to arrive at.

As for the classical Laplacian growth evolution, even though the situation is essentially trivial in the moment coordinates (in view of the explicit solution (57)), it may be nice to put it all in a traditional Hamiltonian framework (with real-valued Hamiltonian function). One possibility then is to choose the phase space to be $\partial\mathbb{D}\times\mathcal{M}_{2N+1}\subset\partial\mathbb{D}\times\mathbb{R}^{2N+1}$, with real coordinates $(\theta, M_0, \mathrm{Re}\, M_1, \mathrm{Im}\, M_1 \ldots, \mathrm{Re}\, M_N, \mathrm{Im}\, M_N)$, symplectic form

$$\begin{aligned}\omega =& \frac{1}{2}d\theta\wedge dM_0 + d\,\mathrm{Re}\, M_1\wedge d\,\mathrm{Im}\, M_1 + \cdots + d\,\mathrm{Re}\, M_N\wedge d\,\mathrm{Im}\, M_N\\ =& \frac{1}{2\mathrm{i}}\left[\mathrm{i}d\theta\wedge dM_0 + d\bar{M}_1\wedge dM_1 + \cdots + d\bar{M}_N\wedge dM_N\right]\end{aligned}$$

and Hamiltonian function

$$H(\theta, M_0, \mathrm{Re}\, M_1, \ldots, \mathrm{Im}\, M_N) = \theta.$$

The Hamilton equations are, generally speaking (see [4]),

$$-dH = \omega(\xi,\cdot), \tag{69}$$

where $\xi = \frac{d}{dt}$ is the velocity vector for the evolution, a vector in the tangent space of the phase space. In our case we have (using dot for time derivative)

$$\xi = \dot{\theta}\frac{\partial}{\partial\theta} + \dot{M}_0\frac{\partial}{\partial M_0} + \mathrm{Re}\,\dot{M}_1\frac{\partial}{\partial\,\mathrm{Re}\, M_1} + \cdots + \mathrm{Im}\,\dot{M}_N\frac{\partial}{\partial\,\mathrm{Im}\, M_N},$$

giving in (69)

$$\dot{\theta} = 0, \quad \dot{M}_0 = 2,$$

$$\operatorname{Re} \dot{M}_j = \operatorname{Im} \dot{M}_j = 0 \quad (1 \le j \le N),$$

as expected (cf. (57)).

Note that the first term in ω, with $dM_0 = 2dt$, can be identified with $-dx \wedge dy$, in the notation of Proposition 8.1.

Acknowledgments This work has been performed within the framework of the European Science Foundation Research Networking Programme HCAA and has been supported by the Swedish Research Council and the Göran Gustafsson Foundation.

References

1. D. Aharonov, H.S. Shapiro, Domains on which analytic functions satisfy quadrature identities. J. Analyse Math. **30**, 39–73 (1976)
2. L.V. Ahlfors, *Complex analysis: An introduction of the theory of analytic functions of one complex variable*, 2nd edn (McGraw-Hill Book, New York, 1966)
3. H. Alexander, J. Wermer, *Several complex variables and Banach algebras*, vol. 35 of Graduate Texts in Mathematics, 3rd edn. (Springer, New York, 1998)
4. V.I. Arnold, *Mathematical Methods of Classical Mechanics* (Springer, New York, 1978). Translated from the Russian by K. Vogtmann and A. Weinstein, Graduate Texts in Mathematics, 60
5. R.W. Carey, J.D. Pincus, An exponential formula for determining functions. Indiana Univ. Math. J. **23**, 1031–1042 (1973/74)
6. V.G. Cherednichenko, *Inverse Logarithmic Potential Problem*, Inverse and Ill-posed Problems Series (VSP, Utrecht, 1996)
7. P.J. Davis, *The Schwarz Function and Its Applications* (The Mathematical Association of America, Buffalo, NY, 1974). The Carus Mathematical Monographs, No. 17
8. H.M. Farkas, I. Kra, *Riemann Surfaces*, vol. 71 of Graduate Texts in Mathematics, 2nd edn. (Springer, New York, 1992)
9. J.D. Fay, *Theta Functions on Riemann Surfaces*, Lecture Notes in Mathematics, vol. 352 (Springer, Berlin, 1973)
10. L.A. Galin, Unsteady filtration with a free surface. C. R. (Doklady) Acad. Sci. URSS (N.S.) **47**, 246–249 (1945)
11. B. Gustafsson, On a differential equation arising in a Hele-Shaw flow moving boundary problem. Ark. Mat. **22**, 251–268 (1984)
12. B. Gustafsson, On quadrature domains and an inverse problem in potential theory. J. Analyse Math. **55**, 172–216 (1990)
13. B. Gustafsson, M. Putinar, An exponential transform and regularity of free boundaries in two dimensions. Ann. Scuola Norm. Sup. Pisa Cl. Sci. (4) **26**, 507–543 (1998)
14. B. Gustafsson, M. Putinar, Linear analysis of quadrature domains. II. Isr. J. Math. **119**, 187–216 (2000)
15. B. Gustafsson, M. Putinar, *Linear Analysis of Quadrature Domains. IV*, in Quadrature domains and their applications, vol. 156 of Oper. Theory Adv. Appl., (Birkhäuser, Basel, 2005), pp. 173–194
16. B. Gustafsson, A. Sebbar, Critical points of Green's function and geometric function theory. Indiana Univ. J. **61**, 939–1017 (2012)

17. B. Gustafsson, H.S. Shapiro, *What Is a Quadrature Domain?* in Quadrature domains and their applications, vol. 156 of Oper. Theory Adv. Appl. (Birkhäuser, Basel, 2005), pp. 1–25
18. B. Gustafsson, V. Tkachev, On the Jacobian of the harmonic moment map. Complex Anal. Oper. Theory **3**, 399–417 (2009)
19. B. Gustafsson, V.G. Tkachev, The resultant on compact Riemann surfaces. Comm. Math. Phys. **286**, 313–358 (2009)
20. B. Gustafsson, A. Vasil′ev, *Conformal and Potential Analysis in Hele-Shaw Cells*, Advances in Mathematical Fluid Mechanics (Birkhäuser Verlag, Basel, 2006)
21. F.R. Harvey, H.B. Lawson, Jr., On boundaries of complex analytic varieties. I. Ann. Math. (2) **102**, 223–290 (1975)
22. D. Khavinson, M. Mineev-Weinstein, M. Putinar, Planar elliptic growth. Complex Anal. Oper. Theory **3**, 425–451 (2009)
23. I.K. Kostov, I. Krichever, M. Mineev-Weinstein, P.B. Wiegmann, A. Zabrodin, *The τ-function for analytic curves*, in Random matrix models and their applications, vol. 40 of Math. Sci. Res. Inst. Publ. (Cambridge University Press, Cambridge, 2001), pp. 285–299
24. I. Krichever, A. Marshakov, A. Zabrodin, Integrable structure of the Dirichlet boundary problem in multiply-connected domains. Comm. Math. Phys. **259**, 1–44 (2005)
25. O.S. Kuznetsova, V.G. Tkachev, Ullemar's formula for the Jacobian of the complex moment mapping. Complex Var. Theory Appl. **49**, 55–72 (2004)
26. Y.-L. Lin, Perturbation theorems for Hele-Shaw flows and their applications. Ark. Mat. **49**, 357–382 (2011)
27. A. Marshakov, P. Wiegmann, A. Zabrodin, Integrable structure of the Dirichlet boundary problem in two dimensions. Comm. Math. Phys. **227**, 131–153 (2002)
28. M. Martin, M. Putinar, *Lectures on Hyponormal Operators*, vol. 39 of Operator Theory: Advances and Applications (Birkhäuser Verlag, Basel, 1989)
29. M. Mineev-Weinstein, A. Zabrodin, Whitham-Toda hierarchy in the Laplacian growth problem. J. Nonlinear Math. Phys. **8**, 212–218 (2001). Nonlinear evolution equations and dynamical systems (Kolimbary, 1999)
30. J.D. Pincus, D. Xia, J.B. Xia, The analytic model of a hyponormal operator with rank one self-commutator. Integr. Equat. Oper. Theory **7**, 516–535 (1984)
31. P.Y. Polubarinova-Kochina, On a problem of the motion of the contour of a petroleum shell. Dokl. Akad. Nauk USSR **47**, 254–257 (1945)
32. M. Putinar, A two-dimensional moment problem. J. Funct. Anal. **80**, 1–8 (1988)
33. M. Putinar, The L problem of moments in two dimensions. J. Funct. Anal. **94**, 288–307 (1990)
34. M. Putinar, On a class of finitely determined planar domains. Math. Res. Lett. **1**, 389–398 (1994)
35. M. Putinar, Linear analysis of quadrature domains. Ark. Mat. **33**, 357–376 (1995)
36. M. Putinar, Extremal solutions of the two-dimensional L-problem of moments. J. Funct. Anal. **136**, 331–364 (1996)
37. M. Putinar, Extremal solutions of the two-dimensional L-problem of moments. II. J. Approx. Theory **92**, 38–58 (1998)
38. M. Reissig, L. von Wolfersdorf, A simplified proof for a moving boundary problem for Hele-Shaw flows in the plane. Ark. Mat. **31**, 101–116 (1993)
39. S. Richardson, Hele-Shaw flows with a free boundary produced by the injection of fluid into a narrow channel. J. Fluid Mech. **56**, 609–618 (1972)
40. J. Ross, D. Witt Nyström, The Hele-Shaw flow and moduli of holomorphic discs. arXiv:1212.2337 [math.CV] (2012)
41. M. Sakai, A moment problem on Jordan domains. Proc. Am. Math. Soc. **70**, 35–38 (1978)
42. M. Sakai, *Quadrature Domains*, vol. 934 of Lecture Notes in Mathematics (Springer, Berlin, 1982)
43. M. Schiffer, D.C. Spencer, *Functionals of Finite Riemann Surfaces* (Princeton University Press, Princeton, NJ, 1954)

44. H.S. Shapiro, *The Schwarz Function and Its Generalization to Higher Dimensions*, University of Arkansas Lecture Notes in the Mathematical Sciences, vol. 9 (Wiley, New York, 1992). A Wiley-Interscience Publication.
45. F.R. Tian, A Cauchy integral approach to Hele-Shaw problems with a free boundary: the case of zero surface tension. Arch. Rational Mech. Anal. **135**, 175–196 (1996)
46. V.G. Tkachev, Ullemar's formula for the moment map. II. Linear Algebra Appl. **404**, 380–388 (2005)
47. C. Ullemar, *Uniqueness theorem for domains satisfying a quadrature identity for analytic functions*, Research Bulletin TRITA-MAT-1980-37, Royal Institute of Technology, Department of Mathematics, Stockholm, 1980
48. A.N. Varchenko, P.I. Etingof, *Why the Boundary of a Round Drop Becomes a Curve of Order Four*, American Mathematical Society AMS University Lecture Series, 3rd edn. (Providence, Rhode Island, 1992)
49. A. Vasil′ev, From the Hele-Shaw experiment to integrable systems: a historical overview. Complex Anal. Oper. Theory **3**, 551–585 (2009)
50. Y.P. Vinogradov, P.P. Kufarev, On a problem of filtration. Akad. Nauk SSSR. Prikl. Mat. Meh. **12**, 181–198 (1948)
51. A. Weil, *Scientific Works. Collected papers. Vol. I (1926–1951)* (Springer, New York, 1979)
52. H. Weyl, *Die Idee der Riemannschen Fläche*, Vierte Auflage. Unveränderter Nachdruck der dritten, Vollständig umgearbeiteten Auflage, B. G. Teubner Verlagsgesellschaft, Stuttgart, 1964
53. P.B. Wiegmann, A. Zabrodin, Conformal maps and integrable hierarchies. Comm. Math. Phys. **213**, 523–538 (2000)
54. D.V. Yakubovich, Real separated algebraic curves, quadrature domains, Ahlfors type functions and operator theory. J. Funct. Anal. **236**, 25–58 (2006)
55. L. Zalcman, *Some inverse problems of potential theory*, in Integral geometry (Brunswick, Maine, 1984), vol. 63 of Contemp. Math., (Amer. Math. Soc., Providence, RI, 1987), pp. 337–350.

From the Virasoro Algebra to Krichever–Novikov Type Algebras and Beyond

Martin Schlichenmaier

Abstract Starting from the Virasoro algebra and its relatives the generalization to higher genus compact Riemann surfaces was initiated by Krichever and Novikov. The elements of these algebras are meromorphic objects which are holomorphic outside a finite set of points. A crucial and non-trivial point is to establish an almost-grading replacing the honest grading in the Virasoro case. Such an almost-grading is given by splitting the set of points of possible poles into two non-empty disjoint subsets. Krichever and Novikov considered the two-point case. Schlichenmaier studied the most general multi-point situation with arbitrary splittings. Here we will review the path of developments from the Virasoro algebra to its higher genus and multi-point analogs. The starting point will be a Poisson algebra structure on the space of meromorphic forms of all weights. As sub-structures the vector field algebras, function algebras, Lie superalgebras and the related current algebras show up. All these algebras will be almost-graded. In detail almost-graded central extensions are classified. In particular, for the vector field algebra it is essentially unique. The defining cocycle is given in geometric terms. Some applications, including the semi-infinite wedge form representations are recalled. We close by giving some remarks on the Lax operator algebras introduced recently by Krichever and Sheinman.

1 Introduction

Lie groups and Lie algebras are related to symmetries of systems. By the use of the symmetry the system can be better understood, maybe it is even possible to solve it in a certain sense. Here we deal with systems which have an infinite number

M. Schlichenmaier (✉)
Mathematics Research Unit, University of Luxembourg, FSTC, Campus Kirchberg, 6, rue Coudenhove-Kalergi, 1359 Luxembourg-Kirchberg, Luxembourg
e-mail: martin.schlichenmaier@uni.lu

A. Vasil'ev (ed.), *Harmonic and Complex Analysis and its Applications*,
Trends in Mathematics, DOI 10.1007/978-3-319-01806-5_7,

of independent degrees of freedom. They appear, for example, in conformal field theory (CFT), see, e.g., [2,66]. But also in the theory of partial differential equations and at many other places in- and outside of mathematics they play an important role. The appearing Lie groups and Lie algebras are infinite dimensional. Some of the simplest non-trivial infinite dimensional Lie algebras are the Witt algebra and its central extension the Virasoro algebra. We will recall their definitions in Sect. 2. In the sense explained (in particular, in CFT) they are related to what is called the genus zero situation. For CFT on arbitrary genus Riemann surfaces the Krichever–Novikov (KN) type algebras, to be discussed here, will show up as algebras of global symmetry operators.

These algebras are defined via meromorphic objects on compact Riemann surfaces Σ of arbitrary genus with controlled polar behaviour. More precisely, poles are only allowed at a fixed finite set of points denoted by A. The "classical" examples are the algebras defined by objects on the Riemann sphere (genus zero) with possible poles only at $\{0, \infty\}$. This yields, e.g., the Witt algebra, the classical current algebras, including their central extensions the Virasoro, and the affine Kac-Moody algebras [21]. For higher genus, but still only for two points where poles are allowed, they were generalized by Krichever and Novikov [26–28] in 1987. In 1990 the author [36–39] extended the approach further to the general multi-point case.

This extension was not a straightforward generalization. The crucial point is to introduce a replacement of the graded algebra structure present in the "classical" case. Krichever and Novikov found that an almost-grading, see Definition 4.1 below, will be enough to do the usual constructions in representation theory, like triangular decompositions, highest weight modules, Verma modules which are demanded by the applications. In [36, 39] it was realized that a splitting of A into two disjoint non-empty subsets $A = I \cup O$ is crucial for introducing an almost-grading and the corresponding almost-grading was given. In the two-point situation there is only one such splitting (up to inversion), hence there is only one almost-grading, which in the classical case is a honest grading. Similar to the classical situation a Krichever–Novikov algebra should always be considered as an algebra of meromorphic objects with an almost-grading coming from such a fixed splitting.

I would like to point out that already in the genus zero case (i.e. the Riemann sphere case) with more than two points where poles are allowed the algebras will only be almost-graded. In fact, quite a number of interesting new phenomena will show up already there, see [8, 15, 16, 40].

In this review no proofs are supplied. For them I have to refer to the original articles and/or to the forthcoming book [52]. For some applications jointly obtained with Oleg Sheinman, see also [65]. For more on the Witt and Virasoro algebra, see, for example, the book [18].

After recalling the definition of the Witt and Virasoro algebra in Sect. 2 we start with describing the geometric set-up of Krichever–Novikov (KN) type algebras in Sect. 3. We introduce a Poisson algebra structure on the space of meromorphic forms (holomorphic outside of the fixed set A of points where poles are allowed) of all weights (integer and half-integer). Special substructures will yield the function algebra, the vector field algebra and more generally the differential operator algebra.

Moreover, we discuss also the Lie superalgebras of KN type defined via forms of weight $-1/2$. An important example role also is played by the current algebra (arbitrary genus—multi-point) associated with a finite-dimensional Lie algebra.

In Sect. 4 we introduce the almost-grading induced by the splitting of A into "incoming" and "outgoing" points, $A = I \cup O$.

In Sect. 5 we discuss central extensions for our algebras. Central extensions appear naturally in the context of quantization and regularization of actions. We give for all our algebras geometrically defined central extensions. The defining cocycle for the Virasoro algebra obviously does not make any sense in the higher genus and/or multi-point case. For the geometric description we use projective and affine connections. In contrast to the classical case there are many inequivalent cocycles and central extensions. If we restrict our attention to the cases where we can extend the almost-grading to the central extensions, the author obtained complete classification and uniqueness results. They are described in Sect. 5.3.

In Sect. 6 we present further results. In particular, we discuss how from the representation of the vector field algebra (or more general of the differential operator algebra) on the forms of weight λ one obtains semi-infinite wedge representations (fermionic Fock space representations) of the centrally extended algebras. These representations have ground states (vacua), creation and annihilation operators. We add some words about $b - c$ systems, Sugawara construction, Wess–Zumino–Novikov–Witten (WZNW) models, Knizhnik–Zamolodchikov (KZ) connections, and deformations of the Virasoro algebra.

Recently, a new class of current type algebras the Lax operator algebras were introduced by Krichever and Sheinman [25, 29]. I will report on them in Sect. 7.

In the closing Sect. 8 some historical remarks (also on related works) on Krichever–Novikov type algebras and some references are given. More references can be found in [52].

2 The Witt and Virasoro Algebra

2.1 The Witt Algebra

The *Witt algebra* $\mathcal{W}$, also sometimes called Virasoro algebra without central term,[1] is the complex Lie algebra generated as vector space by the elements $\{e_n \mid n \in \mathbb{Z}\}$ with Lie structure

$$[e_n, e_m] = (m - n)e_{n+m}, \quad n, m \in \mathbb{Z} . \tag{1}$$

[1] In the book [18] arguments are given why it is more appropriate just to use Virasoro algebra, as Witt introduced "his" algebra in a characteristic p context. Nevertheless, I decided to stick here to the most common convention.

It can be realized as complexification of the Lie algebra of polynomial vector fields $\mathrm{Vect}_{pol}(S^1)$ on the circle S^1. The latter is a subalgebra of $\mathrm{Vect}(S^1)$, the Lie algebra of all C^∞ vector fields on the circle. In this realization

$$e_n := -\mathrm{i}\,\exp \mathrm{i}\,n\varphi\,\frac{d}{d\varphi}, \qquad n \in \mathbb{Z}\,. \tag{2}$$

The Lie product is the usual Lie bracket of vector fields.

If we extend these generators to the whole punctured complex plane, we obtain

$$e_n = z^{n+1}\frac{d}{dz}, \qquad n \in \mathbb{Z}\,. \tag{3}$$

This gives another realization of the Witt algebra as the algebra of those meromorphic vector fields on the Riemann sphere $\mathbb{P}^1(\mathbb{C})$ which are holomorphic outside $\{0\}$ and $\{\infty\}$.

Let z be the (quasi) global coordinate z (quasi, because it is not defined at ∞). Let $w = 1/z$ be the local coordinate at ∞. A global meromorphic vector field v on $\mathbb{P}^1(\mathbb{C})$ will be given on the corresponding subsets, where z, resp. w are defined as

$$v = \left(v_1(z)\frac{d}{dz}\,,\ v_2(w)\frac{d}{dw}\right), \qquad v_2(w) = -v_1(z(w))w^2. \tag{4}$$

The function v_1 will determine the vector field v. Hence, we will usually just write v_1 and in fact identify the vector field v with its local representing function v_1, which we will denote by the same letter.

For the bracket we calculate

$$[v,u] = \left(v\frac{d}{dz}u - u\frac{d}{dz}v\right)\frac{d}{dz}. \tag{5}$$

The space of all meromorphic vector fields constitutes a Lie algebra. The subspace of those meromorphic vector fields which are holomorphic outside of $\{0,\infty\}$ is a Lie subalgebra. Its elements can be given as

$$v(z) = f(z)\frac{d}{dz} \tag{6}$$

where f is a meromorphic function on $\mathbb{P}^1(\mathbb{C})$, which is holomorphic outside $\{0,\infty\}$. Those are exactly the Laurent polynomials $\mathbb{C}[z,z^{-1}]$. Consequently, this subalgebra has the set $\{e_n, n \in \mathbb{Z}\}$ as basis elements. The Lie product is the same and it can be identified with the Witt algebra $\mathcal{W}$.

The subalgebra of global holomorphic vector fields is $\langle e_{-1}, e_0, e_1\rangle_{\mathbb{C}}$. It is isomorphic to the Lie algebra $\mathfrak{sl}(2,\mathbb{C})$.

The algebra $\mathcal{W}$ is more than just a Lie algebra. It is a graded Lie algebra. If we set for the degree $\deg(e_n) := n$ then $\deg([e_n,e_m]) = \deg(e_n) + \deg(e_m)$ and we obtain the degree decomposition

$$\mathcal{W} = \bigoplus_{n\in\mathbb{Z}} \mathcal{W}_n, \qquad \mathcal{W}_n = \langle e_n \rangle_{\mathbb{C}} \,. \tag{7}$$

Note that $[e_0, e_n] = n\, e_n$, which says that the degree decomposition is the eigen-space decomposition with respect to the adjoint action of e_0 on $\mathcal{W}$.

Algebraically $\mathcal{W}$ can also be given as Lie algebra of derivations of the algebra of Laurent polynomials $\mathbb{C}[z, z^{-1}]$.

2.2 *The Virasoro Algebra*

In the process of quantizing or regularization one is often forced to modify an action of a Lie algebra. A typical example is given by the product of infinite sums of operators. Quite often they are only well defined if a certain "normal ordering" is introduced. In this way the modified action will only be a projective action. This can be made to an honest Lie action by passing to a suitable central extension of the Lie algebra.

For the Witt algebra the universal one-dimensional central extension is the *Virasoro algebra* $\mathcal{V}$. As vector space it is the direct sum $\mathcal{V} = \mathbb{C} \oplus \mathcal{W}$. If we set for $x \in \mathcal{W}$, $\hat{x} := (0, x)$, and $t := (1, 0)$, then its basis elements are $\hat{e}_n$, $n \in \mathbb{Z}$ and t with the Lie product

$$[\hat{e}_n, \hat{e}_m] = (m-n)\hat{e}_{n+m} - \frac{1}{12}(n^3 - n)\delta_n^{-m}\, t, \qquad [\hat{e}_n, t] = [t, t] = 0\,, \tag{8}$$

for[2] all $n, m \in \mathbb{Z}$. If we set $\deg(\hat{e}_n) := \deg(e_n) = n$ and $\deg(t) := 0$, then $\mathcal{V}$ becomes a graded algebra. The algebra $\mathcal{W}$ will only be a subspace, not a subalgebra of $\mathcal{V}$. It will be a quotient. In some abuse of notation we identify the element $\hat{x} \in \mathcal{V}$ with $x \in \mathcal{W}$. Up to equivalence and rescaling the central element t, this is besides the trivial (splitting) central extension the only central extension.

3 The Krichever–Novikov Type Algebras

3.1 *The Geometric Set-Up*

For the whole article let Σ be a compact Riemann surface without any restriction for the genus $g = g(\Sigma)$. Furthermore, let A be a finite subset of Σ. Later we will

[2] Here δ_k^l is the Kronecker delta which is equal to 1 if $k = l$, otherwise zero.

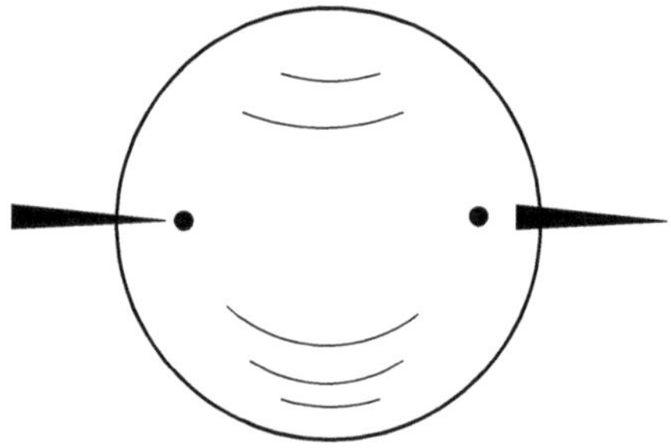

Fig. 1 Riemann surface of genus zero with one incoming and one outgoing point

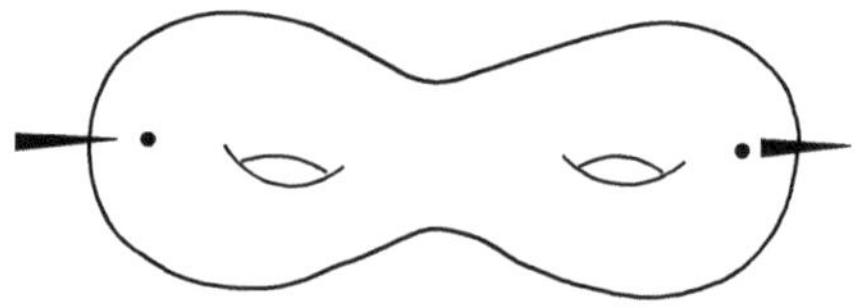

Fig. 2 Riemann surface of genus two with one incoming and one outgoing point

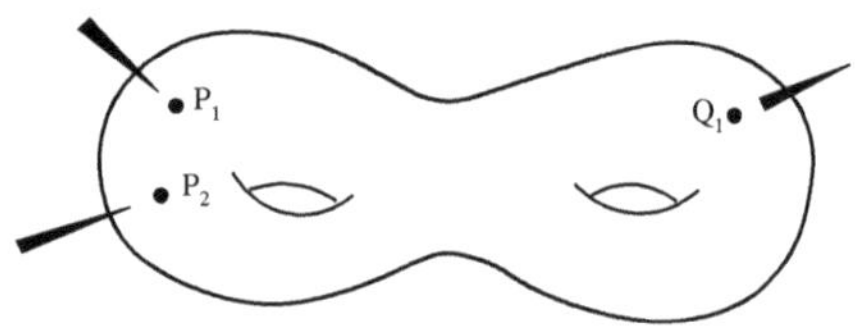

Fig. 3 Riemann surface of genus two with two incoming points and one outgoing point

need a splitting of A into two non-empty disjoint subsets I and O, i.e. $A = I \cup O$. Set $N := \#A$, $K := \#I$, $M := \#O$, with $N = K + M$. More precisely, let

$$I = (P_1, \ldots, P_K), \quad \text{and} \quad O = (Q_1, \ldots, Q_M) \tag{9}$$

be disjoint ordered tuples of distinct points ("marked points", "punctures") on the Riemann surface. In particular, we assume $P_i \neq Q_j$ for every pair (i, j). The points in I are called the *in-points*, the points in O the *out-points*. Sometimes we consider I and O simply as sets.

In the article we sometimes refer to the classical situation. By this we understand

$$\Sigma = \mathbb{P}^1(\mathbb{C}) = S^2, \quad I = \{z = 0\}, \quad O = \{z = \infty\} \tag{10}$$

The following Figs. 1–3 exemplify the different situations:

Our objects, algebras, structures, ... will be meromorphic objects defined on Σ which are holomorphic outside of the points in A. To introduce the objects let $\mathcal{K} = \mathcal{K}_\Sigma$ be the canonical line bundle of Σ, resp. the locally free canonically sheaf. The local sections of the bundle are the local holomorphic differentials. If $P \in \Sigma$ is a point and z a local holomorphic coordinate at P, then a local holomorphic differential can be written as $f(z)dz$ with a local holomorphic function f defined in a neighbourhood of P. A global holomorphic section can be described locally with respect to a covering by coordinate charts $(U_i, z_i)_{i \in J}$ by a system of local holomorphic functions $(f_i)_{i \in J}$, which are related by the transformation rule induced by the coordinate change map $z_j = z_j(z_i)$ and the condition $f_i dz_i = f_j dz_j$ yielding

$$f_j = f_i \cdot \left(\frac{dz_j}{dz_i}\right)^{-1} . \tag{11}$$

Moreover, a meromorphic section of $\mathcal{K}$ is given as a collection of local meromorphic functions $(h_i)_{i \in J}$ for which the transformation law (11) still is true.

In the following λ is either an integer or a half-integer. If λ is an integer, then

1. $\mathcal{K}^\lambda = \mathcal{K}^{\otimes \lambda}$ for $\lambda > 0$,
2. $\mathcal{K}^0 = \mathcal{O}$, the trivial line bundle, and
3. $\mathcal{K}^\lambda = (\mathcal{K}^*)^{\otimes(-\lambda)}$ for $\lambda < 0$.

Here as usual $\mathcal{K}^*$ denotes the dual line bundle to the canonical line bundle. The dual line bundle is the holomorphic tangent line bundle, whose local sections are the holomorphic tangent vector fields $f(z)(d/dz)$. If λ is a half-integer, then we first have to fix a "square root" of the canonical line bundle, sometimes called a *theta-characteristics*. This means we fix a line bundle L for which $L^{\otimes 2} = \mathcal{K}$.

After such a choice of L is done we set $\mathcal{K}^\lambda = \mathcal{K}_L^\lambda = L^{\otimes 2\lambda}$. In most cases we will drop the mentioning of L, but we have to keep the choice in mind. Also the fine-structure of the algebras we are about to define will depend on the choice. But the main properties will remain the same.

Remark 3.1. A Riemann surface of genus g has exactly 2^{2g} non-isomorphic square roots of $\mathcal{K}$. For $g = 0$ we have $\mathcal{K} = \mathcal{O}(-2)$, and $L = \mathcal{O}(-1)$, the tautological bundle, is the unique square root. Already for $g = 1$ we have 4 non-isomorphic ones. As in this case $\mathcal{K} = \mathcal{O}$ one solution is $L_0 = \mathcal{O}$. But we have also other bundles L_i, $i = 1, 2, 3$. Note that L_0 has a non-vanishing global holomorphic section, whereas this is not the case for L_1, L_2, L_3. In general, depending on the parity of $\dim \mathrm{H}(\Sigma, L)$, one distinguishes even and odd theta characteristics L. For $g = 1$ the bundle $\mathcal{O}$ is an odd, the others are even theta characteristics.

We set

$$\begin{aligned} \mathcal{F}^\lambda := \mathcal{F}^\lambda(A) := \{ f \text{ is a global meromorphic section of } K^\lambda \mid \\ \text{such that } f \text{ is holomorphic over } \Sigma \setminus A \} . \end{aligned} \tag{12}$$

We will drop the set A in the notation. Obviously, $\mathcal{F}^\lambda$ is an infinite dimensional $\mathbb{C}$-vector space. Recall that in the case of half-integer λ everything depends on the theta characteristic L.

The elements of the space $\mathcal{F}^\lambda$ we call *meromorphic forms of weight* λ (with respect to the theta characteristic L). In local coordinates z_i we can write such a form as $f_i dz_i^\lambda$, with f_i a local holomorphic, resp. meromorphic form.

Special important cases of the weights are the functions ($\lambda = 0$), the space is also denoted by $\mathcal{A}$, the vector fields ($\lambda = -1$), denoted by $\mathcal{L}$, the differentials ($\lambda = 1$), and the quadratic differentials ($\lambda = 2$).

Next we introduce algebraic operations on the space of all weights

$$\mathcal{F} := \bigoplus_{\lambda \in \frac{1}{2}\mathbb{Z}} \mathcal{F}^\lambda. \tag{13}$$

These operations will allow us to introduce the algebras we are heading for.

3.2 Associative Structure

The natural map of the locally free sheaves of rang one

$$\mathcal{K}^\lambda \times \mathcal{K}^\nu \to \mathcal{K}^\lambda \otimes \mathcal{K}^\nu \cong \mathcal{K}^{\lambda+\nu}, \quad (s,t) \mapsto s \otimes t, \tag{14}$$

defines a bilinear map

$$\cdot : \mathcal{F}^\lambda \times \mathcal{F}^\nu \to \mathcal{F}^{\lambda+\nu}. \tag{15}$$

With respect to local trivializations this corresponds to the multiplication of the local representing meromorphic functions

$$(s\, dz^\lambda, t\, dz^\nu) \mapsto s\, dz^\lambda \cdot t\, dz^\nu = s \cdot t\ dz^{\lambda+\nu}. \tag{16}$$

If there is no danger of confusion, then we will mostly use the same symbol for the section and for the local representing function.

The following is obvious

Proposition 3.2. *The vector space $\mathcal{F}$ is an associative and commutative graded (over $\frac{1}{2}\mathbb{Z}$) algebras. Moreover, $\mathcal{A} = \mathcal{F}^0$ is a subalgebra.*

Definition 3.3. The associative algebra $\mathcal{A}$ is the Krichever–Novikov function algebra (associated with (Σ, A)).

Of course, it is the algebra of meromorphic functions on Σ which are holomorphic outside of A. The spaces $\mathcal{F}^\lambda$ are modules over $\mathcal{A}$. In the classical situation $\mathcal{A} = \mathbb{C}[z, z^{-1}]$, the algebra of Laurent polynomials.

3.3 Lie Algebra Structure

Next we define a Lie algebra structure on the space $\mathcal{F}$. The structure is induced by the map

$$\mathcal{F}^\lambda \times \mathcal{F}^\nu \to \mathcal{F}^{\lambda+\nu+1}, \qquad (s,t) \mapsto [s,t], \tag{17}$$

which is defined in local representatives of the sections by

$$(s\,dz^{\lambda}, t\,dz^{\nu}) \mapsto [s\,dz^{\lambda}, t\,dz^{\nu}] := \left((-\lambda)s\frac{dt}{dz} + \nu\, t\frac{ds}{dz}\right) dz^{\lambda+\nu+1}, \qquad (18)$$

and bilinearly extended to $\mathcal{F}$.

Proposition 3.4 ([42, 52]).

(a) The bilinear map $[.,.]$ *defines a Lie algebra structure on* $\mathcal{F}$.
(b) The space $\mathcal{F}$ *with respect to* $\cdot$ *and* $[.,.]$ *is a Poisson algebra.*

Next we consider certain important substructures.

3.4 The Vector Field Algebra and the Lie Derivative

For $\lambda = \nu = -1$ in (17) we end up in $\mathcal{F}^{-1}$ again. Hence,

Proposition 3.5. *The subspace* $\mathcal{L} = \mathcal{F}^{-1}$ *is a Lie subalgebra, and the* $\mathcal{F}^{\lambda}$*'s are Lie modules over* $\mathcal{L}$.

As forms of weight -1 are vector fields, $\mathcal{L}$ could also be defined as the Lie algebra of those meromorphic vector fields on the Riemann surface Σ which are holomorphic outside of A. The product (18) gives the usual Lie bracket of vector fields and the Lie derivative for their actions on forms. Due to its importance let us specialize this. We obtain (naming the local functions with the same symbol as the section)

$$[e, f]_{|}(z) = \left[e(z)\frac{d}{dz}, f(z)\frac{d}{dz}\right] = \left(e(z)\frac{df}{dz}(z) - f(z)\frac{de}{dz}(z)\right)\frac{d}{dz}, \qquad (19)$$

$$\nabla_e(g)_{|}(z) = L_e(g)_{|} = e\,.\,g_{|} = \left(e(z)\frac{dg}{dz}(z) + \lambda g(z)\frac{de}{dz}(z)\right)(dz)^{\lambda}. \qquad (20)$$

Definition 3.6. The algebra $\mathcal{L}$ is called *Krichever–Novikov type vector field algebra* (associated with (Σ, A)).

In the classical case this gives the Witt algebra.

3.5 The Algebra of Differential Operators

In $\mathcal{F}$, considered as Lie algebra, $\mathcal{A} = \mathcal{F}^0$ is an abelian Lie subalgebra and the vector space sum $\mathcal{F}^0 \oplus \mathcal{F}^{-1} = \mathcal{A} \oplus \mathcal{L}$ is also a Lie subalgebra of $\mathcal{F}$. In an equivalent way it can also be constructed as semi-direct sum of $\mathcal{A}$ considered as abelian Lie algebra and $\mathcal{L}$ operating on $\mathcal{A}$ by taking the derivative.

Definition 3.7. This Lie algebra is called the *Lie algebra of differential operators of degree* ≤ 1 of KN type (associated with (Σ, A)) and is denoted by $\mathcal{D}^1$.

In more direct terms $\mathcal{D}^1 = \mathcal{A} \oplus \mathcal{L}$ as vector space direct sum and endowed with the Lie product

$$[(g,e),(h,f)] = (e\,.\,h - f\,.\,g\,,\,[e,f]). \tag{21}$$

The spaces $\mathcal{F}^\lambda$ will be Lie-modules over $\mathcal{D}^1$.

Its universal enveloping algebra will be the algebra of all differential operators of arbitrary degree [39, 41, 46].

3.6 The Superalgebra of Half Forms

Next we consider the associative product

$$\cdot\ \mathcal{F}^{-1/2} \times \mathcal{F}^{-1/2} \to \mathcal{F}^{-1} = \mathcal{L}. \tag{22}$$

We introduce the vector space and the product

$$\mathcal{S} := \mathcal{L} \oplus \mathcal{F}^{-1/2}, \quad [(e,\varphi),(f,\psi)] := ([e,f] + \varphi \cdot \psi, e\,.\,\varphi - f\,.\,\psi). \tag{23}$$

Usually we will denote the elements of $\mathcal{L}$ by $e, f, \ldots$, and the elements of $\mathcal{F}^{-1/2}$ by $\varphi, \psi, \ldots$.

The definition (23) can be reformulated as an extension of $[.,.]$ on $\mathcal{L}$ to a "super-bracket" (denoted by the same symbol) on $\mathcal{S}$ by setting

$$[e,\varphi] := -[\varphi,e] := e\,.\,\varphi = \left(e\frac{d\varphi}{dz} - \frac{1}{2}\varphi\frac{de}{dz}\right)(dz)^{-1/2} \tag{24}$$

and

$$[\varphi,\psi] := \varphi \cdot \psi\,. \tag{25}$$

We call the elements of $\mathcal{L}$ elements of even parity, and the elements of $\mathcal{F}^{-1/2}$ elements of odd parity. For such elements x we denote by $\bar{x} \in \{\bar{0}, \bar{1}\}$ their parity.

The sum (23) can also be described as $\mathcal{S} = \mathcal{S}_{\bar{0}} \oplus \mathcal{S}_{\bar{1}}$, where $\mathcal{S}_{\bar{i}}$ is the subspace of elements of parity $\bar{i}$.

Proposition 3.8 ([51]). *The space $\mathcal{S}$ with the above introduced parity and product is a Lie superalgebra.*

Definition 3.9. The algebra $\mathcal{S}$ is the Krichever–Novikov type Lie superalgebra (associated with (Σ, A)).

Classically this Lie superalgebra corresponds to the Neveu–Schwarz superalgebra. See in this context also [3, 6, 10].

3.7 Jordan Superalgebra

Leidwanger and Morier-Genoux introduced in [30] a Jordan superalgebra in the Krichever–Novikov setting, i.e.

$$\mathscr{J} := \mathscr{F}^0 \oplus \mathscr{F}^{-1/2} = \mathscr{J}_{\bar{0}} \oplus \mathscr{J}_{\bar{1}}. \tag{26}$$

Recall that $\mathscr{A} = \mathscr{F}^0$ is the associative algebra of meromorphic functions. They define the (Jordan) product $\circ$ via the algebra structures for the spaces $\mathscr{F}^\lambda$ by

$$\begin{aligned} f \circ g &:= f \cdot g && \in \mathscr{F}^0, \\ f \circ \varphi &:= f \cdot \varphi && \in \mathscr{F}^{-1/2} \\ \varphi \circ \psi &:= [\varphi, \psi] && \in \mathscr{F}^0. \end{aligned} \tag{27}$$

By rescaling the second definition with the factor 1/2 one obtains a Lie antialgebra. See [30] for more details and additional results on representations.

3.8 Current Algebras

We start with $\mathfrak{g}$ a complex finite-dimensional Lie algebra and endow the tensor product $\overline{\mathfrak{g}} = \mathfrak{g} \otimes_{\mathbb{C}} \mathscr{A}$ with the Lie bracket

$$[x \otimes f, y \otimes g] = [x, y] \otimes f \cdot g, \qquad x, y \in \mathfrak{g}, \quad f, g \in \mathscr{A}. \tag{28}$$

The algebra $\overline{\mathfrak{g}}$ is the higher genus current algebra. It is an infinite dimensional Lie algebra and might be considered as the Lie algebra of $\mathfrak{g}$-valued meromorphic functions on the Riemann surface with poles only outside of A. Note that we allow also the case of $\mathfrak{g}$ an abelian Lie algebra.

Definition 3.10. The algebra $\overline{\mathfrak{g}}$ is called *current algebra* of Krichever Novikov type (associated with (Σ, A)).

Sometimes also the name *loop algebra* is used.

In the classical case the current algebra $\overline{\mathfrak{g}}$ is the standard current algebra $\overline{\mathfrak{g}} = \mathfrak{g} \otimes \mathbb{C}[z^{-1}, z]$ with Lie bracket

$$[x \otimes z^n, y \otimes z^m] = [x, y] \otimes z^{n+m} \qquad x, y \in \mathfrak{g}, \quad n, m \in \mathbb{Z}. \tag{29}$$

To point out the dependence on the geometrical structure we added always "(associated with (Σ, A))" in the definition. For simplicity we will drop it starting from now.

4 Almost-Graded Structure

4.1 Definition of Almost-Gradedness

Recall the classical situation. This is the Riemann surface $\mathbb{P}^1(\mathbb{C}) = S^2$, i.e. the Riemann surface of genus zero, and the points where poles are allowed are $\{0, \infty\})$. In this case the algebras introduced in the last section are graded algebras. In the higher genus case and even in the genus zero case with more than two points where poles are allowed there is no non-trivial grading anymore. As realized by Krichever and Novikov [26] there is a weaker concept, an almost-grading, which to a large extent is a valuable replacement of a honest grading. Such an almost-grading is induced by a splitting of the set A into two non-empty and disjoint sets I and O. The almost-grading is fixed by exhibiting certain basis elements in the spaces $\mathcal{F}^\lambda$ as homogeneous.

Definition 4.1. Let $\mathcal{L}$ be a Lie or an associative algebra such that $\mathcal{L} = \oplus_{n\in\mathbb{Z}}\mathcal{L}_n$ is a vector space direct sum, then $\mathcal{L}$ is called an *almost-graded* (Lie-) algebra if

(i) $\dim \mathcal{L}_n < \infty$,
(ii) There exists constants $L_1, L_2 \in \mathbb{Z}$ such that

$$\mathcal{L}_n \cdot \mathcal{L}_m \subseteq \bigoplus_{h=n+m-L_1}^{n+m+L_2} \mathcal{L}_h, \qquad \forall n, m \in \mathbb{Z}. \tag{30}$$

The elements in $\mathcal{L}_n$ are called *homogeneous* elements of degree n, and $\mathcal{L}_n$ is called *homogeneous subspace* of degree n.

If $\dim \mathcal{L}_n$ is bounded with a bound independent of n, we call $\mathcal{L}$ *strongly almost-graded*. If we drop the condition that $\dim \mathcal{L}_n$ is finite, we call $\mathcal{L}$ *weakly almost-graded*.

In a similar manner almost-graded modules over almost-graded algebras are defined. We can extend in an obvious way the definition to superalgebras, resp. even to more general algebraic structures. This definition makes complete sense also for more general index sets $\mathbb{J}$. In fact we will consider the index set $\mathbb{J} = (1/2)\mathbb{Z}$ in the case of superalgebras. The even elements (with respect to the super-grading) will have integer degree, the odd elements half-integer degree.

4.2 Separating Cycle and Krichever–Novikov Duality

Let C_i be positively oriented (deformed) circles around the points P_i in I, $i = 1, \ldots, K$ and C_j^* positively oriented ones around the points Q_j in O, $j = 1, \ldots, M$.

A cycle C_S is called a separating cycle if it is smooth, positively oriented of multiplicity one and if it separates the in- from the out-points. It might have multiple components. In the following we will integrate meromorphic differentials on Σ without poles in $\Sigma \setminus A$ over closed curves C. Hence, we might consider the C and C' as equivalent if $[C] = [C']$ in $\mathrm{H}(\Sigma \setminus A, \mathbb{Z})$. In this sense we can write for every separating cycle

$$[C_S] = \sum_{i=1}^{K}[C_i] = -\sum_{j=1}^{M}[C_j^*]. \tag{31}$$

The minus sign appears due to the opposite orientation. Another way for giving such a C_S is via level lines of a "proper time evolution", for which I refer to [36].

Given such a separating cycle C_S (resp. cycle class) we define a linear map

$$\mathcal{F}^1 \to \mathbb{C}, \qquad \omega \mapsto \frac{1}{2\pi \mathrm{i}} \int_{C_S} \omega. \tag{32}$$

As explained above the map will not depend on the separating line C_S chosen, as two of such will be homologous and the poles of ω are only located in I and O.

Consequently, the integration of ω over C_S can also be described over the special cycles C_i or equivalently over C_j^*. This integration corresponds to calculating residues

$$\omega \quad \mapsto \quad \frac{1}{2\pi \mathrm{i}} \int_{C_S} \omega \;=\; \sum_{i=1}^{K} \mathrm{res}_{P_i}(\omega) \;=\; -\sum_{l=1}^{M} \mathrm{res}_{Q_l}(\omega). \tag{33}$$

Definition 4.2. The pairing

$$\mathcal{F}^\lambda \times \mathcal{F}^{1-\lambda} \to \mathbb{C}, \quad (f, g) \mapsto \langle f, g\rangle := \frac{1}{2\pi \mathrm{i}} \int_{C_S} f \cdot g, \tag{34}$$

between λ and $1-\lambda$ forms is called *Krichever–Novikov (KN) pairing*.

Note that the pairing depends not only on A (as the $\mathcal{F}^\lambda$ depend on it) but also critically on the splitting of A into I and O as the integration path will depend on it. Once the splitting is fixed the pairing will be fixed too.

By exhibiting dual basis elements further down we will see that it is non-degenerate.

4.3 The Homogeneous Subspaces

Depending on whether λ is integer or half-integer we set $\mathbb{J}_\lambda = \mathbb{Z}$ or $\mathbb{J}_\lambda = \mathbb{Z} + 1/2$. For $\mathcal{F}^\lambda$ we introduce for $m \in \mathbb{J}_\lambda$ subspaces $\mathcal{F}^\lambda_m$ of dimension K, where $K = \#I$, by exhibiting certain elements $f^\lambda_{m,p} \in \mathcal{F}^\lambda$, $p = 1, \ldots, K$ which constitute a basis of $\mathcal{F}^\lambda_m$. Recall that the spaces $\mathcal{F}^\lambda$ for $\lambda \in \mathbb{Z} + 1/2$ depend on the chosen square root L (the theta characteristic) of $\mathcal{K}$. The elements are the elements of degree m. As explained in the following, the degree is in an essential way related to the zero orders of the elements at the points in I.

Let $I = \{P_1, P_2, \ldots, P_K\}$ then we have for the zero-order at the point $P_i \in I$ of the element $f^\lambda_{n,p}$

$$\operatorname{ord}_{P_i}(f^\lambda_{n,p}) = (n + 1 - \lambda) - \delta_i^p, \quad i = 1, \ldots, K\,. \tag{35}$$

The prescription at the points in O is made in such a way that the element $f^\lambda_{m,p}$ is essentially uniquely given. Essentially unique means up to multiplication with a constant.[3] After fixing as additional geometric data a system of coordinates z_l centred at P_l for $l = 1, \ldots, K$ and requiring that

$$f^\lambda_{n,p}(z_p) = z_p^{n-\lambda}(1 + O(z_p))(dz_p)^\lambda \tag{36}$$

the element $f_{n,p}$ is uniquely fixed. In fact, the element $f^\lambda_{n,p}$ only depends on the first jet of the coordinate z_p [54].

Example. Here we will not give the general recipe for the prescription at the points in O, see [36, 39, 52]. Just to give an example which is also an important special case, assume $O = \{Q\}$ is a one-element set. If either the genus $g = 0$, or $g \geq 2$, $\lambda \neq 0,\ 1/2,\ 1$ and the points in A are in generic position, then we require

$$\operatorname{ord}_Q(f^\lambda_{n,p}) = -K \cdot (n + 1 - \lambda) + (2\lambda - 1)(g - 1). \tag{37}$$

In the other cases (e.g., for $g = 1$) there are some modifications at the point in O necessary for finitely many n.

Theorem 4.3 ([36, 39, 52]). *Set*

$$\mathcal{B}^\lambda := \{\, f^\lambda_{n,p} \mid n \in \mathbb{J}_\lambda,\ p = 1, \ldots, K\,\}. \tag{38}$$

Then (a) $\mathcal{B}^\lambda$ is a basis of the vector space $\mathcal{F}^{1-\lambda}$.
(b) The introduced basis $\mathcal{B}^\lambda$ of $\mathcal{F}^\lambda$ and $\mathcal{B}^{1-\lambda}$ of $\mathcal{F}^{1-\lambda}$ are dual to each other with respect to the Krichever–Novikov pairing (34), i.e.

---

[3]Strictly speaking, there are some special cases where some constants have to be added such that the Krichever–Novikov duality (39) is valid, see [36].

$$\langle f_{n,p}^{\lambda}, f_{-m,r}^{1-\lambda}\rangle = \delta_p^r\,\delta_n^m, \quad \forall n,m \in \mathbb{J}_\lambda, \quad r,p = 1,\dots,K. \tag{39}$$

From part (b) of the theorem it follows that the Krichever–Novikov pairing is non-degenerate. Moreover, any element $v \in \mathcal{F}^{1-\lambda}$ acts as linear form on $\mathcal{F}^{\lambda}$ via

$$\Phi_v : \mathcal{F}^{1-\lambda} \mapsto \mathbb{C}, \quad w \mapsto \Phi_v(w) := \langle v, w\rangle. \tag{40}$$

Via this pairing $\mathcal{F}^{1-\lambda}$ can be considered as subspace of $(\mathcal{F}^{\lambda})^*$. But I would like to stress the fact that the identification depends on the splitting of A into I and O as the KN pairing depends on it.

The full space $(\mathcal{F}^{\lambda})^*$ can even be described with the help of the pairing. Consider the series

$$\hat{v} := \sum_{m\in\mathbb{Z}}\sum_{p=1}^{K} a_{m,p} f_{m,p}^{1-\lambda} \tag{41}$$

as a formal series, then $\Phi_{\hat{v}}$ (as a distribution) is a well-defined element of ${\mathcal{F}^{\lambda}}^*$, as it will be only evaluated for finitely many basis elements in $\mathcal{F}^{\lambda}$. Vice versa, every element of ${\mathcal{F}^{\lambda}}^*$ can be given by a suitable $\hat{v}$. Every $\phi \in (\mathcal{F}^{\lambda})^*$ is uniquely given by the scalars $\phi(f_{m,r}^{\lambda})$. We set

$$\hat{v} := \sum_{m\in\mathbb{Z}}\sum_{p=1}^{K} \phi(f_{-m,p}^{\lambda})\, f_{m,p}^{1-\lambda}. \tag{42}$$

Obviously, $\Phi_{\hat{v}} = \phi$. For more information about this "distribution interpretation," see [39, 42].

The dual elements of $\mathcal{L}$ will be given by the formal series (41) with basis elements from $\mathcal{F}^2$ the quadratic differentials, the dual elements of $\mathcal{A}$ correspondingly from $\mathcal{F}^1$ the differentials, and the dual elements of $\mathcal{F}^{-1/2}$ correspondingly from $\mathcal{F}^{3/2}$. The spaces $\mathcal{F}^2$, $\mathcal{F}^1$ and $\mathcal{F}^{3/2}$ themselves can be considered as some kind of restricted duals.

It is quite convenient to use special notations for elements of some important weights:

$$\begin{aligned} e_{n,p} &:= f_{n,p}^{-1}, \quad \varphi_{n,p} := f_{n,p}^{-1/2}, \quad A_{n,p} := f_{n,p}^{0}, \\ \omega^{n,p} &:= f_{-n,p}^{1}, \quad \Omega^{n,p} := f_{-n,p}^{2}. \end{aligned} \tag{43}$$

In view of (39) for the forms of weight 1 and 2 it is convenient to invert the index n and write it as a superscript.

4.4 The Algebras

Theorem 4.4 ([36, 39, 52]). *There exists constants R_1 and R_2 (depending on the genus g, and on the number and splitting of the points in A) independent of $n, m \in \mathbb{J}$ such that for the basis elements*

$$\begin{aligned} f_{n,p}^{\lambda} \cdot f_{m,r}^{\nu} = \quad & f_{n+m,r}^{\lambda+\nu}\delta_p^r \\ & + \sum_{h=n+m+1}^{n+m+R_1}\sum_{s=1}^{K} a_{(n,p)(m,r)}^{(h,s)} f_{h,s}^{\lambda+\nu}, \qquad a_{(n,p)(m,r)}^{(h,s)} \in \mathbb{C}, \\ [f_{n,p}^{\lambda}, f_{m,r}^{\nu}] = \quad & (-\lambda m + \nu n)\, f_{n+m,r}^{\lambda+\nu+1}\delta_p^r \\ & + \sum_{h=n+m+1}^{n+m+R_2}\sum_{s=1}^{K} b_{(n,p)(m,r)}^{(h,s)} f_{h,s}^{\lambda+\nu+1}, \quad b_{(n,p)(m,r)}^{(h,s)} \in \mathbb{C}. \end{aligned} \tag{44}$$

This says in particular that with respect to both the associative and Lie structure the algebra $\mathscr{F}$ is weakly almost-graded. In generic situations and for $N = 2$ points one obtains $R_1 = g$ and $R_2 = 3g$.

The reason why we only have weakly almost-gradedness is that

$$\mathscr{F}^{\lambda} = \bigoplus_{m \in \mathbb{J}_\lambda} \mathscr{F}_m^{\lambda}, \qquad \text{with} \quad \dim \mathscr{F}_m^{\lambda} = K. \tag{45}$$

If we add up for a fixed m all λ we get that our homogeneous spaces are infinite dimensional.

In the definition of our KN type algebra only finitely many λ are involved, hence the following is immediate

Theorem 4.5. *The Krichever–Novikov type vector field algebras $\mathscr{L}$, function algebras $\mathscr{A}$, differential operator algebras $\mathscr{D}^1$, Lie superalgebras $\mathscr{S}$, and Jordan superalgebras $\mathscr{J}$ are all (strongly) almost-graded.*

We obtain

$$\dim \mathscr{L}_n = \dim \mathscr{A}_n = K, \quad \dim \mathscr{S}_n = \dim \mathscr{J}_n = 2K, \quad \dim \mathscr{D}_n^1 = 3K. \tag{46}$$

If $\mathscr{U}$ is one of these algebras, with product denoted by [,] then

$$[\mathscr{U}_n, \mathscr{U}_m] \subseteq \bigoplus_{h=n+m}^{n+m+R_i} \mathscr{U}_h, \tag{47}$$

with $R_i = R_1$ for $\mathscr{U} = \mathscr{A}$ and $R_i = R_2$ otherwise.

For further reference let us specialize the lowest degree term component in (44) for certain special cases.

$$
\begin{aligned}
A_{n,p}\cdot A_{m,r} &= A_{n+m,r}\,\delta_r^p \;+\; \text{h.d.t.}\\
A_{n,p}\cdot f^\lambda_{m,r} &= f^\lambda_{n+m,r}\,\delta_r^p \;+\; \text{h.d.t.}\\
[e_{n,p},e_{m,r}] &= (m-n)\cdot e_{n+m,r}\,\delta_r^p \;+\; \text{h.d.t.}\\
e_{n,p}\cdot f^\lambda_{m,r} &= (m+n\lambda)\cdot f^\lambda_{n+m,r}\,\delta_r^p \;+\; \text{h.d.t.}
\end{aligned}
\tag{48}
$$

Here h.d.t. denote linear combinations of basis elements of degree between $n+m+1$ and $n+m+R_i$,

Finally, the almost-grading of $\mathscr{A}$ induces an almost-grading of the current algebra $\overline{\mathfrak{g}}$ by setting $\overline{\mathfrak{g}}_n=\mathfrak{g}\otimes\mathscr{A}_n$. We obtain

$$
\overline{\mathfrak{g}}=\bigoplus_{n\in\mathbb{Z}}\overline{\mathfrak{g}}_n,\qquad \dim\overline{\mathfrak{g}}_n=K\cdot\dim\mathfrak{g}. \tag{49}
$$

4.5 *Triangular Decomposition and Filtrations*

Let $\mathscr{U}$ be one of the above introduced algebras (including the current algebra). On the basis of the almost-grading we obtain a triangular decomposition of the algebras

$$
\mathscr{U}=\mathscr{U}_{[+]}\oplus\mathscr{U}_{[0]}\oplus\mathscr{U}_{[-]}, \tag{50}
$$

where

$$
\mathscr{U}_{[+]}:=\bigoplus_{m>0}\mathscr{U}_m,\quad \mathscr{U}_{[0]}=\bigoplus_{m=-R_i}^{m=0}\mathscr{U}_m,\quad \mathscr{U}_{[-]}:=\bigoplus_{m<-R_i}\mathscr{U}_m. \tag{51}
$$

By the almost-gradedness the [+] and [−] subspaces are (infinite dimensional) subalgebras. The [0] spaces in general not. Sometimes we will use *critical strip* for $\mathscr{U}_{[0]}$.

With respect to the almost-grading of $\mathscr{F}^\lambda$ we can introduce a filtration

$$
\begin{gathered}
\mathscr{F}^\lambda_{(n)}:=\bigoplus_{m\ge n}\mathscr{F}^\lambda_m,\\
\cdots\;\supseteq\;\mathscr{F}^\lambda_{(n-1)}\;\supseteq\;\mathscr{F}^\lambda_{(n)}\;\supseteq\;\mathscr{F}^\lambda_{(n+1)}\;\cdots
\end{gathered}
\tag{52}
$$

Proposition 4.6.

$$
\mathscr{F}^\lambda_{(n)}:=\{\,f\in\mathscr{F}^\lambda\mid \operatorname{ord}_{P_i}(f)\ge n-\lambda,\forall i=1,\ldots,K\,\}. \tag{53}
$$

This proposition is very important. In case that O has more than one point there are certain choices, e.g. numbering of the points in O, different rules, etc. involved in defining the almost-grading. Hence, if the choices are made differently the subspaces $\mathscr{F}_n^\lambda$ might depend on them, and consequently also the almost-grading. But by this proposition the induced filtration is indeed canonically defined via the splitting of A into I and O.

Moreover, different choices will give equivalent almost-grading. We stress the fact, that under a KN algebra we will always understand one of the introduced algebras (or even some others still to come) together with an almost-grading (resp. equivalence class of almost-grading) introduced by the splitting $A = I \cup O$.

5 Central Extensions

Central extension of our algebras appear naturally in the context of quantization and regularization of actions. In Sect. 6.1 we will see a typical example. Of course they are also of independent mathematical interest.

5.1 Central Extensions and Cocycles

A central extension of a Lie algebra W is a special Lie algebra structure on the vector space direct sum $\hat{W} = \mathbb{C} \oplus W$. If we denote $\hat{x} := (0, x)$ and $t := (1, 0)$, then the Lie structure is given by

$$[\hat{x}, \hat{y}] = \widehat{[x, y]} + \psi(x, y) \cdot t, \quad [t, \hat{W}] = 0, \quad x, y \in W. \tag{54}$$

The map $x \mapsto \hat{x} = (0, x)$ is a linear splitting map. $\hat{W}$ will be a Lie algebra, e.g. will fulfill the Jacobi identity, if and only if ψ is antisymmetric and fulfills the Lie algebra 2-cocycle condition

$$0 = d_2\psi(x, y, z) := \psi([x, y], z) + \psi([y, z], x) + \psi([z, x], y). \tag{55}$$

A 2-cochain ψ is a coboundary if there exists a $\varphi : W \to \mathbb{C}$ such that

$$\psi(x, y) = \varphi([x, y]). \tag{56}$$

One easily shows that a coboundary is a cocycle. Hence, we can define the second Lie algebra cohomology $\mathrm{H}^2(W, \mathbb{C})$ of W with values in the trivial module $\mathbb{C}$ as this quotient.

There is the notion of equivalence of central extensions. For the definition in terms of short exact sequences, I refer to the standard textbooks. Equivalently, they can be described by a different choice of the linear splitting map. Instead of $x \mapsto$

$\hat{x} = (0, x)$ one chooses $x \mapsto \hat{x}' = (\varphi, x)$ with $\varphi : W \to \mathbb{C}$ a linear form. For $\varphi \neq 0$ the Lie structure corresponding to (54) will be different, but equivalent. In fact in (54) we obtain a different 2-cocycle ψ'.

From this definition follows that that two central extensions are equivalent if and only if the difference of their defining 2-cocycles ψ and ψ' is a coboundary. In this way the second Lie algebra cohomology $\mathrm{H}^2(W, \mathbb{C})$ classifies equivalence classes of central extensions. The class [0] corresponds to the trivial central extension. In this case the splitting map is a Lie homomorphism. To construct central extensions of our algebras we have to find such Lie algebra 2-cocycles.

Clearly, equivalent central extensions are isomorphic. The opposite is not true. Furthermore, in our case we can always rescale the central element by multiplying it with a non-zero scalar. This is not an equivalence of central extensions but nevertheless an irrelevant modification. Hence we will mainly be interested in central extensions modulo equivalence and rescaling. They are classified by [0] and the elements of the projectivized cohomology space $\mathbb{P}(\mathrm{H}^2(W, \mathbb{C}))$.

In the classical case we have $\dim \mathrm{H}^2(\mathscr{W}, \mathbb{C}) = 1$, hence there are only two essentially different central extensions, the splitting one given by the direct sum $\mathbb{C} \oplus \mathscr{W}$ of Lie algebras and the up to equivalence and rescaling unique non-trivial one, the Virasoro algebra $\mathscr{V}$.

Remark 5.1. Given a vector space bases $\{e_\rho \mid \rho \in \mathscr{R}\}$ of W, a vector space basis of $\hat{W}$ will be given by $\{\hat{e}_\rho := (0, e_\rho) \mid \rho \in \mathscr{R}, t := (1, 0)\ \}$. An equivalent central extension can be described as a change of basis and rescaling of the form

$$\hat{e}'_\rho = \hat{e}_\rho + \varphi(e_\rho)\, t, \qquad t' = \alpha \cdot t,\ \alpha \in \mathbb{C}^*. \tag{57}$$

5.2 *Geometric Cocycles*

The defining cocycle for the Virasoro algebra obviously does not make any sense in the higher genus and/or multi-point case. We need a geometric description. For this we have first to introduce connections.

5.2.1 Projective and Affine Connections

Let $(U_\alpha, z_\alpha)_{\alpha \in J}$ be a covering of the Riemann surface Σ by holomorphic coordinates with transition functions $z_\beta = f_{\beta\alpha}(z_\alpha)$.

Definition 5.2. (a) A system of local (holomorphic, meromorphic) functions $R = (R_\alpha(z_\alpha))$ is called a (holomorphic, meromorphic) *projective connection* if it transforms as

$$R_\beta(z_\beta) \cdot (f'_{\beta,\alpha})^2 = R_\alpha(z_\alpha) + S(f_{\beta,\alpha}), \qquad \text{with} \quad S(h) = \frac{h'''}{h'} - \frac{3}{2}\left(\frac{h''}{h'}\right)^2, \tag{58}$$

the Schwartzian derivative. Here $'$ denotes differentiation with respect to the coordinate z_α.

(b) A system of local (holomorphic, meromorphic) functions $T = (T_\alpha(z_\alpha))$ is called a (holomorphic, meromorphic) *affine connection* if it transforms as

$$T_\beta(z_\beta) \cdot (f'_{\beta,\alpha}) = T_\alpha(z_\alpha) + \frac{f''_{\beta,\alpha}}{f'_{\beta,\alpha}}. \tag{59}$$

Every Riemann surface admits a holomorphic projective connection [19, 20]. Given a point P then there exists always a meromorphic affine connection holomorphic outside of P and having maximally a pole of order one there [39].

From their very definition it follows that the difference of two affine (projective) connections will be a (quadratic) differential. Hence, after fixing one affine (projective) connection all others are obtained by adding (quadratic) differentials.

5.2.2 The Function Algebra $\mathscr{A}$

We consider $\mathscr{A}$ as an abelian Lie algebra. Let C be an arbitrary smooth but not necessarily connected curve. We set

$$\psi^1_C(g,h) := \frac{1}{2\pi \mathrm{i}} \int_C g dh, \quad g, h \in \mathscr{A}. \tag{60}$$

5.2.3 The Current Algebra $\overline{\mathfrak{g}}$

For $\overline{\mathfrak{g}} = \mathfrak{g} \otimes \mathscr{A}$ we first have to fix β a symmetric, invariant, bilinear form on $\mathfrak{g}$ (not necessarily non-degenerate). Invariance means that we have $\beta([x,y],z) = \beta(x,[y,z])$ for all $x, y, z \in \mathfrak{g}$. The cocycle is given as

$$\psi^2_{C,\beta}(x \otimes g, y \otimes h) := \beta(x,y) \cdot \frac{1}{2\pi \mathrm{i}} \int_C g dh, \quad x, y \in \mathfrak{g},\ g, h \in \mathscr{A}. \tag{61}$$

5.2.4 The Vector Field Algebra $\mathscr{L}$

Here it is a little bit more delicate. First we have to choose a (holomorphic) projective connection R. We define

$$\psi^3_{C,R}(e,f) := \frac{1}{24\pi \mathrm{i}} \int_C \left(\frac{1}{2}(e'''f - ef''') - R \cdot (e'f - ef') \right) dz\,. \tag{62}$$

Only by the term related to the projective connection it will be a well-defined differential, i.e. independent of the coordinate chosen. It is shown in [39] that it is a cocycle. Another choice of a projective connection will result in a cohomologous one. Hence, the equivalence class of the central extension will be the same.

5.2.5 The Differential Operator Algebra $\mathscr{D}^1$

For the differential operator algebra the cocycles for $\mathscr{A}$ can be extended by zero on the subspace $\mathscr{L}$. The cocycles for $\mathscr{L}$ can be pulled back. In addition there is a third type of cocycles mixing $\mathscr{A}$ and $\mathscr{L}$:

$$\psi^4_{C,T}(e,g) := \frac{1}{24\pi \mathrm{i}} \int_C (eg'' + Teg')dz, \qquad e \in \mathscr{L}, g \in \mathscr{A}, \tag{63}$$

with an affine connection T, with at most a pole of order one at a fixed point in O. Again, a different choice of the connection will not change the equivalence class. For more details on the cocycles, see [46].

5.2.6 The Lie Superalgebra $\mathscr{S}$

Here we have to take into account that it is not a Lie algebra. Hence, the Jacobi identity has to be replaced by the super-Jacobi identity. The conditions for being a cocycle for the superalgebra cohomology will change too. Recall the definition of the algebra from Sect. 3.6, in particular that the even elements (parity 0) are the vector fields and the odd elements (parity 1) are the half-forms. A bilinear form c is a cocycle if the following is true. The bilinear map c will be symmetric if x and y are odd, otherwise it will be antisymmetric.

$$c(x,y) = -(-1)^{\bar{x}\bar{y}} c(x,y). \tag{64}$$

The super-cocycle condition reads as

$$(-1)^{\bar{x}\bar{z}} c(x,[y,z]) + (-1)^{\bar{y}\bar{x}} c(y,[z,x]) + (-1)^{\bar{z}\bar{y}} c(z,[x,y]) \;=\; 0. \tag{65}$$

With the help of c we can define central extensions in the Lie superalgebra sense. If we put the condition that the central element is even, then the cocycle c has to be an even map and c vanishes for pairs of elements of different parity.

By convention we denote vector fields by $e, f, g, \ldots$ and $-1/2$-forms by $\varphi, \psi, \chi, \ldots$ and get

$$c(e,\varphi) = 0, \quad e \in \mathscr{L},\ \varphi \in \mathscr{F}^{-1/2}. \tag{66}$$

The super-cocycle conditions for the even elements is just the cocycle condition for the Lie subalgebra $\mathscr{L}$. The only other non-vanishing super-cocycle condition is for the *(even,odd,odd)* elements and reads as

$$c(e, [\varphi, \psi]) - c(\varphi, e \,.\, \psi) - c(\psi, e \,.\, \varphi) = 0. \tag{67}$$

Here the definition of the product $[e, \psi] := e \,.\, \psi$ was used.

If we have a cocycle c for the algebra $\mathcal{S}$, we obtain by restriction a cocycle for the algebra $\mathcal{L}$. For the mixing term we know that $c(e, \psi) = 0$. A naive try to put just anything for $c(\varphi, \psi)$ (for example, 0) will not work as (67) relates the restriction of the cocycle on $\mathcal{L}$ to its values on $\mathcal{F}^{-1/2}$.

Proposition 5.3 ([51]). *Let C be any closed (differentiable) curve on Σ not meeting the points in A, and let R be any (holomorphic) projective connection then the bilinear extension of*

$$\begin{aligned} \Phi_{C,R}(e, f) &:= \frac{1}{24\pi \mathrm{i}} \int_C \left(\frac{1}{2}(e''' f - e f''') - R \cdot (e' f - e f') \right) dz \\ \Phi_{C,R}(\varphi, \psi) &:= -\frac{1}{24\pi \mathrm{i}} \int_C \left(\varphi'' \cdot \psi + \varphi \cdot \psi'' - R \cdot \varphi \cdot \psi \right) dz \\ \Phi_{C,R}(e, \varphi) &:= 0 \end{aligned} \tag{68}$$

gives a Lie superalgebra cocycle for $\mathcal{S}$, hence defines a central extension of $\mathcal{S}$. A different projective connection will yield a cohomologous cocycle.

A similar formula was given by Bryant in [10]. By adding the projective connection in the second part of (68) he corrected some formula appearing in [3]. He only considered the two-point case and only the integration over a separating cycle. See also [24] for the multi-point case, where still only the integration over a separating cycle is considered.

In contrast to the differential operator algebra case the two parts cannot be prescribed independently. Only with the same integration path (more precisely, homology class) and the given factors in front of the integral it will work. The reason for this that (67) relates both.

5.3 Uniqueness and Classification of Central Extensions

Our cocycles depend on the choice of the connections R and T. But different choices will not change the equivalence class. Hence, this ambiguity does not disturb us. What really matters is that they depend on the integration curve C chosen.

In contrast to the classical situation, for the higher genus and/or multi-point situation there are many essentially different closed curves and hence many non-equivalent central extensions defined by the integration.

But we should take into account that we want to extend the almost-grading from our algebras to the centrally extended ones. This means we take $\deg \hat{x} := \deg x$ and assign a degree $\deg(t)$ to the central element t, and still obtain almost-gradedness.

This is possible if and only if our defining cocycle ψ is "local" in the following sense (the name was introduced in the two-point case by Krichever and Novikov in [26]). There exists $M_1, M_2 \in \mathbb{Z}$ such that

$$\forall n,m: \quad \psi(W_n, W_m) \neq 0 \implies M_1 \leq n+m \leq M_2. \tag{69}$$

Here W stands for any of our algebras (including the supercase). Very important, "local" is defined in terms of the almost-grading, and the grading itself depends on the splitting $A = I \cup O$. Hence what is "local" depends on the splitting too.

We will call a cocycle *bounded* (from above) if there exists $M \in \mathbb{Z}$ such that

$$\forall n,m: \quad \psi(W_n, W_m) \neq 0 \implies n+m \leq M. \tag{70}$$

Similarly bounded from below is defined. Locality means bounded from above and below.

Given a cocycle class we call it bounded (resp. local) if and only if it contains a representing cocycle which is bounded (resp. local). Not all cocycles in a bounded class have to be bounded. If we choose as integration path a separating cocycle C_S, or one of the C_i, then the above introduced geometric cocycles are local, resp. bounded. Recall that in this case integration can be done by calculating residues at the in-points or at the out-points. All these cocycles are cohomologically non-trivial. The theorems in the following concern the opposite direction. They were treated in my works [45, 46, 51].

I start with the vector field case as it will give a model for all the results.

Theorem 5.4 ([46]). *Let $\mathcal{L}$ be the Krichever–Novikov vector field algebra.*

(a) The space of bounded cohomology classes is K-dimensional ($K = \#I$). A basis is given by setting the integration path in (62) to C_i, $i = 1, \ldots, K$ the little (deformed) circles around the points $P_i \in I$.
(b) The space of local cohomology classes is one-dimensional. A generator is given by integrating (62) over a separating cocycle C_S.
(c) Up to equivalence and rescaling there is only one non-trivial one-dimensional central extension of the vector field algebra $\mathcal{L}$ which allows an extension of the almost-grading.

Part (c) says that the result is completely analogous to the case of the Witt algebra. Here I would like to repeat again the fact that for $\mathcal{L}$ depending on the set A and its possible splittings into two disjoint subsets there are different almost-gradings. Hence, the "unique" central extension finally obtained will also depend on the splitting. Only in the two-point case there is only one splitting possible.

The above theorem is a model for all other classification results. We will always obtain a statement about the bounded (from above) cocycles and then for the local cocycles.

As $\mathscr{A}$ is an abelian Lie algebra every skew-symmetric bilinear form will be a non-trivial cocycle. Hence, there is no hope of uniqueness. But if we add the condition of $\mathscr{L}$-invariance, i.e.

$$\psi(e.g, h) + \psi(g, e.h) = 0, \quad \forall e \in \mathscr{L},\ g, h \in \mathscr{A} \tag{71}$$

things will change.

Let us denote the subspace of local cohomology classes by $\mathrm{H}^2_{\mathrm{loc}}$, and the subspace of local and $\mathscr{L}$-invariant by $\mathrm{H}^2_{\mathscr{L},\mathrm{loc}}$. Note that the condition is only required for at least one representative in the cohomology class. We collect a part of the results for the other algebras in the following theorem.

Theorem 5.5.

(1) $\dim \mathrm{H}^2_{\mathscr{L},\mathrm{loc}}(\mathscr{A}, \mathbb{C}) = 1,$
(2) $\dim \mathrm{H}^2_{\mathrm{loc}}(\mathscr{S}, \mathbb{C}) = 1,$
(3) $\dim \mathrm{H}^2_{\mathrm{loc}}(\mathscr{D}^1, \mathbb{C}) = 3,$
(4) $\dim \mathrm{H}^2_{\mathrm{loc}}(\overline{\mathfrak{g}}, \mathbb{C}) = 1$ *for* $\mathfrak{g}$ *a simple finite-dimensional Lie algebra,*

A basis of the cohomology spaces are given by taking the cohomology classes of the cocycles (60), (61), (62), (63), (68) obtained by integration over a separating cycle C_S.

Correspondingly, we obtain also for these algebras the corresponding result about uniqueness of almost-graded central extensions. For the differential operator algebra we got three independent cocycles. This generalizes results of [1] for the classical case.

For the bounded cocycle classes we have to multiply the dimensions above by K. For the supercase with odd central elements the bounded cohomology vanishes.

For $\mathfrak{g}$ a reductive Lie algebra and the cocycle $\mathscr{L}$-invariant if restricted to the abelian part, a complete classification of local cocycle classes for $\overline{\mathfrak{g}}$ can be found in [45]. Note that in the case of a simple Lie algebra every symmetric, invariant bilinear form β is a multiple of the Cartan-Killing form.

I would like to mention that in all the applications I know of, the cocycles coming from representations, regularizations, etc. are local. Hence, the uniqueness or classification can be used.

6 Further Results

Above the basic concepts, results about the structure of these Krichever–Novikov type algebras and their central extensions were treated. Of course, this does not close the story. I will add some further important constructions and applications but due to space limitations only in a very condensed manner.

6.1 Semi-infinite Forms and Fermionic Fock Space Representations

Our Krichever–Novikov vector field algebras $\mathscr{L}$ have as Lie modules the spaces $\mathscr{F}^\lambda$. These representations are not of the type physicists are usually interested in, as there is no ground state (no vacuum). There are neither creation nor annihilation operators which can be used to construct the full representation out of a vacuum state.

To obtain such desired representations the almost-grading comes into play. First, using the grading of $\mathscr{F}^\lambda$ it is possible to construct starting from $\mathscr{F}^\lambda$, the forms of weight $\lambda \in 1/2\mathbb{Z}$, the *semi-infinite wedge forms* $\mathscr{H}^\lambda$.

The vector space $\mathscr{H}^\lambda$ is generated by basis elements which are formal expressions

$$\Phi = f^\lambda_{(i_1)} \wedge f^\lambda_{(i_2)} \wedge f^\lambda_{(i_3)} \wedge \cdots , \tag{72}$$

where $(i_1) = (m_1, p_1)$ is a double index indexing our basis elements. The indices are in strictly increasing lexicographical order. They are stabilizing in the sense that they will increase exactly by one starting from a certain point, depending on Φ. The action of $\mathscr{L}$ can be extended by Leibniz rule from $\mathscr{F}^\lambda$ to $\mathscr{H}^\lambda$. But a problem arises. For elements of the critical strip $\mathscr{L}_{[0]}$ of the algebra $\mathscr{L}$ it might happen that it produces infinitely many contributions. The action has to be regularized (as physicists like to call it, but it is a well-defined mathematical procedure).

Here the almost-grading has its second appearance. By the (strong) almost-graded module structure of $\mathscr{F}^\lambda$ the algebra $\mathscr{L}$ can be imbedded into the Lie algebra of both-sided infinite matrices

$$\overline{gl}(\infty) := \{A = (a_{ij})_{i,j\in\mathbb{Z}} \mid \exists r = r(A), \text{ such that } a_{ij} = 0 \text{ if } |i-j| > r\}, \tag{73}$$

with "infinitely many diagonals". The embedding will depend on the weight λ. For $\overline{gl}(\infty)$ there exists a procedure for the regularization of the action on the semi-infinite wedge forms [11, 22], see also [23] for a nice pedagogical treatment. In particular, there is a unique non-trivial central extension $\widehat{gl}(\infty)$. If we pull-back the defining cocycle for the extension, we obtain a central extension $\widehat{\mathscr{L}}_\lambda$ of $\mathscr{L}$ and the required regularization of the action of $\widehat{\mathscr{L}}_\lambda$ on $\mathscr{H}^\lambda$. As the embedding of $\mathscr{L}$ depends on the weight λ the cocycle will do so. The pull-back cocycle will be local. Hence, by the classification results of Sect. 5.3 it is the unique central extension class defined by (62) integrated over C_S (up to a λ dependent rescaling).

In $\mathscr{H}^\lambda$ there are invariant subspaces, which are generated by a certain "vacuum vectors". Such a vacuum Φ_T is given by an element of the form (72) which starts with the element $f^\lambda_{(T,1)}$, and the indices for the following ones increase always by one. The subalgebra $\mathscr{L}_{[+]}$ annihilates the vacuum, the central element and the other elements of degree zero act by multiplication with a constant on the vacuum and the whole representation state is generated by $\mathscr{L}_{[-]} \oplus \mathscr{L}_{[0]}$ from the vacuum.

As the function algebra $\mathscr{A}$ operates as multiplication operators on $\mathscr{F}^\lambda$ the above representation can be extended to the algebra $\mathscr{D}^1$ (see details in [39]) after one passes over to central extensions. The cocycle again is local and hence, up to coboundary, it will be a certain linear combination of the 3 generating cocycles for the differential operator algebra. In fact it will be

$$c_\lambda[\psi^3_{C_S}] + \frac{2\lambda - 1}{2}[\psi^4_{C_S}] - [\psi^1_{C_S}], \qquad c_\lambda := -2(6\lambda^2 - 6\lambda + 1). \tag{74}$$

Recall that ψ^3 is the cocycle for the vector field algebra, ψ^1 the cocycle for the function algebra, and ψ^4 the mixing cocycle, see [46] for details. Note that the expression for c_λ appears also in Mumford's formula [35] relating divisors on the moduli space of curves.

Also the representation on $\mathscr{H}^\lambda$ gives a projective representation of the algebra of $\mathscr{D}_\lambda$ of differential operators of all orders. It is exactly the combination (74) which lifts to a cocycle for $\mathscr{D}_\lambda$ and gives a central extension $\hat{\mathscr{D}}_\lambda$. For $\mathscr{L}$ we could rescale the central element. Hence the central extension $\hat{\mathscr{L}}$ did not depend essentially on the weight. Here this is different. The central extension $\widehat{\mathscr{D}^1}_\lambda$ depends on it.

For the centrally extended current algebras $\hat{\mathfrak{g}}$, the affine algebra of KN type, in a similar way fermionic Fock space representations can be constructed, see [54, 63].

6.2 b–c Systems

Related to the above there are other quantum algebra systems which can be realized on $\mathscr{H}^\lambda$. On the space $\mathscr{H}^\lambda$ the forms $\mathscr{F}^\lambda$ act by wedging elements $f^\lambda \in \mathscr{F}^\lambda$ in front of the semi-infinite wedge form, i.e.

$$\Phi \mapsto f^\lambda \wedge \Phi. \tag{75}$$

Using the Krichever–Novikov duality pairing (34) to contract in the semi-infinite wedge form the entries with the form $f^{1-\lambda} \in \mathscr{F}^{1-\lambda}$ the latter form will act $\mathscr{H}^\lambda$. For Φ a basis element (72) of $\mathscr{H}^\lambda$ the contraction defines via

$$i(f^{1-\lambda})\Phi = \sum_{l=1}^{\infty}(-1)^{l-1}\langle f^{1-\lambda}, f^\lambda_{i_l}\rangle \cdot f^\lambda_{(i_1)} \wedge f^\lambda_{(i_2)} \wedge \cdots \check{f}^\lambda_{(i_l)} \cdots . \tag{76}$$

Here $\check{f}^\lambda_{(i_l)}$ indicates as usual that this element will not be there anymore.

Both operations together create a Clifford algebra structure, which is sometimes called a $b - c$ system, see [39, 42, 52].

6.3 Sugawara Representation

Given an admissible representation of the centrally extended current algebra $\hat{\mathfrak{g}}$ we can construct the so-called *Sugawara operators*. Here admissible means, that the central element operates as constant $\times$ identity, and that every element v in the representation space will be annihilated by the elements in $\hat{\mathfrak{g}}$ of sufficiently high degree (the degree depends on the element v). The Sugawara operator is an infinite formal sum of operators and is constructed as the product of the current operators which are again formal infinite sum of operators. To make the product well defined a normal ordering has to be set, which moves the annihilation operators to the right to act first. It turns out that after some rescaling the operators appearing in the formal sum of the Sugawara operators give a representation of a centrally extended vector field algebra $\mathscr{L}$. The central extension is due to the appearance of the normal ordering. Again the defining cocycle is local and we know that the central extension defined by the representation is the central extension given by our geometric cocycle $\psi^3_{C_S}$. See [42, 43, 53] for details.

6.4 Wess–Zumino–Novikov–Witten Models and Knizhnik–Zamolodchikov Connection

Despite the fact that it is a very important application, the following description is extremely condensed. More can be found in [44, 54, 55]. See also [52, 65]. WZNW models are defined on the basis of a fixed finite-dimensional Lie algebra $\mathfrak{g}$. One considers families of representations of the affine KN algebras $\hat{\mathfrak{g}}$ (which is an almost-graded central extension of the current algebra $\overline{\mathfrak{g}}$ of KN type) defined over the moduli space of Riemann surfaces of genus g with $K + 1$ marked points and splitting of type $(K, 1)$. The single point in O will be a reference point. The data of the moduli of the Riemann surface and the marked points enter the definition of the algebra $\hat{\mathfrak{g}}$ and the representation. The construction of certain co-invariants yields a special vector bundle of finite rank over moduli space, called the vector bundle of conformal blocks. With the help of the Krichever Novikov vector field algebra, and using the Sugawara construction, the *Knizhnik–Zamolodchikov (KZ) connection* is given. It is projectively flat. An essential fact is that certain elements in the critical strip $\mathscr{L}_{[0]}$ of the vector field algebra correspond to infinitesimal deformations of the moduli and to moving the marked points. This gives a global operator approach in contrast to the semi-local approach of Tsuchiya et al. [66].

6.5 Geometric Deformations of the Witt and Virasoro Algebra

As the second Lie algebra cohomology of the Witt and Virasoro algebra with values in their adjoint module vanishes [14, 15, 50] both are formally and infinitesimally rigid. This means that all formal (and infinitesimal) families where the special fiber is one of these algebras are equivalent to the trivial one. Nevertheless, we showed in [15] that there exists naturally defined families of Krichever–Novikov vector field algebras defined for the torus with two marked points [12,33,40]. These families are obtained by a geometric degeneration process. The families have as special element the Witt algebra (resp. Virasoro algebra). All other fibers are non-isomorphic to it. Hence, these families are even locally non-trivial. This is a phenomena which can only be observed for infinite dimensional algebras. See also the case of affine Lie algebra and some general treatment in [16, 17, 48].

7 Lax Operator Algebras

Recently, a new class of current type algebras appeared, the Lax operator algebras. As the naming indicates, they are related to infinite dimensional integrable systems [64]. The algebras were introduced by Krichever [25], and Krichever and Sheinman [29]. Here I will report on their definition.

Compared to the KN current type algebra we will allow additional singularities which will play a special role. The points where these singularities are allowed are called *weak singular points*. The set is denoted by

$$W = \{\gamma_s \in \Sigma \setminus A \mid s = 1, \ldots, R\}. \tag{77}$$

Let $\mathfrak{g}$ be one of the classical matrix algebras $\mathfrak{gl}(n)$, $\mathfrak{sl}(n)$, $\mathfrak{so}(n)$, $\mathfrak{sp}(2n)$. We assign to every point γ_s a vector $\alpha_s \in \mathbb{C}^n$ (resp. $\in \mathbb{C}^{2n}$ for $\mathfrak{sp}(2n)$). The system

$$\mathscr{T} := \{(\gamma_s, \alpha_s) \in \Sigma \times \mathbb{C}^n \mid s = 1, \ldots, R\} \tag{78}$$

is called *Tyurin data*.

Remark 7.1. In case that $R = n \cdot g$ and for generic values of (γ_s, α_s) with $\alpha_s \neq 0$ the tuples of pairs $(\gamma_s, [\alpha_s])$ with $[\alpha_s] \in \mathbb{P}^{n-1}(\mathbb{C})$ parameterize semi-stable rank n and degree $n\,g$ framed holomorphic vector bundles as shown by Tyurin [67]. Hence, the name Tyurin data.

We consider $\mathfrak{g}$-valued meromorphic functions

$$L : \Sigma \to \mathfrak{g}, \tag{79}$$

which are holomorphic outside $W \cup A$, have at most poles of order one (resp. of order two for $\mathfrak{sp}(2n)$) at the points in W, and fulfill certain conditions at W depending on $\mathscr{T}$. To describe them let us fix local coordinates w_s centred at γ_s, $s = 1, \ldots, R$. For **$\mathfrak{gl}(n)$** the conditions are as follows. For $s = 1, \ldots, R$ we require that there exist $\beta_s \in \mathbb{C}^n$ and $\kappa_s \in \mathbb{C}$ such that the function L has the following expansion at $\gamma_s \in W$

$$L(w_s) = \frac{L_{s,-1}}{w_s} + L_{s,0} + \sum_{k>0} L_{s,k} w_s^k, \tag{80}$$

with

$$L_{s,-1} = \alpha_s \beta_s^t, \quad \operatorname{tr}(L_{s,-1}) = \beta_s^t \alpha_s = 0, \quad L_{s,0}\,\alpha_s = \kappa_s \alpha_s. \tag{81}$$

In particular, if $L_{s,-1}$ is non-vanishing, then it is a rank 1 matrix, and if $\alpha_s \neq 0$, then it is an eigenvector of $L_{s,0}$. The requirements (81) are independent of the chosen coordinates w_s.

It is not at all clear that the commutator of two such matrix functions fulfills again these conditions. But it is shown in [29] that they indeed close to a Lie algebra (in fact in the case of $\mathfrak{gl}(n)$ they constitute an associative algebra under the matrix product). If one of the $\alpha_s = 0$, then the conditions at the point γ_s correspond to the fact that L has to be holomorphic there. If all α_s's are zero or $W = \emptyset$, we obtain back the current algebra of KN type. In some sense the Lax operator algebras generalize them. But their definition is restricted to the case that our finite-dimensional Lie algebra has to be one from the list. In the bundle interpretation of the Tyurin data the KN case corresponds to the trivial rank n bundle.

For $\mathfrak{sl}(n)$ the only additional condition is that in (80) all matrices $L_{s,k}$ are trace-less. The conditions (81) remain unchanged.

In the case of $\mathfrak{so}(n)$ one requires that all $L_{s,k}$ in (80) are skew-symmetric. In particular, they are trace-less. Following [29] the set-up has to be slightly modified. First only those Tyurin parameters α_s are allowed which satisfy $\alpha_s^t \alpha_s = 0$. Then the 1 relation in (81) is changed to obtain

$$L_{s,-1} = \alpha_s \beta_s^t - \beta_s \alpha_s^t, \quad \operatorname{tr}(L_{s,-1}) = \beta_s^t \alpha_s = 0, \quad L_{s,0}\,\alpha_s = \kappa_s \alpha_s. \tag{82}$$

For $\mathfrak{sp}(2n)$ we consider a symplectic form $\hat{\sigma}$ for $\mathbb{C}^{2n}$ given by a non-degenerate skew-symmetric matrix σ. The Lie algebra $\mathfrak{sp}(2n)$ is the Lie algebra of matrices X such that $X^t\sigma + \sigma X = 0$. The condition $\operatorname{tr}(X) = 0$ will be automatic. At the weak singularities we have the expansion

$$L(w_s) = \frac{L_{s,-2}}{w_s^2} + \frac{L_{s,-1}}{w_s} + L_{s,0} + L_{s,1} w_s + \sum_{k>1} L_{s,k} w_s^k. \tag{83}$$

The condition (81) is modified as follows (see [29]): there exist $\beta_s \in \mathbb{C}^{2n}$, $\nu_s, \kappa_s \in \mathbb{C}$ such that

$$L_{s,-2} = \nu_s \alpha_s \alpha_s^t \sigma, \quad L_{s,-1} = (\alpha_s \beta_s^t + \beta_s \alpha_s^t)\sigma, \quad {\beta_s}^t \sigma \alpha_s = 0, \quad L_{s,0}\,\alpha_s = \kappa_s \alpha_s. \tag{84}$$

Moreover, we require $\alpha_s^t \sigma L_{s,1} \alpha_s = 0$. Again under the point-wise matrix commutator the set of such maps constitute a Lie algebra.

The next step is to introduce an almost-graded structure for these Lax operator algebras induced by a splitting of the set $A = I \cup O$. This is done for the two-point case in [29] and for the multi-point case in [57]. From the applications there is again a need to classify almost-graded central extensions. The author obtained this jointly with Sheinman in [56, 57], see also [49] for an overview. For the Lax operator algebras associated with the simple algebras $\mathfrak{sl}(n), \mathfrak{so}(n), \mathfrak{sp}(n)$ it will be unique (meaning: given a splitting of A there is an almost-grading and with respect to this there is up to equivalence and rescaling only one non-trivial almost-graded central extension). For $\mathfrak{gl}(n)$ we obtain two independent local cocycle classes if we assume $\mathscr{L}$-invariance on the reductive part.

8 Some Historical Remarks

In this section I will give some historical remarks (also on related works) on Krichever–Novikov type algebras. Space limitations do not allow to give a complete list of references. For this I have to refer to [52].

In 1987 the ground-breaking work of Krichever and Novikov [26–28] in the two-point case initiated the subject. They introduced the vector field algebra, the function algebra and the affine algebra with their almost-graded structure. To acknowledge their work these algebras are nowadays called Krichever–Novikov (KN) type algebras. Sheinman joint in by investigating the affine algebras and their representations [58–63].

As it should have become clear from this review what should be called a KN type algebra is not the algebra alone but the algebra together with a chosen almost-grading. It is exactly the step of introducing such an almost-grading which is not straightforward.

From the application in CFT, string theory, etc. there was clearly the need to pass over to a multi-point picture. In the multi-point case this was done 1989 by the author in [36–39]. The almost-grading is induced by splitting of the set A of points where poles are allowed into two non-empty disjoint subsets I and O. In the applications quite often such a splitting is naturally given. In the context of fields and strings the points correspond to incoming and outgoing fields and strings, respectively. Without considering an almost-grading Dick [13] gave also a generalization of the vector field algebra. He obtained results similar to [37]. Only in the work of Sadov [34] 1990 an almost-grading is also discussed.

Note that in the two-point case there is only one splitting. Hence, quite often one does not mention explicitly the grading for the Witt and Virasoro algebra, respectively, the almost-grading for the KN type algebras. Nevertheless, the grading is heavily used.

The genus zero and two-point case is the classical well-studied case. But already for genus zero and more than two points there are interesting things so study, see [8, 9, 15, 16, 38, 47]. For genus one, the complex torus case, there is [7, 12, 15, 16, 33, 38].

After the work of Krichever and Novikov appeared physicists got very much interested in these algebras and the possibilities of using these objects for a global operator approach. Especially I would like to mention the work of the people around Bonora [3–6] and by Bryant [10]. This includes also the superversions. A lot of more names could be given.

It is not only the possible applications in physics which makes the KN type algebras so interesting. From the mathematical point general infinite dimensional Lie algebras are hard to approach. KN type algebras supply examples of them which are given in a geometrical context, hence (hopefully) better to understand. A typical example of this are the families of geometric deformation of the Witt algebra which I mentioned in Sect. 6.5 obtained by degenerations of tori. Quite recently also in the context of Jordan Superalgebras and Lie antialgebras [32] examples were constructed on the basis of KN type algebras [24, 30, 31]

In this review we discussed extensively the case of 2nd Lie algebra cohomology with values in the trivial module. But we did not touch the question of the general Lie algebra cohomology of Krichever–Novikov type algebras. Here I refer, e.g., to the work of Wagemann [68, 69].

Acknowledgements Partial support by the Internal Research Project GEOMQ11, University of Luxembourg, is acknowledged.

References

1. E. Arbarello, C. deConcini, V. Kac, C. Procesi, Moduli spaces of Curves and representation theory. Comm. Math. Phys. **117**, 1–36 (1988)
2. A.A. Belavin, A.M. Polyakov, A.B. Zamolodchikov, Infinite conformal symmetry in two-dimensional quantum field theory. Nucl. Phys. B **241**, 333–380 (1084)
3. L. Bonora, M. Martellini, M. Rinaldi, J. Russo, Neveu-Schwarz- and Ramond-type Superalgebras on genus g Riemann surfaces. Phys. Lett. B **206**(3), 444–450 (1988)
4. L. Bonora, M. Rinaldi, J. Russo, K. Wu, The Sugawara construction on genus g Riemann surfaces. Phys. Lett. **B208**, 440–446 (1988)
5. L. Bonora, A. Lugo, M. Matone, J. Russo, A global operator formalism on higher genus Riemann surfaces: $b - c$ systems. Comm. Math. Phys. **123**, 329–352 (1989)
6. L. Bonora, M. Matone, F. Toppan, K. Wu, Real weight $b - c$ systems and conformal field theories in higher genus. Nuclear Phys. B **334**(3), 716–744 (1990)
7. M.R. Bremner, On a Lie algebra of vector fields on a complex torus. J. Math. Phys. **31**, 2033–2034 (1990)
8. M.R. Bremner, Structure of the Lie algebra of polynomial vector fields on the Riemann sphere with three punctures. J. Math. Phys. **32**, 1607–1608 (1991)
9. M.R. Bremner, Four-point affine Lie algebras. Proc. Am. Math. Soc. **123**, 1981–1989 (1995)
10. P. Bryant, Graded Riemann surfaces and Krichever–Novikov algebras. Lett. Math. Phys. **19**, 97–108 (1990)
11. E. Date, M. Jimbo, T. Miwa, M. Kashiwara, Transformation groups for soliton equations. Publ. RIMS, Kyoto 394 (1982)
12. Th. Deck, Deformations from Virasoro to Krichever–Novikov algebras. Phys. Lett. B **251**, 335–540 (1990)
13. R. Dick, Krichever–Novikov-like bases on punctured Riemann surfaces. Lett. Math. Phys. **18**, 255–265 (1989)
14. A. Fialowski, Formal Rigidity of the Witt and Virasoro Algebra, arXiv:1202.3132
15. A. Fialowski, M. Schlichenmaier, Global deformations of the Witt algebra of Krichever–Novikov type. Comm. Contemp. Math. **5**(6), 921–946 (2003)
16. A. Fialowski, M. Schlichenmaier, Global geometric deformations of current algebras as Krichever–Novikov type algebras. Comm. Math. Phys. **260**, 579–612 (2005)

17. A. Fialowski, M. Schlichenmaier, Global geometric deformations of the Virasoro algebra, current and affine algebras by Krichever–Novikov type algebras. Int. J. Theor. Phys. **46**(11), 2708–2724 (2007)
18. L. Guieu, C. Roger, *L'algèbre et le groupe de Virasoro* (Les publications CRM, Montreal, 2007)
19. R.C. Gunning, *Lectures on Riemann Surfaces* (Princeton Math. Notes, NJ, 1966)
20. N.S. Hawley, M. Schiffer, Half-order differentials on Riemann surfaces. Acta Math. **115**, 199–236 (1966)
21. V.G. Kac, *Infinite Dimensional Lie Algebras* (Cambridge University Press, Cambridge, 1990)
22. V.G. Kac, D.H. Petersen, Spin and wedge representations of infinite-dimensional Lie algebras and groups. Proc. Nat. Acad. Sci. USA **78**, 3308–3312 (1981)
23. V. Kac, A. Raina, *Highest Weight Representations of Infinite Dimensional Lie Algebras* (World Scientific, Singapore, 1987)
24. M. Kreusch, Extensions of Superalgebras of Krichever–Novikov Type, arXiv:1204.4338v2
25. I.M. Krichever, Vector bundles and Lax equations on algebraic curves. Comm. Math. Phys. **229**, 229–269 (2002)
26. I.M. Krichever, S.P. Novikov, Algebras of Virasoro type, Riemann surfaces and structures of the theory of solitons. Funktional Anal. i. Prilozhen. **21**(2), 46–63 (1987)
27. I.M. Krichever, S.P. Novikov, Virasoro type algebras, Riemann surfaces and strings in Minkowski space. Funktional Anal. i. Prilozhen. **21**(4), 47–61 (1987)
28. I.M. Krichever, S.P. Novikov, Algebras of Virasoro type, energy-momentum tensors and decompositions of operators on Riemann surfaces. Funktional Anal. i. Prilozhen. **23**(1), 46–63 (1989)
29. I.M. Krichever, O.K. Sheinman, Lax operator algebras. Funct. Anal. Appl. **41**, 284–294 (2007)
30. S. Leidwanger, S. Morier-Genoud, Superalgebras associated to Riemann surfaces: Jordan algebras of Krichever–Novikov type. Int. Math. Res. Not. doi:10.1093/imrn/rnr196
31. S. Leidwanger, S. Morier-Genoud, Universal enveloping algebras of Lie antialgebras. Algebras Represent. Theor. **15**(1), 1–27 (2012)
32. V. Ovsienko, Lie antialgebras: prémices. J. Algebra **325**(1), 216–247 (2011)
33. A. Ruffing, Th. Deck, M. Schlichenmaier, String Branchings on complex tori and algebraic representations of generalized Krichever–Novikov algebras. Lett. Math. Phys. **26**, 23–32 (1992)
34. V.A. Sadov, Bases on multipunctured Riemann surfaces and interacting string amplitudes. Comm. Math. Phys. **136**, 585–597 (1991)
35. M. Schlichenmaier, *An Introduction to Riemann Surfaces, Algebraic Curves and Moduli Spaces, 2nd enlarged edition* (Springer, Berlin, 2007). 1st edition published 1989
36. M. Schlichenmaier, Central extensions and semi-infinite wedge representations of Krichever–Novikov algebras for more than two points. Lett. Math. Phys. **20**, 33–46 (1990)
37. M. Schlichenmaier, Krichever–Novikov algebras for more than two points. Lett. Math. Phys. **19**, 151–165 (1990)
38. M. Schlichenmaier, Krichever–Novikov algebras for more than two points: explicit generators. Lett. Math. Phys. **19**, 327–336 (1990)
39. M. Schlichenmaier, *Verallgemeinerte Krichever - Novikov Algebren und deren Darstellungen*. Ph.D. thesis, Universität Mannheim, 1990
40. M. Schlichenmaier, Degenerations of generalized Krichever–Novikov algebras on tori. J. Math. Phys. **34**, 3809–3824 (1993)
41. M. Schlichenmaier, Differential operator algebras on compact Riemann surfaces, in *Generalized Symmetries in Physics*, ed. by H.-D. Doebner, V.K. Dobrev, A.G. Ushveridze (Clausthal, 1993) (World Scientific, Singapore, 1994), pp. 425–434
42. M. Schlichenmaier, Zwei Anwendungen algebraisch-geometrischer Methoden in der theoretischen Physik: Berezin-Toeplitz-Quantisierung und globale Algebren der zweidimensionalen konformen Feldtheorie. Habilitation Thesis, University of Mannheim, June 1996
43. M. Schlichenmaier, Sugawara operators for higher genus Riemann surfaces. Rep. Math. Phys. **43**, 323–339 (1999)

44. M. Schlichenmaier, Elements of a global operator approach to Wess–Zumino–Novikov–Witten models, in *Lie Theory and Its Applications in Physics III*, ed. by H.D. Doebner, V.K. Dobrev, J. Hilgert (Clausthal, July 1999) (World Scientific, Singapore, 2000), pp. 204–220. arXiv:math/0001040
45. M. Schlichenmaier, Higher genus affine Lie algebras of Krichever–Novikov type. Moscow Math. J. **3**, 1395–1427 (2003)
46. M. Schlichenmaier, Local cocycles and central extensions for multi-point algebras of Krichever–Novikov type. J. reine angew. Math. **559**, 53–94 (2003)
47. M. Schlichenmaier, Higher genus affine Lie algebras of Krichever–Novikov type, in *Proceedings of the International Conference on Difference Equations, Special Functions, and Applications*, Munich (World-Scientific, Singapore, 2007)
48. M. Schlichenmaier, Deformations of the Witt, Virasoro, and current algebra, in *Generalized Lie Theory in Mathematics, Physics and Beyond*, ed. by Silvestrov et al. (Springer, Berlin, 2009), pp. 219–234
49. M. Schlichenmaier, Almost-graded central extensions of Lax operator algebras. Banach Center Publ. **93**, 129–144 (2011)
50. M. Schlichenmaier, An Elementary Proof of the Vanishing of the Second Cohomology of the Witt and Virasoro Algebra with Values in the Adjoint Module, arXiv:1111.6624v1, 2011. doi:10.1515/forum-2011–0143, February 2012
51. M. Schlichenmaier, Lie Superalgebras of Krichever–Novikov Type. Almost-Grading and Central Extensions, arXiv:1301.0484
52. M. Schlichenmaier, *Krichever–Novikov Algebras. Theory and Applications* (forthcoming)
53. M. Schlichenmaier, O.K. Sheinman, Sugawara construction and Casimir operators for Krichever–Novikov algebras. J. Math. Sci. **92**, 3807–3834 (1998). Translated from Itoki Nauki i Tekhniki **38**, q-alg/9512016 (1996)
54. M. Schlichenmaier, O.K. Sheinman, Wess-Zumino-Witten-Novikov theory, Knizhnik-Zamolodchikov equations, and Krichever-Novikov algebras, I. Russian Math. Surv. (Uspekhi Math. Nauk.) **54**, 213–250 (1999), math.QA/9812083
55. M. Schlichenmaier, O.K. Sheinman, Knizhnik–Zamolodchikov equations for positive genus and Krichever–Novikov algebras. Russ. Math. Surv. **59**(4), 737–770 (2004)
56. M. Schlichenmaier, O.K. Sheinman, Central extensions of Lax operator algebras. Uspheki Math. Mauk. **63**(4), 131–172 (2008); Russ. Math. Surv. **63**(4), 727–766 (2008)
57. M. Schlichenmaier, Multipoint Lax operator algebras. Almost-graded structure and central extensions. arXiv:1304.3902 (2013)
58. O.K. Sheinman, Elliptic affine Lie algebras. Funktional Anal. i Prilozhen. **24**(3), 210–219 (1992)
59. O.K. Sheinman, Highest Weight modules over certain quasi-graded Lie algeebras on elliptic curves. Funktional Anal. i Prilozhen. **26**(3), 203–208 (1992)
60. O.K. Sheinman, Affine Lie algebras on Riemann surfaces. Funktional Anal. i Prilozhen. **27**(4), 54–62 (1993)
61. O.K. Sheinman, Modules with highest weight for affine Lie algebras on Riemann surfaces. Funktional Anal. i Prilozhen. **29**(1), 44–55 (1995)
62. O.K. Sheinman, Second order casimirs for the affine Krichever–Novikov algebras $\widehat{\mathfrak{gl}}_{g,2}$ and $\widehat{\mathfrak{sl}}_{g,2}$. Moscow Math. J. **1**, 605–628 (2001)
63. O.K. Sheinman, The fermion model of representations of affine Krichever–Novikov algebras. Funktional Anal. i Prilozhen. **35**(3), 209–219 (2001)
64. O.K. Sheinman, Lax operator algebras and Hamiltonian integrable hierarchies. Russ. Math. Surv. **66**, 145–171 (2011)
65. O.K. Sheinman, *Current Algebras on Riemann Surfaces*. Expositions in Mathematics, vol. 58 (De Gruyter, Berlin, 2012)
66. A. Tsuchiya, K. Ueno, Y. Yamada, Conformal field theory on universal family of stable curves with gauge symmetries. Adv. Stud. Pure Math. **19**, 459–566 (1989)

67. A.N. Tyurin, Classification of vector bundles on an algebraic curve of an arbitrary genus. Soviet Izvestia, ser. Math. **29**, 657–688 (1969)
68. F. Wagemann, Some remarks on the cohomology of Krichever–Novikov algebras. Lett. Math. Phys. **47**, 173–177 (1999). Erratum: Lett. Math. Phys. **52**, 349 (2000)
69. F. Wagemann, More remarks on the cohomology of Krichever–Novikov algebras. Afr. Mat. **23**(1), 41–52 (2012)

MIX
Papier aus verantwortungsvollen Quellen
Paper from responsible sources
FSC® C105338
FSC
www.fsc.org

If you have any concerns about our products,
you can contact us on
ProductSafety@springernature.com

In case Publisher is established outside the EU,
the EU authorized representative is:
Springer Nature Customer Service Center GmbH
Europaplatz 3, 69115 Heidelberg, Germany

Printed by Libri Plureos GmbH
in Hamburg, Germany